Laser Control and Manipulation of Molecules

Laser Control and Manipulation of Molecules

A. D. Bandrauk, Editor
Université de Sherbrooke

Y. Fujimura, Editor
Tohoku University

R. J. Gordon, Editor
University of Illinois at Chicago

American Chemical Society, Washington, DC

Library of Congress Cataloging-in-Publication Data

Laser control and manipulation of molecules / A. D. Bandrauk, editor, Y. Fujimura, editor, R. J. Gordon, editor.

p. cm.—(ACS symposium series ; 821)

Includes bibliographical references and index.

ISBN 0–8412–3786–7

1. Lasers in chemistry—Congresses. 2. Laser manipulation (Nuclear physics)—Congresses. 3. Molecular dynamics—Congresses.

I. Bandrauk, André D.- II. Fujimura, Y. (Yuichi)- III. Gordon, R. J. (Robert J.)- 1944 IV. Pacifichem 2000 (2000 : Honolulu, Hawaii) V. Series.

QD461 .L38   2002
542—dc21   2002016464

Foreword

The ACS Symposium Series was first published in 1974 to provide a mechanism for publishing symposia quickly in book form. The purpose of the series is to publish timely, comprehensive books developed from ACS sponsored symposia based on current scientific research. Occasionally, books are developed from symposia sponsored by other organizations when the topic is of keen interest to the chemistry audience.

Before agreeing to publish a book, the proposed table of contents is reviewed for appropriate and comprehensive coverage and for interest to the audience. Some papers may be excluded to better focus the book; others may be added to provide comprehensiveness. When appropriate, overview or introductory chapters are added. Drafts of chapters are peer-reviewed prior to final acceptance or rejection, and manuscripts are prepared in camera-ready format.

As a rule, only original research papers and original review papers are included in the volumes. Verbatim reproductions of previously published papers are not accepted.

ACS Books Department

Contents

Intense Fields

Alignment and Manipulation

Preface

This volume is dedicated to the memory of Kent Wilson, who was a pioneer in the rapidly growing field of coherent control of chemical reactions. The chapters in this volume were presented at a symposium at the 2000 International Chemical Congress of Pacific Basin Societies ("PacifiChem 2000") held in Honolulu, Hawaii, December 14–19, 2000. This symposium dealt with the manipulation of light with matter, a field in which Professor Wilson and his students have made many seminal contributions.

Kent Wilson had an open-mindedness about people and ideas, which led him eventually to cross many intellectual boundaries. He was born in 1937 and was raised in Bryn Gweled Homesteads, a mostly Quaker community near Philadelphia, Pennsylvania. This experiment in communal living helped Kent develop a passion for free thinking. Kent's childhood experiments with moonshine and fireworks eventually led him to do undergraduate research in George Kistiakowsky's laboratory at Harvard College. The greatest influence on Kent's scientific growth was his graduate and post-doctoral mentor, Dudley Herschbach. Kent joined Herschbach's group when the idea of using crossed molecular beams to study the chemical reactions of isolated molecules was in its infancy. Kent's doctoral research dealt with the transition from rebound to stripping reactions in collisions between alkali atoms and halogen containing molecules *(1)*.

Kent's early interest in molecular reaction dynamics was a recurring theme throughout his varied scientific career. One of his first achievements after joining the chemistry faculty at the University of California in San Diego in 1964 was the development of the technique of photofragment spectroscopy. His experiments were among the first applications of laser radiation to probe the structure and dynamics of molecules. His paper with Busch *(2)* on the rebound model for the partitioning of energy among molecular fragments has been cited hundreds of times and is still valuable today.

The years that followed were a period of remarkable creativity in such diverse fields as air pollution, chemical archeology, and the imaging of human brain anatomy. An example of Wilson's interdisciplinary vision is his study of the sources of metal used in ancient Nigerian bronzes. It was previously believed that the lead used in these 15th to 19th century sculptures came from northern Africa, but a statistical, isotopic analysis performed by Goucher et al.

(3) proved that it came from Germany, possibly by a triangular trade route. In the meanwhile, Kent's continuing interest in molecular reaction dynamics led him to study the theory of chemical reactions in the liquid phase.

A hallmark of Kent's scholarship was his merging of science, art, and showmanship. Numerous film productions by his undergraduate *Senses Bureau* inspired a generation of young scientists. His film on protein synthesis, which incorporates modern dance, rock-jazz, and poetry, is still shown in universities today, 30 years after its making. Wilson's conference presentations were usually attended by standing-room-only crowds, many of whom came for the pleasure of experiencing Kent's magical blend of science and art.

Kent's final decade, marked by his entry into the field of coherent control, was perhaps his most creative. At a workshop in Telluride in 1991, he conceived of the idea of using a computer to tailor the properties of a light pulse, which could in turn be used to control the behavior of a wave packet *(4)*. In the following four years, Wilson's group laid much of the theoretical groundwork for "controlling the future of matter." In a landmark paper, Krause, Whitnell, Wilson, Yan, and Mukamel *(5)* developed a method of calculating the optimal weak laser field for producing the best overlap of a wave packet with a given target in phase space. They illustrated the theory by calculating the field needed to produce a "molecular cannon" wave packet focused to a specified position and momentum, an incoming "reflectron" wave packet, and a bound, zero-momentum wave packet focused at a turning point of the trajectory. In later papers the Wilson group extended the theory to strong fields and dissipative systems. Two other groundbreaking theoretical papers describe a method for sculpting Rydberg wave packets *(6)* and the generation of a molecular π-pulse for total inversion of an electronic state population *(7)*.

Concurrent with their theoretical effort, the Wilson group developed the laser tools for controlling matter and applied them to a variety of systems. To produce the bandwidth required for the shaping of wave packets, Yakovlev, Kohler, and Wilson developed a broadly tunable 30 fs amplifier *(8)*. They went on to use this device to generate frequency-chirped laser pulses, which they used to control the evolution of a vibrational wave packet in electronically excited I_2 *(9)*. In succeeding experiments over the next three years they extended their experiments to condensed phases *(10)*, working with molecules as complex as the green fluorescent protein *(11)*. In still another landmark experiment, the Wilson group used a genetic algorithm in a feedback loop to "teach" the laser how to optimize the fluorescence of a laser dye in solution *(12)*.

One of the great engineering accomplishments of the Wilson group was to build an ultrahigh intensity laser, compact enough to fit on a single table and simple enough to be operated by students *(13)*. This instrument was capable of generating intensities in excess of 10^{20} W/cm^2, at which relativistic effects become important. Using this device, Wilson was able to fulfill a lifelong dream of using ultra-short X-ray pulses to monitor atomic and molecular motion in real time. In his final years, Wilson developed an ultra-fast X-ray laser that allowed him to observe atomic motion *(14)* and a laser microscope that allowed him to obtain very high-resolution images of living cells.

In 1998 Kent discovered that he had inoperable prostate cancer. Until the very end of his life, Kent's curiosity about science was unabated. In his final experimental endeavor, he worked with a team of biochemists and physicians to develop new therapies for suppressing the cancer growth in his own body, to which he finally succumbed in March 2000 *(15)*. Shortly before his death, a special issue of *The Journal of Physical Chemistry A* celebrating Kent's work was published. In that issue, Kent wrote a poignant and inspirational account of his values and accomplishments *(16)*.

The work of Kent Wilson has had an enormous impact on both experimental and theoretical methods of controlling matter with light, as evident from an examination of the chapters in this volume. The book is divided into three sections, covering different aspects of the interaction of light with molecules, including laser control in the weak (perturbative) and strong (nonperturbative) regimes and the manipulation of molecules in space. These are new research areas in chemical physics to which Kent Wilson and his group have made notable contributions.

The first section of this book deals with the coherent control of molecular processes in weak fields. The opening chapter by Rabitz discusses the use of closed loop learning algorithms to control a variety of quantum and nonlinear optical phenomena without knowledge of the Hamiltonian, a strategy that was first demonstrated by Wilson's group in controlling the fluorescence of a dye. In the following chapter by Rice and co-workers, a multi-pulse generalization of STIRAP (stimulated Raman adiabatic passage) is introduced for controlling unimolecular reactions. The use of counterintuitive pulse sequences to transfer 100% of the population to an excited level is conceptually related to Wilson's molecular π-pulse. Generalizations of Wilson's pump–dump method for controlling nuclear motion can be seen in Dantus' use of three-pulse, four-wave mixing to manipulate electronic, vibrational, and rotational coherences, and in Jones' pump–dump control of Rydberg wave packets. The use optimal control theory to design the "best" laser pulse is generalized in the chapter by Cao et al. Chirped pulses, a central idea in much of Wilson's wave packet experiments, are used by Yamashita et al .to control chemical reactions and by Nakamura et al. to control non-adiabatic transitions. The impact of Wilson's pioneering work on vibrational wave packets is evident in the paper by Ohmori et al. on the interaction of two-wave packets in a complex and in the paper by Kobayashi et al., which deals with real-time vibrational spectroscopy. Hoki and Fujimura discuss selective preparation of pure enantiomers from an equal mixture of left-handed and right-handed preoriented enantiomers, using a theory based on the density matrix formalism with a dressed state representation. The photodissociation of preoriented enantiomers has its antecedents in the early work of Busch and Wilson on the photofragmentation of aligned molecules with linearly polarized light. Finally, coherent phase control in the molecular analogue of Young's two-slit experiment discussed by Gordon and co-workers and by Nakajima is complementary to the time domain experiments pioneered by Wilson.

The second section of the book deals with intense field effects. The leading chapters by Gerber and co-workers and by Levis et al. extend the usage of learning algorithms to the strong field regime, where dressed state effects are important. Structural deformation of molecules in intense fields is discussed in the papers by Nguyen-Dang et al., by Yamanouchi, and by Kono and Koseki. Intense field control of the interaction of electron and nuclear dynamics is treated in the chapter by Bandrauk et al.

The final section of the book deals with the manipulation and alignment of molecules with laser fields. Again, the dipole (or induced dipole) interaction of molecules with a linearly polarized laser beam has its antecedents in the early work of Busch and Wilson. The important extension developed by Cai and Friedrich, Sakai et al., and Mathur et al. is that the dipole force of a focused laser beam may be used to *actively* manipulate the motion of molecule. Finally, Lyrra et al. demonstrated Autler–Townes splitting in Li_2, and showed how it could be used to align the angular momentum vector optically.

At the close of his paper on "Controlling the Future of Matter" *(17)*, Kent Wilson wrote: "This has been a quick look at the Holy Grail of using lasers to control matter, and at goals for quantum control to entice us onward into the future. Realistically, some of these dreams will come true, and others will not. But the quest to achieve the goals discussed here will likely lead us to new discoveries, which may in the end be worth more to us than the original goals." Some of these goals have already been fulfilled, and the dreams for the others are still alive.

References

1. Wilson, K. R.; Kwei, G. H.; Norris, J. A.; Herm, R. R.; Birely, J. H.; Herschbach, D. R. *J. Chem. Phys.* **1964**, *41*, 1154.
2. Busch, G. E.; Wilson, K. R. *J. Chem. Phys.* **1972**, *56*, 3626.
3. Goucher, C. L.; Teilhet, J. H.; Wilson, K. R.; Chow, T. J. *Nature (London)* **1976**, *262*, 130.
4. Gordon, R. J. private communication.
5. Krause, J. L.; Whitnell, R. M.; Wilson, K. R.; Yan, Y. J.; Mukamel, S. *J. Chem. Phys.* **1993**, *99*, 6562.

6. Krause, J. L.; Schafer, K. J.; Ben-Nun, M.; Wilson, K. R. *Phys. Rev. Lett.* **1997**, *79,* 4978.
7. Cao, J.; Bardeen, C. J.; Wilson, K. R. *Phys. Rev. Lett.* **1998**, *80,* 1406.
8. Yakovlev, V. V.; Kohler, B.; Wilson, K. R. *Opt. Lett.* **1994,** *19,* 23.
9. Kohler, B.; Yakovlev, V. V.; Che, J.; Krause, J.; Messina, M.; Wilson K. R.; Schwentner, N.; Whitnell, R. W.; Yan. Y. J. *Phys. Rev. Lett.* **1995**, *74,* 3360.
10. Che, J.; Messina, M.; Wilson, K. R.; Apkarian, V. A.; Li, Z.; Martens, C. C.; Zadoyan, R.; Yan. Y. J. *J. Phys. Chem.* **1996**, *100, 7873.
11. Bardeen, C. J.; Yakovlev, V. V.; Squier, J. A.; Wilson, K. R. *J. Am. Chem. Soc.* **1998**, *120,* 13023.
12. Bardeen, C. J.; Yakovlev, V. V.; Wilson, K. R.; Carpenter, S. D.; Weber, P. M.; Warren, W. S. *Chem. Phys. Lett.* **1997**, *280,* 151.
13. Mourou, G. A.; Barty, C. P. J.; Perry, M. D. *Phys.Today* Jan. 1998, 22–28.
14. Rose-Petruck, C.; Jimenez, R.; Guo, T.; Cavalleri, A.; Siders, C. W.; Ráksi, F.; Squier, J. A.; Walker, B. C.; Wilson, K. R.; Barty, C. P. J. *Nature (London)* **1999**, *398,* 310.
15. Herschbach, D. R. *Nature (London)* **2000,** *405,* 902.
16. Wilson, K. R., "Summing Up," *J. Phys. Chem. A* **1999**, *103,* 10022.
17. Kohler, B.; Krause, J.; Ráksi, F.; Wilson, K. R.; Yakovlev, V. V.; Whitnell, R. M.; Yan., Y. J. *Acc. Chem. Res.* **1995**, *28,* 133

A. D. Bandrauk
Laboratorie de Chimie Théorique
Faculté des Science
Université de Sherbrooke
Québec, Québec J1K 2R1, Canada

Y. Fujimura
Department of Chemistry
Graduate School of Science
Tohoku University
Sendai 980–8578, Japan

R. J. Gordon
Department of Chemistry
Mail Code 111
University of Illinois at Chicago
845 West Taylor Street
Chicago, IL 60607–7061

Coherent Optimal Control

Chapter 1

Controlling Molecular Motion:
The Molecule Knows Best

Herschel Rabitz

Department of Chemistry, Princeton University, Princeton, NJ 08544

Over the past decade, an intense effort has brought together theoretical and laboratory tools for controlling molecular motion with tailored laser pulses. Various means for designing laser pulses are available, including a new procedure, discussed here, for carrying out the effort when there is uncertainty in the Hamiltonian. Presently, the most viable general procedure for achieving successful control over quantum systems in the laboratory is through the use of closed loop learning algorithms. The logic behind the operation of such algorithms is discussed, along with a summary of several recent laboratory achievements exploiting closed loop learning to control quantum and nonlinear optical phenomena.

Interest in the control of quantum systems has grown since the original suggestion, in the 1960's, of using lasers to manipulate chemical reactivity*(1)*. This latter goal still stands as an important challenge, and there are very promising recent experimental results*(2,3)*. Many more objectives are now of interest, including the manipulation of electron transport in semiconductors, excitons in solids, quantum optics, quantum computers, and high harmonic generation, amongst others. In addition, there is the prospect of using similar quantum system control techniques to invert the observed dynamics data and learn high quality information about the underlying atomic-scale interactions. The primary means of control in all of these areas is through the use of tailored laser fields, whose formation is an emerging laboratory technology exhibiting considerable flexibility for practical applications.

Regardless of the application, the essence of the underlying control concept is captured by the goal

$$|\psi_i\rangle \rightarrow |\psi_f\rangle \tag{1}$$

of steering a quantum system from a specified initial state $|\psi_i\rangle$ to a desired final state $|\psi_f\rangle$. In the laboratory, the actual objective is an expectation value $\langle\psi_i|O|\psi_f\rangle$ over a suitable observable operator O. These statements may also be generalized to include the density matrix, rather than the wavefunction. As a problem in quantum system control, these goals are typically expressed in terms of seeking a tailored laser electric field $\varepsilon(t)$ that couples into the Schrödinger equation

$$i\hbar\frac{\partial}{\partial t}|\psi\rangle = \left[H_0 - \mu\cdot\varepsilon(t)\right]|\psi\rangle \tag{2}$$

through the dipole μ. By assumption, the dynamics under the free Hamiltonian H_0 does not evolve the system in the desired way expressed in Eq. (1). Regardless of the physical application, the general mechanism for achieving quantum control is through the manipulation of constructive and destructive quantum wave interferences. The goal is to create maximum constructive interference in the state $|\psi_f\rangle$ according to Eq. (1), while simultaneously achieving maximum destructive interference in all other states $|\psi_{f'}\rangle$, $f' \neq f$ at the desired target time T. A simple analogy to this process is the traditional double slit experiment. However, a wave interference experiment with two slits will lead to only minimal resolution; in the context of quantum control, two pathways can produce limited selectivity when there are many accessible final states for discrimination. Thus, a multitude of effective slits needs to be created at the molecular scale in order to realize high quality control into a single state, while eliminating the flux into all other states, as best as possible. This logic leads to the need for introducing a control field $\varepsilon(t)$ having sufficiently rich structure to simultaneously manipulate the phases and amplitudes of all of the pathways connecting the initial and final states.

The physical picture above prescribes a general mechanism for successful quantum control of any type: High quality control calls for the external field to fully cooperate with all of the dynamical capabilities of the quantum system. As quantum systems are generally capable of complex dynamical behavior, a simple conclusion is that the most successful control fields, in turn, will reflect that complex structure. This conclusion is borne out in the recent successful control experiments, especially those involving strong fields encompassing nonlinear quantum dynamical responses to the fields*(2-6)*.

Recognizing that the general means of achieving control is through tailored optical fields $\varepsilon(t)$ lays the groundwork for the task of identifying successful control fields. This task may be accomplished either by computational design or direct discovery of the control field in the laboratory. The state of these two approaches has evolved considerably in recent years. It is natural to express control as an optimization process, as we always desire to produce the *best* outcome in the laboratory. The search for an optimal control field $\varepsilon(t)$ poses a nonlinear problem, in terms of either field design or a direct laboratory search for the field. The nonlinearity of the problem is important to appreciate, as it underlies the techniques and concepts being developed to determine the optimal fields. The origin of the nonlinearity lies in the control field $\varepsilon(t)$ being dependent on the current state $\left|\psi(t)\right\rangle$ of the evolving quantum system, as well as the future desired state $\left|\psi_f(T)\right\rangle$ or the physical objective. The relationship between the control field and the evolving state of the system can be highly complex in many cases. Nonlinearity is inherent in all areas of temporal system control, but it is an unusual perspective in quantum mechanics, which is normally thought of as involving linear dynamics: given the Hamiltonian, solve the linear Schrödinger equation. However, in the present context of quantum system control, we are seeking to find a piece of the Hamiltonian (i.e., $\varepsilon(t)$ in the term $-\mu \cdot \varepsilon(t)$), which is initially unknown by definition of the physical control problem.

The next section will first discuss theoretical design by optimal control theory techniques, along with some suggestions on future directions for development. This will be followed by a presentation of the case for closed loop laboratory learning techniques, as presently, they provide the only viable generic means of achieving control over many quantum system objectives, especially in those cases where the systems and their dynamics may be highly complex. A discussion of some recent laboratory results in this area will also be presented. Finally, the last section will give some brief concluding remarks.

Computational Design of Controls to Optimally Manipulate Quantum Systems

Control field design might be achieved by a variety of means, starting with intuition, on through perturbation theory techniques, and ultimately, with *optimal* design*(7)*. Given the complexities of quantum dynamics phenomena, intuition is likely to work only in the simplest of cases, and perturbation theory (i.e., the weak field regime) is operative under special circumstances. Even in the perturbation theory regime, the control field may have subtle structure reflecting the need to discriminate amongst may attainable physical objectives. Thus, in general, the best means of approaching computational control field design is through optimization techniques. Optimal control theory entails first prescribing a cost functional containing the desired physical objective, as well as costs against any undesirable physical behavior (e.g., breaking the wrong chemical bond in a dissociation control experiment). In addition, the cost functional may contain penalties against any unattainable laser field structure or other laboratory limitations. The goal is maximization of the cost functional to balance all of these criteria, subject to the dynamics described by the Schrödinger equation. There are many ways of expressing cost functionals for quantum control design, but the simplest approach will lead to the following Euler-Lagrange equations whose solution prescribes the desired optimal field $\varepsilon(t)$:

$$i\hbar \frac{\partial}{\partial t}\left|\psi(t)\right\rangle = \left[H_0 - \mu \cdot \varepsilon(t)\right]\left|\psi(t)\right\rangle \ , \ \left|\psi(0)\right\rangle = \left|\psi_i\right\rangle \tag{3a}$$

$$i\hbar \frac{\partial}{\partial t}\left|\lambda(t)\right\rangle = \left[H_0 - \mu \cdot \varepsilon(t)\right]\left|\lambda(t)\right\rangle \ , \ \left|\lambda(T)\right\rangle = \sigma O\left|\psi(T)\right\rangle \tag{3b}$$

$$\varepsilon(t) = \Im\left\langle\lambda(t)\left|\mu\right|\psi(t)\right\rangle \tag{3c}$$

$$\sigma = \left\langle\psi(T)\left|O\right|\psi(T)\right\rangle - O^* \tag{3d}$$

The electric field in Eq. (3c) may be substituted into Eqs. (3a) and (3b), to produce a coupled set of cubically nonlinear Schrödinger-type equations, where the goal is to drive the eigenvalue of this system σ to be as small as possible, with O^* being the target for the expectation value $\left\langle\psi(T)\left|O\right|\psi(T)\right\rangle$ at the time T.

In general, the solution of these equations calls for iteration techniques due to their nonlinearity, which may be embodied into the collective equations (3) being rewritten as

$$i\hbar \frac{\partial}{\partial t}\left|\psi(t)\right\rangle = \left[H_0 - \mu \cdot \varepsilon(\psi,\sigma)\right]\left|\psi(t)\right\rangle \tag{4}$$

where the control field $\varepsilon(\psi,\sigma)$ is explicitly shown to depend on the current state ψ of the evolving quantum system, as well as the final state through the eigenvalue σ. This reexpression of the field shows the nonlinear nature of the

6

quantum control field design equations, as well as the essential need for iteration, as the state at the final time plays a role in guiding the current evolution of the physical system. Many control field designs have been performed in recent years*(1)*, with the conclusions that (a) fields can be designed to manipulate rotational, vibrational, and electronic degrees of freedom, and (b) such fields typically have structure reflecting the complexity of the dynamics placed under control, especially when high yields are required (e.g., unit population in a target state).

A basic question is whether a means can be found to circumvent the need for iterative calculations in the above variational formulation. Tracking theory provides an alternative*(8)*, whereby a track $f(t) = \left(\psi(t) | O | \psi(t) \right)$ is specified *a priori,* covering the control interval $O \leq t \leq T$ out to the target $f(T) = O^*$. The goal is to find the field $\varepsilon(t)$, such that the dynamics follows the track. A rigorous non-iterative tracking procedure exists for meeting this goal, but the process demands good physical intuition for the specified track $f(t)$, in order for the field $\varepsilon(t)$ to be free of undesirable features.

A variety of alternative computational design formulations may be developed that avoid iteration. The origin of the iterative operations needed to solve Eq. (3) is the hard demand that Schrödinger's equation be precisely satisfied in the design process. In fact, the Hamiltonian for most realistic quantum systems is not precisely known. It is possible to take advantage of this apparent limitation through the use of design cost functionals of the following form*(9)*

$$J = \int_0^T \alpha(t) \left(\left\langle \psi(t) | O | \psi(t) \right\rangle - O^* \right)^2 dt + \int_0^T \beta(t) \varepsilon^2(t) dt +$$

$$\int_0^T \gamma(t) \left\langle \left(-i\hbar \frac{\partial}{\partial t} - H \right) \psi(t) \Big| \left(i\hbar \frac{\partial}{\partial t} - H \right) \psi(t) \right\rangle dt \tag{5}$$

The positive definite weights $\alpha(t)$, $\beta(t)$, and $\gamma(t)$ manage the contributions of the three integral terms in Eq. (5). Importantly, the last term introduces a quadratic cost, to relax the hard demand that Schrödinger's equation be precisely followed*(10)*. As an illustration of employing J for control field design, the numerical study shown in Figure 1 indicated that reasonable physical intuition could be applied to specify the qualitative behavior of $\alpha(t)$, $\beta(t)$, and $\gamma(t)$, to balance the goals of meeting the target O^*, while satisfying Schrödinger's equation to an acceptable degree and keeping the laser fluence to a minimum. The goal in such calculations is to provide a reasonable trial control field $\varepsilon(t)$ (i.e., a field that takes the physical system at least to the vicinity of the target

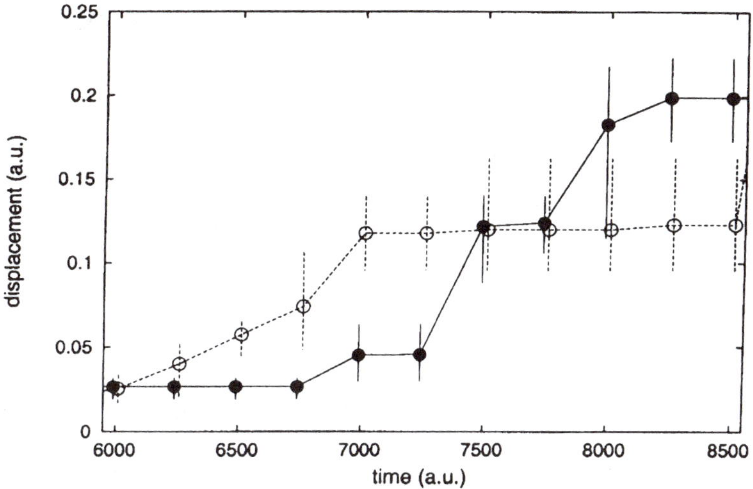

(Reproduced with permission from reference 9. Copyright 2001 European Physical Journal D.)

Figure 1: Results from the control of a coupled harmonic chain of five bonds. The radiatively active dipole is in bond 1 (dotted line) and the target for stretching is bond 5 (solid line) at the other end of the molecule. The vertical lines span the distribution of the maximum stretch achieved from testing the field design with an ensemble of Hamiltonian. The control shows robustness over reasonable variations of the Hamiltonian, and significant excitation is seen at 8000 a.u., when the target bond achieves a displacement larger than the dipole bond.

$O*$), which may then be refined in the laboratory, working with the actual physical system. Thus, in Figure 1, the test shows the quality of the design achieved through Eq. (5), by employing the field with an ensemble of Hamiltonians having ± 10% variation around the one employed in the design process.

The reasonable statistical behavior implies that such a field design would permit efficient laboratory refinement, by starting with a significant signal in the target state. Reliable approximate field design techniques could have an increasingly important role in facilitating laboratory control studies.

Laboratory Closed Loop Control of Quantum Systems

A possible approach to achieving successful control over quantum systems would be through the route of (a) theoretical control field design, followed by (b) implementation of the design in the laboratory using modern laser pulse-shaping techniques. Although the viability of this design-based route will surely improve in the coming years [c.f., the concept of adequate approximate design techniques utilizing Eq. (5)], at present, the overall situation may be succinctly summarized by the following: If the design can be carried out reliably, then the physical system will likely not be of much interest, while for interesting physical systems, reliable designs can not be performed. This conundrum arises for three reasons:

1. The Hamiltonians for the most interesting polyatomic and solid-state systems are not known accurately.
2. Solving the quantum design equations is a heavy computational task, especially in the interesting strong field regime.
3. Precise execution of any particular design likely would be difficult to achieve in the laboratory.

The importance of point 1 cannot be overemphasized. For example, laser control over molecular systems was originally, and continues to be, motivated by the desire to manipulate large polyatomic molecules. The only means of successfully controlling complex quantum systems in the foreseeable future is through the introduction of closed loop learning techniques in the laboratory*(11)*. This realization led to a proposed practical procedure to take advantage of the following circumstances:

A. The molecule "knows" its own Hamiltonian.
B. When exposed to a laboratory field $\varepsilon(t)$, a molecule will solve its Schrödinger equation in real ultrafast molecular-scale time, with absolute fidelity.
C. Laser pulse shapers with duty cycles of up to hundreds or more distinct pulses per second are becoming available under full computer control.
D. Many physical objectives may be expressed in terms of easily detectable outcomes calling for little or no refined data analysis.
E. Fast algorithms exist to recognize patterns in the emerging relationships between the fields and their observable molecular impacts to automatically suggest new (better) control fields.

The synthesis of steps A,…,E produces an efficient closed loop learning procedure for teaching lasers to control quantum systems and a schematic of this process is shown in Figure 2. The quantum system, upon each cycle of the loop, is replaced by a new one, thereby avoiding the need for ultrafast excursions around the loop, and eliminating any concerns about the observation process disturbing the dynamics.

Laser fields and the general laboratory environment will inherently be contaminated with noise, and an essential requirement for successful quantum system control is the achievement of robustness to realistic disturbances. Theoretical studies have shown that computational field designs, without

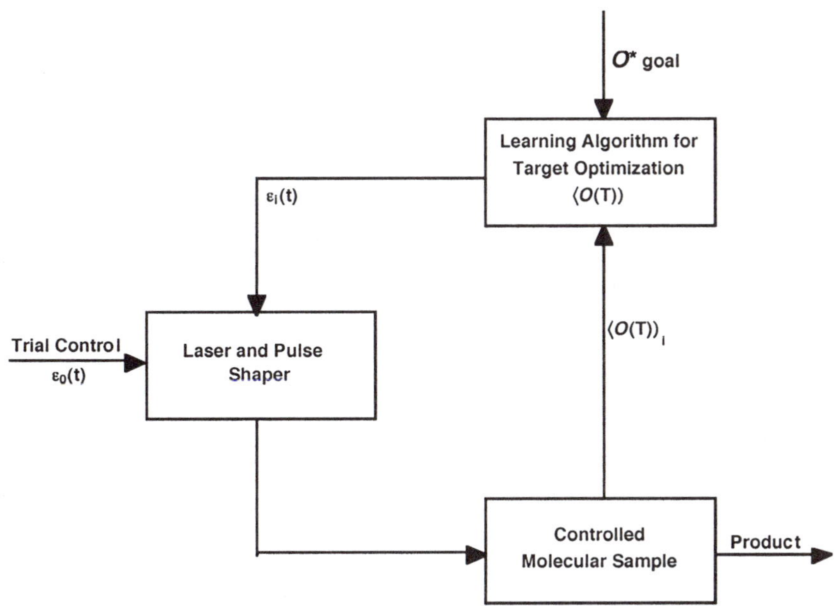

Figure 2: Closed loop learning process for performing optimal control experiments. In a sequence of excursions i = 1,2,... around the loop, the learning algorithm guides the shaped laser pulse to steer $\langle O(T) \rangle$ toward the product goal O by observing the patterns of behavior evident in the laser control settings for $\varepsilon_i(t)$ and their molecular actions $\langle O(T) \rangle_i$.*

imposing robustness requirements, at first sight, may appear to produce successful results; however, the physical outcome from the designs can be very sensitive to laboratory noise. A reasonable speculation is that closed loop learning control in the laboratory will inherently introduce a degree of robustness by virtue of the need to work within the true environment, including all of its exigencies: Only those controls that are robust will survive the learning process in Figure 2. However, the most effective approach should be through the performance of closed loop laboratory optimal control with the inclusion of explicit robustness criteria in the cost functional. Significantly, little, if any, additional effort will be necessary in the laboratory to assure such robustness, as the natural performance of signal averaging provides the statistical information to determine the variance of the physical observation. Figure 3 presents a simulation of how random disturbances may influence designs*(12)*. The four cases show (A) learning to produce nearly unit population in the target state

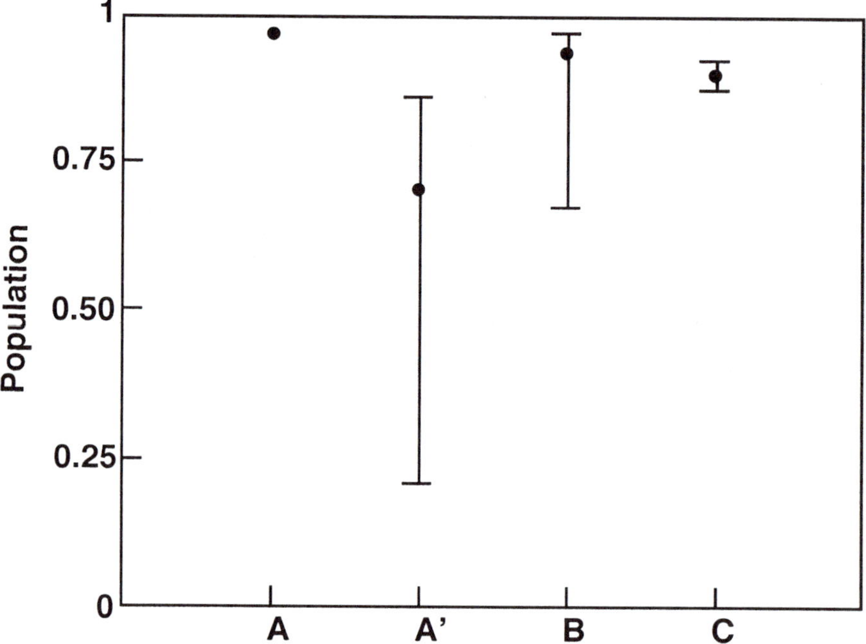

Figure 3: Reducing the influence of laboratory noise on quantum control by introducing robustness demands into the learning algorithm. The results (see the text for details) are from a simulation of a closed loop control experiment for steering a model 10-level quantum system from state 1 to state 8.

with no field noise (an idealized situation unattainable in the laboratory), (A′) case A post-assayed for robustness to ± 5% noise (shows that the idealized control in case A is highly unstable), (B) learning in the presence of ± 5% noise (this procedure is the current laboratory practice), and (C) robust learning in the presence of ± 5% noise. The results in case B vs. that of A and A′ show that some degree of robustness is naturally achieved in the learning control process operating in the presence of noise. Most significant is case C, where imposing robustness criteria *during* the laboratory learning process is shown to greatly enhance the degree of robustness attained while costing very little in terms of the achieved target population.

The capabilities of closed loop learning techniques to control quantum systems can be appreciated by considering the increasing number of successful laboratory experiments for a broad variety of applications. The first successful experiment was carried out by Wilson, et al., in 1997 in the control of laser dye fluorescence*(13)*. A summary of the some recent applications is given below.

Optimally Controlled Dissociation and Rearrangement of Molecules

Selective chemical bond breaking was put forth as a challenge in the laser community some 40 years ago. With the pioneering work of Gerber. et al.*(2)*, and now, extended to organic molecules by Levis*(3)*, et al., truly chemically interesting systems are finally succumbing to tailored strong field closed loop laser control. A most important feature of these closed loop experiments is the use of high intensity shaped laser pulses, where dynamic power broadening can range up to ~ 10 eV. In this regime, the bonding electrons throughout the molecule are disturbed significantly and coherently, to dynamically bring in and out resonance, the necessary molecular vibronic levels to achieve control. Simple considerations based on spectroscopic resonance criteria are no longer operative. At first sight, it may appear that little control could be achieved under such extreme conditions, but the sensitivity of the molecular dynamics to the nature of the shaped laser fields has led to several clear demonstrations of control over molecular reactivity. One significant case is the selective manipulation of acetophenone

$$\phi + CH_3 - C = O \qquad (6a)$$

$$CH_3 + \phi - C = O \qquad (6b)$$

$$\phi - CH_3 + CO \qquad (6c)$$

where selectivity towards any of the three products was observed through mass spectrometric detection of the associated ion yields*(3)*. The first two product channels (6a) and (6b) correspond to the classic case of breaking one bond versus another in competition with each other, and the experiment showed that dynamic selectivity over a range of at least a factor of 4 could be achieved for these two channels. Perhaps the most dramatic result is observed in the third product channel (6c) producing toluene, which requires breaking two bonds and forming a third, under laser-guided control. The detailed mechanism of such processes is not known at this time, but in general terms, the shaped laser pulse creates a suitable vibronic wave packet to steer the system out the desired channel.

Although earlier arguments suggested that carefully operating with the resonant states of the free Hamiltonian H_0 should provide the means of achieving molecular control, the present successful experiments on complex molecules imply that operating in the strong field regime may actually provide the best opportunity for achieving control for a broad variety of molecules. The utilization of significant dynamic power broadening allows for creating a generic apparatus, into which one may place almost any molecule for an attempt at controlled manipulation, with the outcome assessed either through mass spectrometric or optical detection. It is anticipated that the immediate period ahead will witness many such tests and applications of closed loop learning control over molecular reactivity; hence, the title of this chapter.

Control of High Harmonic Generation

The passage of a sufficiently high intensity laser pulse through many gases will result in subsequent re-radiation, not only at the carrier frequency of the driving laser pulse, but also at many high harmonics. For example, a driving infrared or visible laser pulse may yield high harmonics even in the X-ray regime. An important objective is the enhancement of a particular high harmonic at the expense of other harmonics. This is a very demanding goal, as the high harmonics are all strongly coupled together. Control of second harmonic generation was easily demonstrated under closed loop operation*(14)*, and most significantly, Bartels, et al., were able to achieve closed loop high harmonic enhancement*(4)*. They showed that the 23rd harmonic of argon in the soft X-ray region could be enhanced by a factor of ~ 10, using a 19-element phase modulator. Although the processes involved were physically complex, subsequent to the experiments, a full simulation led to a phase modulation pattern similar to the optimal form discovered in the laboratory*(15)*. It was shown that the successful enhancement of a single harmonic occurred through the creation of a complex electron wave packet that acted to constructively focus the radiation intensity in the desired high harmonic, while minimizing the intensity in other harmonic channels. Selective high harmonic generation as a source of tailored radiation may have many practical applications.

Learning Control of Ultrafast Semiconductor Optical Switching

A need exists in the optical communications and computing communities is for ultrafast solid-state switches for propagating laser pulses. Keller, et al., demonstrated that a closed loop apparatus operating with a solid-state sample could be used to teach a laser pulse how to manipulate an electron wave packet*(5)*, to act as a differential transmission switch for a second laser pulse.

An ultrafast switch operating this way was created in AlGaAs to function on a time scale of nearly 50 femtoseconds, over a positive and negative differential transmission range of a factor of ~ 4. This initial significant demonstration opens the way to further refinement for possible practical applications.

Adaptive Control of Nonlinear Femtosecond Pulse Propagation in Optical Fibers

Telecommunications with optical fibers is typically carried out in the weak field regime, thereby avoiding nonlinear dispersion effects. However, a number of industrial and laboratory applications, with transmission distances on the order of meters, involve high intensity pulses in the femtosecond regime. Clean pulses, such as a Gaussian, entering a fiber under these conditions will generally undergo severe distortion upon transmission through the fiber. Omenetto, et al., demonstrated that closing the loop around such a fiber transmission experiment with a phase modulator to shape the input pulse was capable of producing very clean transmitted pulses *(6)*. The quality enhancement was dramatic, and there may be a number of practical applications of this technology.

The examples cited above are likely only the beginning of many laboratory illustrations of closed loop control over quantum systems. In some cases, this capability may be primarily of academic interest for studying the underlying dynamics phenomena, while in other cases, practical industrial or biological applications may ensue. The period ahead should be exciting, as the patterns found in the evolving experiments should provide a basis for establishing the scientific principles, mechanisms, and rules of thumb for controlling complex quantum phenomena.

Conclusion

Seeking to control quantum phenomena has been a long-term objective, extending for some 40 years since the originally posed goal of using lasers to control chemical reactivity. The persistence of the scientific community involved in the long trek speaks to the importance of this subject, which also is attested to by the rapid increase in the type and number of applications under current exploration. Most importantly, after many years of effort, the conceptual, theoretical, and experimental tools are finally coming together to show success in the laboratory on systems of true significance.

For relatively simple systems (e.g., those where only a few quantum levels are important for achieving the control objective), computational techniques should be capable of producing viable designs for laboratory execution. However, complex systems such as polyatomic molecules in the strong field

14

regime [c.f., the applications in Eq. (6)] are not currently amenable to reliable computational control field design. It is anticipated that the immediate period ahead will see a rapid expansion in the number of closed loop learning control experiments with molecules of ever more complexity, including some possibly of biological significance. Although computational design remains wanting in such applications, theoretical tools should become increasingly important as a means to create better laboratory algorithms and subsequently to analyze the mechanistic and physical behavior demonstrated in the experiments. The proper balance of theory, computation, and laboratory tools should draw on the best features of each of these techniques, in a synergistic fashion, to meet the control objectives.

References

1. Rabitz, H; de Vivie-Riedle, R; Motzkus, M; Kompa, K. *Science.* **2000**, *288*, 824-828; Rice, S.A; Zhao, M. *Optical Control of Molecular Dynamics*; John Wiley & Sons: New York, NY, 2000.
2. Assion, A.; Baumert, T.; Bergt, M.; Brixner, T., Keifer, B.; Seyfried, V.; Strehle, M.; Gerber, G. *Science.* **1998,** *282*, 919-922.
3. Levis, R.J.; Menkir, G.; Rabitz, H. *Science.* **2001**, *292*, 709-713.
4. Bartels, R.; Backus, S.; Zeek, E.; Misoguti, L.; Vdovin, G.; Christov, I.P.; Murnane, M.M.; Kapteyn, H.C. *Nature.* **2000**, *406*, 164-166.
5. Kunde, J.; Baumann, B.; Arlt, S.; Morier-Genoud, F.; Siegner, U.; Keller, U. *Appl. Phys. Lett.* **2000**, *77*, 924-926.
6. Omenetto, F.G.; Moores, M.D.; Luce, B.P.; Reitze, D.H.; Taylor, A.J. *Appl. Optics.* **2001**, in press.
7. Shi, S.; Rabitz, H. *J. Chem. Phys.* **1990**, *92*, 364-376.
8. Zhao, Y.; Kühn, O. *J. Phys. Chem. A.* **2000**, *104*, 4882-4888; Gross, P., Singh, H.; Rabitz, H.; Mease, K.; Huang, G.M. *Phys. Rev. A.* **1993**, *47*, 4593-4604.
9. Murtha, Z.; Rabitz, H. *Eur. Phys. J. D.* **2001**, *14*, 141-145.
10. Shen, H.; Dussault, J.P.; Bandrauk, A.S. *Chem. Phys. Lett.* **1994**, *221,* 498-506.
11. Judson, R.S.; Rabitz, H. *Phys. Rev. Lett.* **1992**, *68*, 1500-1503; Rabitz, H.; Shi, S. In *Advances in Molecular Vibrations and Collision Dynamics*; Bowman, J., Ed.; JAI Press Inc.; 1991; Vol. 1, Part A, pp. 187 – 214.
12. Geremia, J.M.; Zhu, W.; Rabitz, H. *J. Chem. Phys.* **2000**, *113*, 10841-10848.
13. Bardeen, C.J.; Yakovlev, V.V.; Wilson, K.R.; Carpenter, S.D.; Weber, P.M.; Warren, W.S. *Chem. Phys. Lett.* **1997**, *280*, 151-158.
14. Meshulach, D.; Silberberg, Y. *Nature.* **1998**, *396*, 239-242.
15. Bartels, R.; Backus, S.; Christov, I.; Kapteyn, H.; Murnane, M. *Chem. Phys.* **2001**, in press.

Acknowledgments

The author acknowledges support from the National Science Foundation and the Department of Defense, and would like to thank Robert Levis for his constructive comments on the chapter.

Chapter 2

Variations on the Theme of Stimulated Raman Adiabatic
Passagr: Control of Chemical Reactions

Suhail P. Shah, Vandana-Kurkal and Stuart A. Rice

**The Department of Chemistry and The James Franck Institute,
The University of Chicago, Chicago, IL 60637**

We describe the exploitation of adiabatic transfer of population between selected states of a molecule as a means for the control of product selectivity in a branching unimolecular reaction. The particular procedures employed to achieve that adiabatic transfer of population depend on the character of the spectrum of states of the molecule; the ones we consider are either extensions of, or are illuminated by, the STIRAP method. The results of studies of efficient selective population of one of a pair of nearly degenerate states in the thiophosgene molecule, and of efficient promotion of the HCN $\rightarrow$ CNH isomerization reaction, are reported. In the former case the adiabatic population transfer is achieved by use of the Kobrak-Rice extended STIRAP method, while in the latter case it is achieved by use of consecutive STIRAP processes. In both cases the efficiency of selective population transfer is shown to be insensitive to radiative coupling between background states and the subset of active states, and to radiative coupling between the background states. We also report the results of a study of adiabatic population transfer to one of a pair of degenerate states when the STIRAP condition (existence of a field-matter state with zero eigenvalue) cannot be satisfied by the subset of active states in the molecular spectrum of states. The criteria for efficient and selective population transfer in this case are discussed.

Introduction

Active control of product selectivity in a chemical reaction with optical fields, first suggested on the basis of theoretical arguments, has been experimentally demonstrated for several small molecule fragmentation and ionization reactions [1]. The key step in achieving active control is the exploitation of quantum interference effects to generate sensibly complete selective transfer of population between specified states of a molecule. Of the currently available methods for efficiently transferring population within a subset of quantum states selected from the full manifold of states of a molecule, those that rely on adiabatic passage are very appealing because they satisfy important criteria for experimental implementation. For example, the stimulated Raman adiabatic passage (STIRAP) process [2] has been shown to be an efficient means of generating population transfer in three state and multi-state systems. An important feature of the STIRAP process is that the "counter-intuitive" ordering of pulses used (see below) is remarkably robust with respect to variation in the overlap between the pulses and to their respective shapes, provided the criteria for adiabatic passage are met.

The STIRAP process involves the creation of stationary states of a coupled coherent field-matter system. In its simplest form this scheme relies on two suitably timed coherent pulses of light coupling a three state system that has non vanishing transition moments that connect state $|1>$ with state $|2>$ and state $|2>$ with state $|3>$, but has zero transition moment between states $|1>$ and $|3>$. The fields associated with the pump pulse Ω_p that couples the initial state $|1>$ to the intermediate state $|2>$, and the Stokes pulse Ω_s that couples $|2>$ to the final state $|3>$, are large enough to generate many cycles of Rabi oscillation between $|1>$ and $|2>$, and between $|2>$ and $|3>$. The coherent field-matter eigenstates can be represented as linear superpositions of the states $|1>$, $|2>$ and $|3>$, and the system wavefunction as a linear combination of the coherent field-matter eigenstates. Selective population transfer from state $|1>$ to state $|3>$ is generated by varying the contributions to the system wavefunction of the coherent field-matter eigenstates; this is accomplished through control of the ratio of Rabi frequencies. Adiabatic transfer of population from state $|1>$ to state $|3>$ accompanies this variation when there is counterintuitive ordering of the pulses, with the Stokes pulse preceding but overlapping the pump pulse. Criteria for variation of the Rabi frequencies that satisfy the condition of adiabatic transfer of population can be found in the literature; they contain constraints on both the strengths and the rates of change of the Stokes and pump fields.

Several extensions of the original STIRAP method of population transfer have been reported [3-11]. Thus, it is now possible to design efficient population transfers between initial and final states in systems in which the intermediate states are autoionizing or replaced by unstructured continua. It is now accepted that in a ladder-like distribution of energy states, provided the lifetimes of the target states are longer than the pulse durations, complete population transfer from an arbitrary initial state to a selected target state is possible using a suitable set of overlapping pulses. And, it is possible to use an extended STIRAP method for selective population transfer from an initial state to one of two degenerate final states, which thereby permits control of product selectivity in a branching unimolecular reaction.

There are several different ways of implementing optical control of molecular dynamics, all derived from the same basic principles. Although these method are independent of the size of the system to be controlled, accurate prediction of the character of the control field becomes more and more difficult as the number of degrees of freedom increases. For example, the multidimensionality and complicated topography of the potential energy surfaces of a large polyatomic molecule makes it very difficult to use pulse timing control of product selection in a unimolecular reaction because the evolution of a wavepacket on a complicated potential energy surface is difficult to visualize and follow. The multipath interference control of product selection in a unimolecular reaction of a large polyatomic molecule becomes more difficult as the number of states per unit energy interval increases because identifying the suitable pairs of pathways between the same initial and the final states requires detailed knowledge of the character of all the states involved. Hence it is desirable to have a reduced states formalism of control process. Amongst its other attributes, we regard STIRAP and extended STIRAP methods of control of population transfer to be examples of reduced states descriptions of molecular dynamics.

In this paper we briefly discuss three issues:
(1) How sensitive is the selectivity of adiabatic population transfer between specified initial and final states to the presence of background states that are radiatively coupled to the subset of active states and are radiatively coupled to each other?
(2) Can adiabatic population transfer be used for control of product formation in an isomerization reaction involving large atomic displacements?
(3) Can adiabatic population transfer be used to selectively populate one of a pair of degenerate states when the conditions for STIRAP are not met?

To address (1) we have examined the Kobrak-Rice [10,11] extended STIRAP method for selective population of one of a pair of nearly degenerate states in thiophosgene, taking account of radiative transitions between the active subset of states and a large set of background states, and radiative transitions between the background states. The selectivity of population transfer is shown to be insensitive to the presence of those background states. To address (2) we have studied the HCN $\rightarrow$ CNH isomerization reaction. The use of straightforward STIRAP excitation with the ground vibrational state of HCN the initial state, a state just above the barrier to isomerization the intermediate state, and the ground vibrational state of HNC the final state is not feasible, as it involves very high intensity pulses due to the very small transition dipole moments between the states mentioned. However, the use of successive STIRAP excitations to generate the product CNH is feasible, and is shown to be very efficient even when account is taken of radiative transitions between the active subset of states and a large set of background states, and radiative transitions between the background states. To address (3) we have studied selective adiabatic population transfer to one of a pair of degenerate states in a subsystem of states with the following characteristics: The ground state $|1\rangle$ is radiatively coupled to two degenerate states, $|2\rangle$ and $|3\rangle$, and these are in turn radiatively coupled to a fourth (branch) state $|4\rangle$. Conditions for complete transfer of population from state $|1\rangle$ to state $|2\rangle$ or state $|3\rangle$ are determined for a wide range of ratios of the transition moments between the states. It is shown that both selectivity and efficiency of population transfer can be achieved even when the STIRAP condition is not satisfied.

Effect of background coupling on the selectivity of population transfer in thiophosgene

The thiophosgene molecule ($SCCl_2$) has three stretching ($v_1v_2v_3$) and three bending ($v_4v_5v_6$) vibrational degrees of freedom. The vibrational spectrum has been very carefully studied by Gruebele and co-workers [12]. Consequently the frequencies, deviations from harmonic behaviour and the energies of the states are known, which facilitates the calculation of transition dipole moments between all states up to 21000 cm^{-1} above the vibrationless ground state. The data we used for the energies and the vibrational states were taken from reference 12.

The Kobrak-Rice extended STIRAP technique for controlled population transfer is based on a subset of five states: an initial state, an intermediate state, a branch state and a pair of degenerate target states. The initial state and the intermediate states are coupled by a resonant pump pulse and a Stokes pulse resonantly couples the intermediate state and the degenerate target states. A third pulse, longer than the pump and Stokes pulses and on throughout their duration, resonantly couples the degenerate target states and the branch state. These active states are represented by thick lines Fig. 1. The rotating wave approximation to the Hamiltonian for this system of five states and three fields is

$$\mathbf{H} = \frac{1}{2}\begin{pmatrix} 0 & \Omega_p & 0 & 0 & 0 \\ \Omega_p & 0 & \Omega_{S(3)} & \Omega_{S(4)} & 0 \\ 0 & \Omega_{S(3)} & 0 & 0 & \Omega_{b(3)} \\ 0 & \Omega_{S(4)} & 0 & 0 & \Omega_{b(4)} \\ 0 & 0 & \Omega_{b(3)} & \Omega_{b(4)} & 0 \end{pmatrix}$$

where $\Omega_p = 2\pi\mu_{12}\varepsilon_p(t)/h$, $\Omega_{S(i)} = 2\pi\mu_{2i}\varepsilon_S(t)/h$ and $\Omega_{b(i)} = 2\pi\mu_{5i}\varepsilon_b(t)/h$ are the time dependent Rabi frequencies coupling the states indicated. The field intensities chosen to provide sufficient splitting of the bare states manifold to

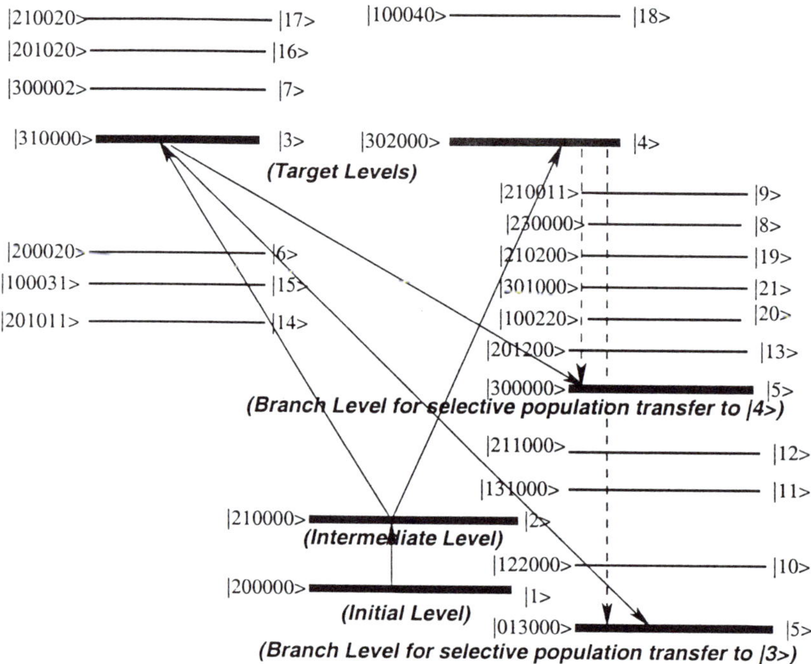

Figure 1 : Schematic diagram of all the vibrational states of thiophosgene molecule considered in the present work.

facilitate the adiabatic transfer of population from the ground state to the selected target state are shown below

	Intensity (W/cm^2)	t_0 (ps)	FWHM (ps)
Pump pulse	1.21×10^9	85	85
Stokes pulse	1.46×10^8	0	143
Branch pulse	9.30×10^{11}	215	215

The Schrödinger equation was solved in a basis of bare matter eigenstates using a fourth order Runge-Kutta integrator. The results [13] indicate that almost 100% of the population of the initial state is transferred to the desired target state using this subset of five states. It is important to note that the vibrational spectrum of thiophosgene is rich in the energy range from which the five levels are chosen. We now show that the selectivity in the population transfer to the desired target states is not affected by radiative coupling of the subset of active states to the background states and radiative coupling between the background states. The background states considered for this work are displayed as light lines in Fig. 1. The selective population transfer achieved with counterintuitive and intuitive ordering of the pump and Stokes pulses is displayed in Figs. 2a and 3b as a function of detuning of the branch frequency.

As expected, our results show that the ratio of populations transferred to the target states with counterintuitive excitation has an inverse dependence on the ratio of the magnitudes of the transition moments connecting the branch state and the target states. Our calculations also show that the selectivity and magnitude of the population transfer to a target state are sensibly independent of the nonresonant coupling with the background states. The effect of detuning of the branch frequency from resonance is of the same magnitude as when the background states are not considered for the case of counterintuitive ordering of the pump and Stokes pulses. When intuitive ordering of the pump and Stokes pulses is employed, there is an oscillatory dependence of the population transfer on detuning, and a reduction in the absolute yield in the target state. These oscillations in the population arise from interference effects.

Successive STIRAP excitation control of the HCN $\rightarrow$ CNH reaction

Active control of product formation in an isomerization reaction is usually very difficult because of the occurrence of large atomic displacements in the reaction and because there is non negligible non-linear coupling of the molecular degrees of freedom along the reaction path. However, because of these complications, isomerization reactions provide a useful vehicle for testing the control fields calculated using different reduced representations of molecular dynamics. We have used the HCN $\rightarrow$ CNH isomerization for this purpose, since the path of the

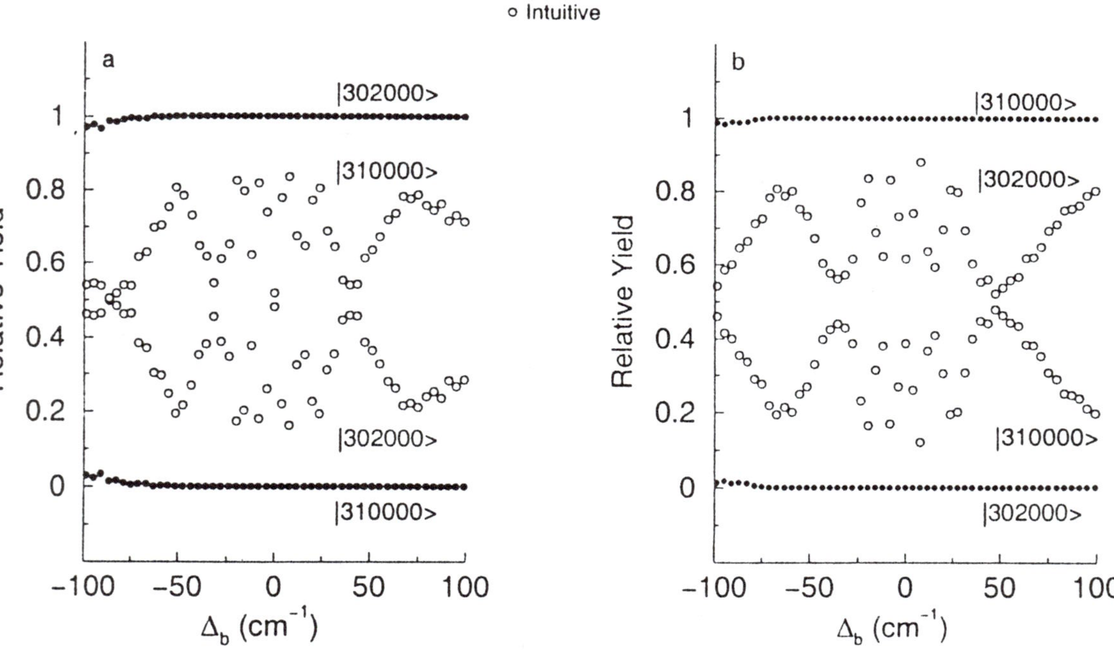

Figure 2 : (a) Variation in the population transfer to target states |310000> and |302000> with |300000> as the branch state, calculated with inclusion of all of the radiative couplings between the 21 states shown in Fig. 1 ; (b) Same as (a), but with |013000> as the branch state.

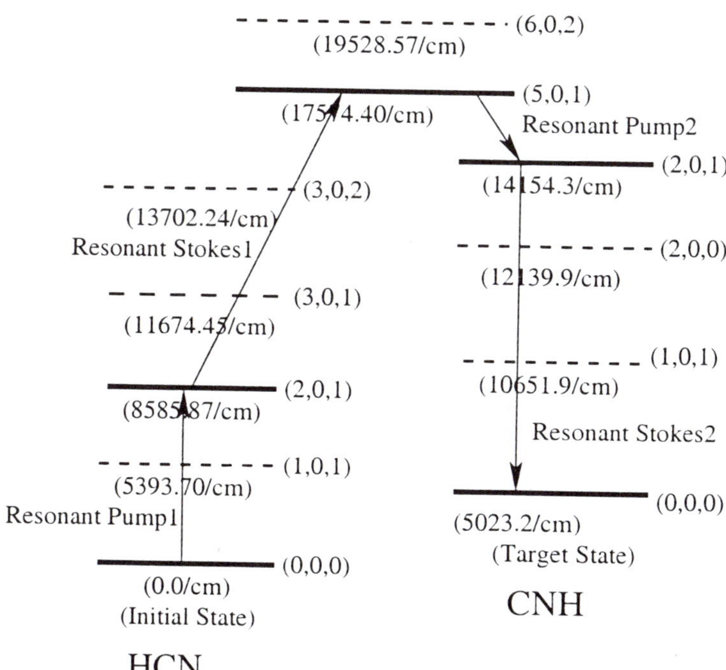

Figure 3 : A schematic diagram of the vibrational states of HCN and CNH considered in this work.

reaction is simple (largely determined by the bending motion). The barrier to this isomerization reaction is large, with the obvious consequence that the Franck-Condon factors that connect the ground vibrational state of HCN with its vibrationally excited states decrease rapidly with increasing vibrational excitation and are sensibly zero for a transition from the ground vibrational level to a delocalized state of the molecule with energy just above the barrier. Hence the population in the ground vibrational state of HCN cannot be easily transferred to the ground vibrational state of CNH by, say, delayed pulses (as in the Tannor-Rice control method). A different approach to controlling the isomerization is based on the use of successive vibrational transitions. One possible path of that type is invoked when an optimal shaped field is used to maximize the formation of CNH. The optimal field is found to generate successive transitions up and down the ladder of vibrational states of HCN and CNH respectively. Shah and Rice [14,15] reported a study of the accuracy of an optimal field for the HCN $\rightarrow$ CNH isomerization calculated using the Zhao–Rice reduced space representation [16]. They showed that at each level of approximation to the reaction path (determined by the number of degrees of freedom included in the calculation), an optimal field that generates high product yield can be found. However, the field calculated at a particular level of approximation does not give any insight for the calculation of the optimal field at the next level of approximation, or to the optimal field when all the degrees of freedom of the HCN molecule are taken into account.

As already indicated in the Introduction, the use of straightforward STIRAP excitation to control the isomerization reaction by using the ground vibrational state of HCN for the initial state, a state just above the barrier to isomerization for the intermediate state, and the ground vibrational state of HNC for the final state is not feasible, as it involves very high intensity pulses [17] due to the very small transition dipole moments between the states mentioned. We now show that an alternative approach, that uses successive STIRAP excitations to generate the product CNH, is feasible and very efficient.

The three dimensional potential energy surface for non rotating HCN/CNH is very well studied [18,19]. The molecular degrees of freedom (dominated by the C-H, N-H, C-N stretching motions and the CNH bending motion) are strongly coupled. Experimental values for the vibrational energy levels of HCN used in the present study were taken from reference 18. The transition dipole moments associated with the transitions between these states were obtained from the IR intensities of the vibrational transitions reported by Bostchwina *et al* [19].

We seek two successive efficient transfers of population, $g \rightarrow i_1 \rightarrow i_2$ and $i_2 \rightarrow i_3 \rightarrow f$, that have the property that the transfer of population from the ground state of HCN to the ground state of CNH is sensibly complete without the use of excessively large Stokes and pump fields. We refer to the first set of pulses as pump1 (p1) and Stokes 1 (S1), and to the second set of pulses as pump 2 (p2) and Stokes (S2). The states that we considered for the successive STIRAP excitations are shown in Fig. 4.

The first STIRAP process is associated with the states HCN states $g \equiv (0,0,0)$ and $i_1 \equiv (2,0,1)$ and the delocalized state $i_2 \equiv (5,0,1)$. The Stokes 1 pulse is in resonance with the transition between $(2,0,1)$ of HCN and the delocalized state $(5,0,1)$; the pump 1 pulse is resonant with the transition $(0,0,0) \rightarrow (2,0,1)$.

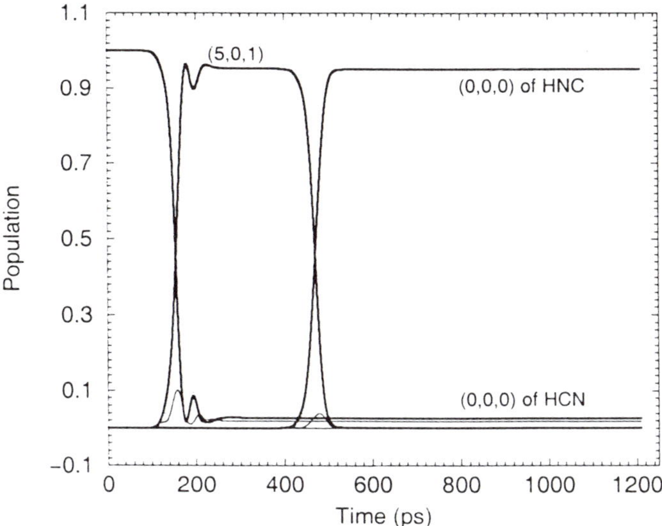

Figure 4 : Variation of population with respect to time in the presence of background states

The second set of pulses involves transitions between the delocalized state $i_2 \equiv$ (5,0,1), and the CNH states $i_3 \equiv$ (2,0,1) and $f \equiv$ (0,0,0). The Stokes 1 pulse precedes the pump1 pulse and, after the population transfer to state i_2, the Stokes 2 pulse and the pump 2 pulse are applied.

The Hamiltonian for this system is, using the rotating wave approximation,

$$\mathbf{H} = \frac{1}{2}\begin{pmatrix} 0 & \Omega_{p1} & 0 & 0 & 0 \\ \Omega_{p1} & 0 & \Omega_{S1} & 0 & 0 \\ 0 & \Omega_{S1} & 0 & \Omega_{p2} & 0 \\ 0 & 0 & \Omega_{p2} & 0 & \Omega_{S2} \\ 0 & 0 & 0 & \Omega_{S2} & 0 \end{pmatrix}$$

Ω_{p1}, Ω_{S1}, Ω_{p2} and Ω_{S2} are the Rabi frequencies associated with the pump 1, Stokes 1, pump 2 and Stokes 2 pulses, respectively. The field parameters associated with these pulses are shown below.

	Intensity (W/cm^2)	t_0 (ps)	Frequency (a.u.)	FWHM (ps)
Stokes 1	8.40×10^{11}	133	0.04095	85
Pump 1	9.30×10^{11}	194	0.03912	85
Stokes 2	9.30×10^{10}	423	0.04160	85
Pump 2	3.38×10^{11}	484	0.01558	85

The states considered in the above discussion are now coupled to a set of background states taken from the manifold of vibrational states. The radiative couplings between all the states (shown in Fig. 3) are incorporated in our study the effect of background coupling on the conversion of HCN to CNH. The variation of population with respect to time in the presence of the two sets of pulses is shown in Fig. 4.

We find, from this calculation, that 95.06% of the population initially in the (0,0,0) state of HCN is transferred to the (0,0,0) state of CNH with negligible transfer to the intermediate state and to the background states. Based on these results, we infer that even in the presence of a substantial number of background states that are radiatively coupled to the active states of the subset and to each other, very efficiently transfer of population to a selected target state is possible.

Optimized Coherent Population Transfer to Degenerate Final States in Multi-Level Systems

The efficiency and completeness of the population transfer generated by a STIRAP process suggests that it may be related to population transfer generated by an optimal control process. Although the efforts to show that the STIRAP sequence of Stokes and pump pulses arise naturally from optimal control theory have been unsuccessful thus far, there are intriguing indications of a theoretical analog between the two methodologies. For example, Malinovsky and Tannor [5] have shown that for N-state systems with a ladder-like distribution of states, the STIRAP two pulse field can be obtained from an optimization algorithm. Using a variant of optimal control theory, they studied the effect of "local optimization" of an initial field guess to maximize the ratio of the population of a selected target state to that of a fully populated initial state. The control field obtained exactly matched the STIRAP process field, which in the N-level case consists of $N-1$ overlapping pulses, each resonant with transitions between adjacent states of the distribution of states.

This section addresses the question whether optimization algorithms that, in principle, yield the set of all control fields that guide the transfer of population from the initial state to the target state, are able to generate optimal fields in systems in which the final states are degenerate. In particular, we shall focus our attention on a simple four-level system for which the conditions for a STIRAP process are not met. Specifically, we examine a four state system of states with a "diamond" arrangement of nonzero transition moments [20]. This system has a manifold of dressed eigenstates that does not contain a state whose projections to intermediate states are identically zero.

The optimization scheme we use is derived from techniques developed previously in the theory of control of product formation in a chemical reaction. Consider a system with a pair of degenerate target states, $|\xi\rangle$ and $|\eta\rangle$, embedded within a subsystem of N states. These states are radiatively coupled by M ($M < N$) radiation fields. In the interaction representation, the Hamiltonian for the system represents the interaction energy of the field-matter system, and the effect of the field is fully characterized by the set of Rabi frequencies of the radiation fields, $\{\Omega\}$ and their respective detunings from resonance $\{\Delta\}$. Assuming the field interacts with the system between an initial time $t = 0$ and final time t_{max}, the condition for optimized population transfer may be written as

$$\nabla_{\{\Omega\},\{\Delta\}}\left[|a_\xi(t)|^2 - |a_\eta(t)|^2 - \sum_{i=1}^{N-2}|a_i(t)|^2\right]\Bigg|_{t=t_{max}} = 0$$

where $|a_\xi(t)|^2$ denotes the time-dependent population of the desired target state, $|a_\eta(t)|^2$ denotes the time-dependent population of the state degenerate to the target state, and $\sum_{i=1}^{N-2}|a_i(t)|^2$ is the sum of populations of other states in the system. The computational algorithm performs a gradient search for the incident fields that maximize the population of the target state while simultaneously minimizing the population of the other states in the system [13,14], where the populations taken are those at the end of the interaction with the field.

The four state system we have examined is shown in Fig. 5. In a molecule, these states may be an initially fully populated ground vibrational state of the ground electronic potential energy surface, a pair of degenerate vibrational (target) states near the continua of different exit channels of an electronically excited potential energy surface, and a fourth (branch) state which may be a ro-vibronic excited state. A pump pulse radiatively couples the ground state to the two degenerate states and a Stokes pulse radiatively couples the two degenerate states to the branch state. The Schrödinger equation for the four state system may then be written in matrix form as

$$\begin{bmatrix} \dot{a}_1(t) \\ \dot{a}_2(t) \\ \dot{a}_3(t) \\ \dot{a}_4(t) \end{bmatrix} = \frac{1}{2} \begin{pmatrix} 0 & i\alpha\Omega_p e^{-i\Delta_p t} & i\beta\Omega_p e^{-i\Delta_p t} & 0 \\ i\alpha\Omega_p e^{i\Delta_p t} & 0 & 0 & i\gamma\Omega_p e^{-i\Delta_s t} \\ i\beta\Omega_p e^{i\Delta_p t} & 0 & 0 & i\delta\Omega_p e^{-i\Delta_s t} \\ 0 & i\gamma\Omega_p e^{i\Delta_s t} & i\delta\Omega_p e^{i\Delta_s t} & 0 \end{pmatrix} \begin{bmatrix} a_1(t) \\ a_2(t) \\ a_3(t) \\ a_4(t) \end{bmatrix}$$

The differences between the transition moments and the corresponding differences in the Rabi frequencies have been subsumed into four parameters α, β, δ and γ that play the role of normalized transition dipole moments; they are multipliers which permit representation of the several Rabi frequencies connecting states $|1\!>$ and $|2\!>$, $|1\!>$ and $|3\!>$, $|2\!>$ and $|4\!>$ and $|3\!>$ and $|4\!>$ in the form $\alpha\Omega_p$, $\beta\Omega_p$, $\gamma\Omega_s$ and $\delta\Omega_s$, respectively, subject to the constraints $\alpha^2 + \beta^2 = 1$ and $\delta^2 + \gamma^2 = 1$.

We seek to selectively populate one of the two degenerate target states using relatively simple pulses. To be useful, the population transfer scheme must be efficient even when the initial state and the target state are coupled by an extremely small transition dipole moment. In our calculations we used a variety of different ratios of transition dipole moments connecting the ground state to each of the two degenerate target states. The field is optimized to maximize the population in the selected target state while simultaneously minimizing the population in the other target state and the branch state. For a two-pulse field, the Rabi frequencies of the pump and Stokes pulses are unknowns. The optimization criterion leaves a large number of combinations of Rabi frequencies which achieve the population transfer goal [20]. We find that it is sufficient to fix the shapes of the pump and Stokes pulses to be Gaussian with optimizable time delays. This choice provides a direct comparison with the locally optimized fields, studied previously, as well as being simpler to generate in the laboratory. In our calculations we assumed the Gaussian envelope of the

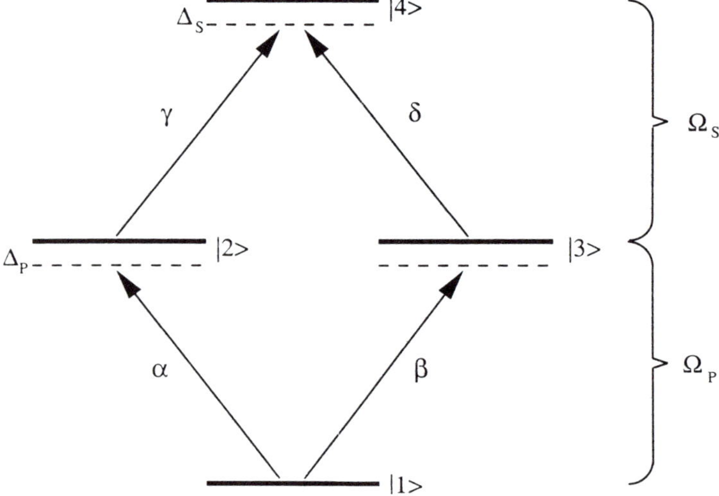

Figure 5 : Four-level diamond system with pump and Stokes pulses. The dipole moment factors α, β, γ and δ are related by $\alpha^2 + \beta^2 = 1$ and $\gamma^2 + \delta^2 = 1$.

α	β	γ	δ	Optimal Target Yield	$t_p\text{-}t_s$
1.0000	0.0100	0.6690	0.7433	0.9995	1.270
0.1961	0.9806	0.1961	0.9806	0.9610	-2.560
0.9806	0.1961	0.1961	0.9806	0.9806	-0.810
0.1961	0.9806	0.9806	0.1961	0.5359	-3.290
0.9806	0.1961	0.9806	0.1961	0.0384	-2.560

Rabi frequencies of each pulse to have a maximum at 32GHz, which represents the minimum required for the algorithm to converge upon the optimum solution. Our calculations have covered a considerable range of values of the ratios α/β and γ/δ. The results are summarized in the table below.

In the case that α/β = 100/1 and γ/δ = 9/10, the target state is practically "dark" with respect to transitions from the initial state. The "counter-intuitive" ordering of pulses ($t_p - t_s > 0$) when full transfer of population to a target state occurs bears a resemblance to the mechanism of population transfer via adiabatic passage. However, the population dynamics indicate that with the optimized pulses incident on the system, there are often large fluctuations of the populations of states |2> and |4> which eventually vanish at the end of the interaction with the field. The level-switching behavior of STIRAP [2] is clearly absent here.

In the case where α/β = 1/5 and γ/δ = 1/5, the optimal sequence of pulses is "intuitive", with the pump pulse preceding the Stokes pulse, and with a significant overlap between them. The large population transferred is a consequence of the fact that the magnitudes of the transition dipole moments connecting the target state to the initial and branch states are much larger than those that connect state |2> with the initial and branch states. Conversely, when α/β = 5/1 and γ/δ = 5/1, the maximum population transferable is found to be no greater than about 4 percent. Again, the optimal sequence of pulses is in the intuitive order, indicating that the mechanism of population transfer resembles a stimulated emission pumping from the initial to the final states.

When α/β = 1/5 and γ/δ = 5/1, we find that the optimal field depends only on the magnitude of the difference between the centers of the pump and Stokes pulses, not their sign. In other words, the *ordering of the pulses is immaterial.* The optimum difference in time of the centers of the pulses listed in the table effects the same population transfer irrespective of whether the pump pulse precedes the Stokes pulse or vice versa [20]. While in this case an optimized pair of pulses transfers nearly 50 percent of the population into the target state, one may opt to use a single pump pulse resonant between the initial state and the two degenerate states to exploit the larger dipole moment transition between the initial state and the final state. Conversely, when α/β = 5/1 and γ/δ = 1/5, the optimum sequence of pulses almost fully populates the target state at the end of the interaction with the pulses. Once again, we find that the factor determining the efficiency of population transfer is the magnitude of the time delay between the pump and Stokes pulses, not their order.

The results we have obtained indicate that in cases where the target state has a close to vanishing dipole moment transition with the initial state, Complete (or nearly complete) population transfer is still achievable provided one finds a suitable branch state in the system that has a dipole moment transition to the target state which is slightly greater than the dipole moment transition to the other degenerate state. Within the range of dipole moments studied, in all but one case one may find a suitable sequence of pulses which may be generated by the experimentalist that effects a population transfer which is STIRAP-like in its efficiency, but different from STIRAP in the dynamics of the transfer process.

Conclusions

One class of methods for overcoming the difficulty that arises in achieving active control of the molecular dynamics of a polyatomic molecule, arising from the dependence of the evolution of a state of the molecule on a large number of coupled degrees of freedom, is based on taking advantage of adiabatic transfer of population between states. This can be accomplished by use of a STIRAP, modified STIRAP, or sequence of STIRAP processes, as well as by use of locally optimized fields. The method of choice depends on details of the structure of the manifold of states of the molecule.

Acknowledgements

The research was supported by a grant from the National Science Foundation (CHE-9807127)

References

1. For a review of the theoretical and experimental status of the field, see; Rice, S. A; Zhao M. *Optical Control of Molecular Dynamics ;* Wiley : New York **2000.**
2. For a review of STIRAP, see: Bergmann, K.; Shore, B. W. *Molecular Dynamics and Spectroscopy by Stimulated Emission Pumping*; Dai, H.C.. Field. R. W., Eds,; World Scientific: Singapore, **1995.**
3. Gaubatz, U.; Rudecki, P.; Schiemann, S.; Bergmann, K. *J. Chem. Phys.* **1990**, 44, 7442
4. Shore, B. W.; Bergmann, K.; Oreg, J.; Rosenwaks, S. *Phys. Rev. A* **1992**, 45, 4888.
5. Malinovsky, V.S,; Tannor, D. J. *Phys. Rev. A* **1997**, 56, 4929.
6. Coulston, G. W.; Bergmann, K. *J. Chem. Phys.* **1992**, 96, 3467.
7. Shore, B. W.; Martin, J.; Fewell, M. P.; Bergmann., K. *Phys. Rev. A* **1995**, 52, 583.
8. Martin, J.; Shore, B. W.; Bergmann, K. *Phys. Rev. A* **1995**, 52, 583.
9. Martin, J.; Shore, B. W.; Bergmann, K. *Phys. Rev. A* **1996**, 54,1556.

10. Kobrak, M.; Rice, S. A. *Phys. Rev. A* **1998**, 57, 2885.
11. Kobrak, M.; Rice, Ś. A. *J. Chem. Phys.* **1998**, 109, 1.
12. Bigwood, R.; Milam, B.; Grucbele, M.; *Chem. Phys. Lett.* **1998**, 287, 333.
13. Kurkal, V.; Rice, S. A. *J. Phys. Chem.* **2000**, xxx.
14. Shah, S. P.; Rice, S. A. *Faraday Transactions*, **1999**, 113, 319.
15. Shah, S. P.; Rice, S. A. *J. Chem. Phys.* **2000**, 113, 6536.
16. Zhao, M.; Rice, S. A. J. W. Hepburn Ed., **1994**, SPIE 2124, 246.
17. Kurkal, V.; Rice, S. A. *Chem. Phys. Lett* (submitted).
18. Bowman, J. W.; Gazdy, B.; Bentley, J. A.; Lee, T. J.; Dateo C. E.; J. *Chem. Phys* **1993**. 99. 308
19. Botschwina, P.; Schulz, B.; Horn, M.; Matuschewski, M.; *Chem. Phys* **1995**, 190, 345
20. Shah, S.P.; Rice, S.A.; (to be submitted)

Chapter 3

Laser Control of Selective Preparation of Preoriented Enantiomers from Their Racemate

K. Hoki and Y. Fujimura

Department of Chemistry, Graduate School of Science, Tohoku University, Sendai 980–8578, Japan

We present a simple treatment of laser control of selective preparation of pre-oriented enantiomers from an equal mixture of R- and L- forms. The control is carried out by using a linearly polarized laser whose polarization direction is determined in a such a way that the laser interacts with L- (R-) enantiomers but not with R-(L-) enantiomers. A four-state model for an enantiomer with one-dimensional, double-well potential is adopted to derive an analytical expression for the time evolution of racemate. The effects of the laser field intensity on the selectivity of enantiomers are discussed. To demonstrate the effectiveness of the simplified treatment of laser control of selective preparation of enantiomers, we considered dynamic chirality control of H_2POSH in their racemate as a model.

Selective preparation of enantiomers is one of the major issues in laser control of chemical reaction dynamics (1, 2). Since the publication of a theoretical paper by Shapiro and Brumer (3), much interest has been shown in the possibility of selective preparation of enantiomers by using coherent properties of lasers (4 - 7), and several types of control scenario have been proposed (8 - 15). The development of quantum control of enantiomers will give a fundamental guide for other interesting fields of research, such as research on the design of effective molecular motor (16). One of the interesting issues in the selective preparation of enantiomers is how to prepare pure enantiomers of R-(L-) form from their racemate, i.e., an equal mixture of the two types of enantiomers. In a previous paper (13), we presented a method for the design of laser fields. An expression for laser pulses for selective preparation was derived based on a locally optimized control theory in the density operator formalism.

In this paper, we present a simple treatment that is an alternative approach to laser control of pre-oriented enantiomers in their racemate. This treatment is not based on optimal control theory. The most important factor in this treatment is the photon polarization direction, which is determined from constraint of optical transitions of R (L)-enantiomers: one of the optical transition paths, i.e., R- or L-enantiomers is optically active, and the other is inactive. It is important to note that pre-orientation of enantiomers is a prerequisite for laser-control of enantiomers within a dipole interaction approximation (13). Orientation of molecules in vapor phase is, in principle, realized by optimal control theory (17,18). We adopted a four-state model to obtain an analytical expression for time-evolution of the system. The simplified treatment is applied to dynamic chirality control of pre-oriented H_2POSH, whose geometrical structure and dipole moment functions were evaluated by an ab initio MO method (9). The results are compared with those previously obtained by using the locally designed pulse- shaping method.

Theoretical

Four-state Model

Consider pre-oriented enantiomers in a mixture of their racemates. We carry out laser control of a selective preparation of the R-form (L-form) from their racemates by using lasers. The reaction potential of the selective preparation is assumed to be characterized by a one-dimensional double-well as shown in Fig. 1 (19). The lowest vibrational state (eigen ket, $|0_+\rangle$) and the first excited state ($|0_-\rangle$) are almost degenerated with energy ε_0 because of existing of a high potential barrier. Here, suffix plus (minus) denotes symmetric (anti-symmetric) with respect to the chiral plane that divides enantiomers between L- and R-forms. We take into account two excited states; for example, the second excited state ($|1_+\rangle$) with energy $\varepsilon_1 - \delta$ and the third excited state ($|1_-\rangle$) with $\varepsilon_1 + \delta$. Here 2δ is the separation between these two states. Such a four-state model is commonly used to carrying out laser-selective preparation of enantiomers in a ground electronic state.

Assuming that the energy difference between $|0_+\rangle$ and $|1_+\rangle$ ($|1_-\rangle$) is much larger than kT, where k is Boltzmann constant and T is temperature, the initial density operator $\rho(t_0)$ is expressed in the case of a low temperature limit as

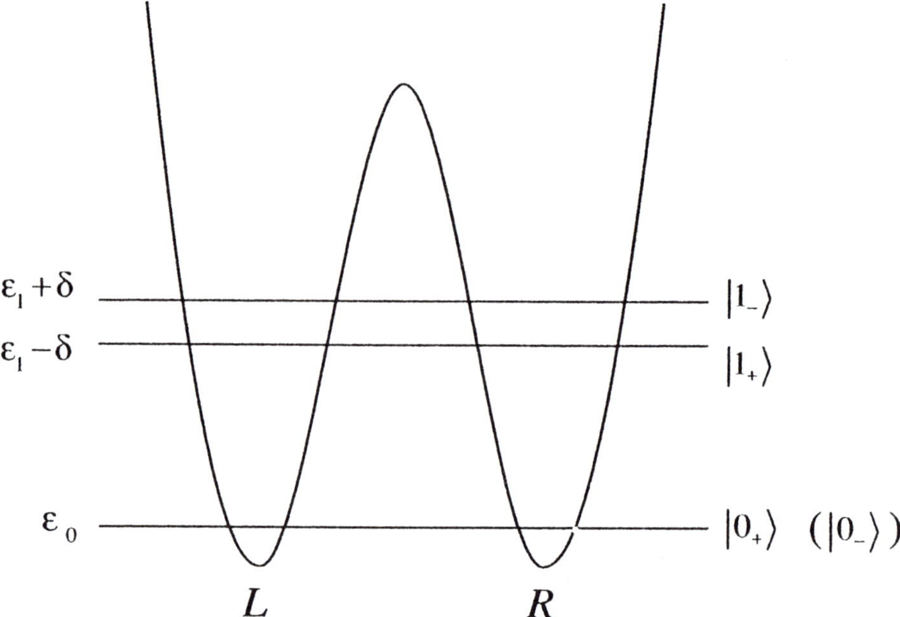

Figure 1. A double-well potential for an enantiomer selective preparation. R (L) denotes the position of the R (L)-enantiomer. A four-state model is also shown.

$$\rho(t_0) = |0_+\rangle \frac{1}{2} \langle 0_+| + |0_-\rangle \frac{1}{2} \langle 0_-|. \tag{1}$$

Two localized states, $|v_R\rangle$ and $|v_L\rangle$ ($v = 0$ and 1), which are localized into a well in the right-hand side and in the left -hand side, are expressed as

$$|v_R\rangle = \frac{1}{\sqrt{2}}\left(|v_+\rangle - |v_-\rangle\right) \tag{2a}$$

and

$$|v_L\rangle = \frac{1}{\sqrt{2}}\left(|v_+\rangle + |v_-\rangle\right), \tag{2b}$$

respectively.

The initial density operator, eq (1), can be rewritten in terms of the localized basis set, eq (2), as

$$\rho(t_0) = |0_R\rangle \frac{1}{2} \langle 0_R| + |0_L\rangle \frac{1}{2} \langle 0_L|. \tag{3}$$

Equation (3) expresses a racemic mixture.

We now specify the target operator for enantiomer selection. As already described in Ref. (13) in detail, the eigenvalues of $\rho(t)$, which are statistical weights, are invariant when the time evolution of $\rho(t)$ is described in terms of a unitary process. Therefore, the final state can be also described in terms of a one-to-one statistical mixture. Let the target population be localized in one of the wells, e.g., to produce the R-form of the enantiomers. In this study, we chose $|0R> \frac{1}{2} <0R| + |1R> \frac{1}{2} <1R|$ as the target operator.

The Hamiltonian of the total system, $H(t)$, is expressed within the semi-classical treatment of the radiation interaction with matter as

$$H(t) = H_0 - \vec{\mu} \cdot \vec{E}(t). \tag{4}$$

Here, H_0 is the molecular Hamiltonian, $\vec{\mu}$ is the dipole moment, and $\vec{E}(t)$ is the laser field:

$$\vec{E}(t) = 2\vec{A}(t)\cos(\omega t),\tag{5}$$

where $\vec{A}(t)$ is the pulse envelope with a photon-polarization vector, and ω is the carrier frequency.

The Hamiltonian matrix, $H(t)$, is given in the eigenstate representation as

$$H(t) = \begin{pmatrix} \varepsilon_0 & 0 & -\langle 0_+|\vec{\mu}|1_+\rangle\cdot\vec{E}(t) & -\langle 0_+|\vec{\mu}|1_-\rangle\cdot\vec{E}(t) \\ 0 & \varepsilon_0 & -\langle 0_-|\vec{\mu}|1_+\rangle\cdot\vec{E}(t) & -\langle 0_-|\vec{\mu}|1_-\rangle\cdot\vec{E}(t) \\ -\langle 1_+|\vec{\mu}|0_+\rangle\cdot\vec{E}(t) & -\langle 1_+|\vec{\mu}|0_-\rangle\cdot\vec{E}(t) & \varepsilon_1-\delta & 0 \\ -\langle 1_-|\vec{\mu}|0_+\rangle\cdot\vec{E}(t) & -\langle 1_-|\vec{\mu}|0_-\rangle\cdot\vec{E}(t) & 0 & \varepsilon_1+\delta \end{pmatrix}\tag{6}$$

where terms such as $\langle 1_+|\vec{\mu}|0_+\rangle$ are the matrix elements of the dipole moment operator $\vec{\mu}$.

The matrix elements in the localized basis set are expressed in a good approximation as

$$\langle 1_L|\vec{\mu}|0_R\rangle = 0 \text{ and } \langle 1_R|\vec{\mu}|0_L\rangle = 0.\tag{7}$$

From eq (7), we can see the following relation between the dipole matrix elements in the eigenstate representation :

$$\langle 1_+|\vec{\mu}|0_+\rangle = \langle 1_-|\vec{\mu}|0_-\rangle \text{ and } \langle 1_+|\vec{\mu}|0_-\rangle = \langle 1_-|\vec{\mu}|0_+\rangle.\tag{8}$$

Determination of the Polarization Direction of the Laser

The ability to selectively prepare pre-oriented enantiomers from their racemate depends on the polarization direction of laser field. This can easily be seen from eq (8). For example, enantiomers of the L-form are optically active between the lowest state $|0_L\rangle$ and excited states $(|1_+\rangle, (|1_-\rangle)$ while enantiomers of the R-form in the lowest state are inactive if the following condition is satisfied:

$$-\langle 1_+|\vec{\mu}|0_+\rangle\cdot\vec{A}(t) = -\langle 1_+|\vec{\mu}|0_-\rangle\cdot\vec{A}(t) \equiv \hbar\Omega(t).\tag{9}$$

This expression is equivalent to

$$\langle 1_+|\vec{\mu}|0_R\rangle\cdot\vec{A}(t) = 0 \text{ and } \langle 1_+|\vec{\mu}|0_L\rangle\cdot\vec{A}(t) = -\sqrt{2}\hbar\Omega(t),\tag{10a}$$

or

$$\langle 1_-|\vec{\mu}|0_R\rangle \cdot \vec{A}(t) = 0 \text{ and } \langle 1_-|\vec{\mu}|0_L\rangle \cdot \vec{A}(t) = -\sqrt{2}\hbar\Omega(t). \tag{10b}$$

If we use linearly polarized laser fields whose polarization direction is determined by eq (9), we can see that the total population of R-enantiomers, $P_R(t)$, is increased compared with that in the initial time $\langle 0_R|\rho(t_0)|0_R\rangle = \dfrac{1}{2}$:

$$P_R(t) = \frac{1}{2} + \langle 1_R|\rho(t)|1_R\rangle \geq \frac{1}{2}. \tag{11}$$

The population of R-enantiomers in the excited state, $|1_R\rangle$, $\langle 1_R|\rho(t)|1_R\rangle$, is transferred from L-enantiomers $\langle 0_L|\rho(t_0)|0_L\rangle = \dfrac{1}{2}$ through a tunneling process after optical excitation.

In a similar way, if we set the laser field polarization to

$$-\langle 1_+|\vec{\mu}|0_+\rangle \cdot \vec{A}'(t) = \langle 1_+|\vec{\mu}|0_-\rangle \cdot \vec{A}'(t) \equiv \hbar\Omega'(t) \tag{12}$$

we obtain

$$\langle 1_+|\vec{\mu}|0_L\rangle \cdot \vec{A}'(t) = 0 \text{ and } \langle 1_+|\vec{\mu}|0_R\rangle \cdot \vec{A}'(t) = -\sqrt{2}\hbar\Omega'(t). \tag{13}$$

Therefore, the total population of L-enantiomers, $P_L(t)$, is increased compared with that in the initial racemate. This is the fundamental principle of the control scenario of selective preparation of pre-oriented enantiomers from their racemate.

Analytical Treatment of Enantiomer Selective Preparation

We now evaluate the time evolution of populations of enantiomers under the laser field condition, eq (9) using the dressed state representation. The time evolution of the density operator $\rho(t)$ obeys the Liouville equation:

$$i\hbar\frac{d}{dt}\rho(t) = [H(t), \rho(t)], \tag{14}$$

where $[\quad,\quad]$ is a commutator.

For simplicity, we restrict ourselves to quantum control of selective preparation of enantiomers by using a stationary laser field. The equation of motion of the density matrix $\boldsymbol{\rho}(t)$ can be expressed in the field-free eigenstate basis set as

$$i\hbar\frac{d}{dt}\boldsymbol{\rho}(t) = \left[\begin{pmatrix} \varepsilon_0 & 0 & 2\hbar\Omega\cos(\omega t) & 2\hbar\Omega\cos(\omega t) \\ 0 & \varepsilon_0 & 2\hbar\Omega\cos(\omega t) & 2\hbar\Omega\cos(\omega t) \\ 2\hbar\Omega\cos(\omega t) & 2\hbar\Omega\cos(\omega t) & \varepsilon_1 - \delta & 0 \\ 2\hbar\Omega\cos(\omega t) & 2\hbar\Omega\cos(\omega t) & 0 & \varepsilon_1 + \delta \end{pmatrix}, \boldsymbol{\rho}(t)\right]$$

(15)

with the initial condition

$$\boldsymbol{\rho}(t_0) = \begin{pmatrix} 1/2 & 0 & 0 & 0 \\ 0 & 1/2 & 0 & 0 \\ 0 & 0 & 0 & 0 \\ 0 & 0 & 0 & 0 \end{pmatrix}.$$

(16)

In eq (15), Rabi frequency Ω is defined as $\hbar\Omega = -\langle 1_+|\vec{\mu}|0_+\rangle \cdot \vec{A}$, where $\vec{A}$ is the amplitude of the laser field with a polarization.

We choose $\hbar\omega = \varepsilon_1 - \varepsilon_0$ as the carrier frequency. Within the rotating wave approximation, eq (15) is rewritten as

$$i\hbar\frac{d}{dt}\tilde{\boldsymbol{\rho}}(t) = \left[\begin{pmatrix} 0 & 0 & 0 & 0 \\ 0 & 0 & 0 & 0 \\ 0 & 0 & -\hbar\tilde{\Omega} & 0 \\ 0 & 0 & 0 & \hbar\tilde{\Omega} \end{pmatrix}, \tilde{\boldsymbol{\rho}}(t)\right],$$

(17)

where

$$\tilde{\boldsymbol{\rho}}(t) = \boldsymbol{\Lambda}(t)\boldsymbol{R}(t)\boldsymbol{\rho}(t)\boldsymbol{R}^{-1}(t)\boldsymbol{\Lambda}^{-1}(t)$$

(18)

with

$$R(t) = \begin{pmatrix} \exp(i\varepsilon_0 t/\hbar) & 0 & 0 & 0 \\ 0 & \exp(i\varepsilon_0 t/\hbar) & 0 & 0 \\ 0 & 0 & \exp(i\varepsilon_1 t/\hbar) & 0 \\ 0 & 0 & 0 & \exp(i\varepsilon_1 t/\hbar) \end{pmatrix}$$

$$(19)$$

and

$$\Lambda(t)^{-1} = \begin{pmatrix} \dfrac{1}{\sqrt{2}} & \dfrac{\delta}{\sqrt{2}\hbar\tilde{\Omega}} & \dfrac{\Omega}{\tilde{\Omega}} & \dfrac{\Omega}{\tilde{\Omega}} \\ -\dfrac{1}{\sqrt{2}} & \dfrac{\delta}{\sqrt{2}\hbar\tilde{\Omega}} & \dfrac{\Omega}{\tilde{\Omega}} & \dfrac{\Omega}{\tilde{\Omega}} \\ 0 & \dfrac{\sqrt{2}\Omega}{\tilde{\Omega}} & \dfrac{-\delta - \hbar\tilde{\Omega}}{2\hbar\tilde{\Omega}} & \dfrac{-\delta + \hbar\tilde{\Omega}}{2\hbar\tilde{\Omega}} \\ 0 & \dfrac{\sqrt{2}\Omega}{\tilde{\Omega}} & \dfrac{\delta - \hbar\tilde{\Omega}}{2\hbar\tilde{\Omega}} & \dfrac{\delta + \hbar\tilde{\Omega}}{2\hbar\tilde{\Omega}} \end{pmatrix}.$$

$$(20)$$

Here,

$$\tilde{\Omega} = \sqrt{\dfrac{\delta^2}{\hbar^2} + 4\Omega^2} .$$

$$(21)$$

The solution of the equation of motion, eq (17), is given as

$$\tilde{\boldsymbol{\rho}}(t) = \begin{pmatrix} 1 & 0 & 0 & 0 \\ 0 & 1 & 0 & 0 \\ 0 & 0 & \exp\left[i\tilde{\Omega}(t-t_0)\right] & 0 \\ 0 & 0 & 0 & \exp\left[-i\tilde{\Omega}(t-t_0)\right] \end{pmatrix} \tilde{\boldsymbol{\rho}}(t_0)$$

$$\bullet \begin{pmatrix} 1 & 0 & 0 & 0 \\ 0 & 1 & 0 & 0 \\ 0 & 0 & \exp\left[-i\tilde{\Omega}(t-t_0)\right] & 0 \\ 0 & 0 & 0 & \exp\left[i\tilde{\Omega}(t-t_0)\right] \end{pmatrix} \tag{22}$$

with $\tilde{\boldsymbol{\rho}}(t_0) = \boldsymbol{\rho}(t_0)$.

Therefore, the time evolution of the population in each localized state can be expressed in an analytical form as

$$P_{0R} = \langle 0_R | \rho(t) | 0_R \rangle = \frac{1}{2},$$

$$P_{0L} = \langle 0_L | \rho(t) | 0_L \rangle = \frac{1}{2} \left[\frac{\delta^2 + 4\hbar^2\Omega^2 \cos\left(\tilde{\Omega}t\right)}{\hbar^2\tilde{\Omega}^2} \right]^2,$$

$$P_{1R} = \langle 1_R | \rho(t) | 1_R \rangle = 2\left[\frac{\delta\Omega\left[1 - \cos\left(\tilde{\Omega}t\right)\right]}{\hbar\tilde{\Omega}^2} \right]^2$$

and

$$P_{1L} = \langle 1_L | \rho(t) | 1_L \rangle = 2\left[\frac{\Omega\sin\left(\tilde{\Omega}t\right)}{\tilde{\Omega}} \right]^2. \tag{23}$$

We can see from eq (23) that, in a stationary laser field, the population in $|1_L\rangle$ is maximized at $\tilde{\Omega}t = \pi/2$ and the population of $|1_R\rangle$ transferred by tunneling from $|1_L\rangle$ is maximized at $\tilde{\Omega}t = \pi$. The maximum population in the localized state, $|1_R\rangle$, is obtained in the case in which Rabi frequency satisfies the condition $|\Omega| = \delta/(2\hbar)$. We notice that the tunneling frequency in the presence of a laser field, eq (21) is different from the tunneling frequency in the absence of a laser field, $2\delta/\hbar$.

Finally, the time evolution of the total population in the right well is expressed as

$$P_R = \frac{1}{2} + 2\left[\frac{\Omega\sin\left(\tilde{\Omega}t\right)}{\tilde{\Omega}}\right]^2. \tag{24}$$

Results and Discussion

In the preceding section, we first presented a simplified treatment of laser selective preparation of pre-oriented enantiomers from their racemate by taking into account photon-polarization direction. Then, we derived analytical expressions for the time evolution of the system by solving the equation of motion of the density matrix in the low temperature limit.

We now demonstrate effects of laser field intensity on time evolution of each component of the localized states in a racemic mixture. Figure 2a shows the time evolution of the enantiomer selective preparation under an optimal condition in which $|\Omega| = \delta/(2\hbar)$, i.e., $\hbar\tilde{\Omega} = \sqrt{2}\delta$ is satisfied. We can see the R-enantiomer production $\left(P_R\right)_{max}$ of nearly 100% from the racemate. This condition means that both the Rabi oscillation and tunneling rate are synchronously matched; i.e., the entire population of the $1L$ localized state (L-form in the first excited state) created from the ground state is transferred to the target state, the $1R$ state (R-form in the first excited state), by tunneling processes. The tunneling rate depends on the laser intensity applied. This is one of the features in the present treatment while the tunneling rate is independent of laser intensity in a previous treatment, which was based on a locally optimized control theory (9,13).

In Fig.2b, a strong field case compared with the optimal field case is shown. Here, $|\Omega| = \delta/\hbar$, i.e., $\hbar\tilde{\Omega} = \sqrt{5}\delta$ is adoted as Rabi frequency. In this strong field case, a large population transfer to the $1L$ localized state from the $0L$ (L-form in the ground state) can be seen compared with the other two cases (Figs. 2a and

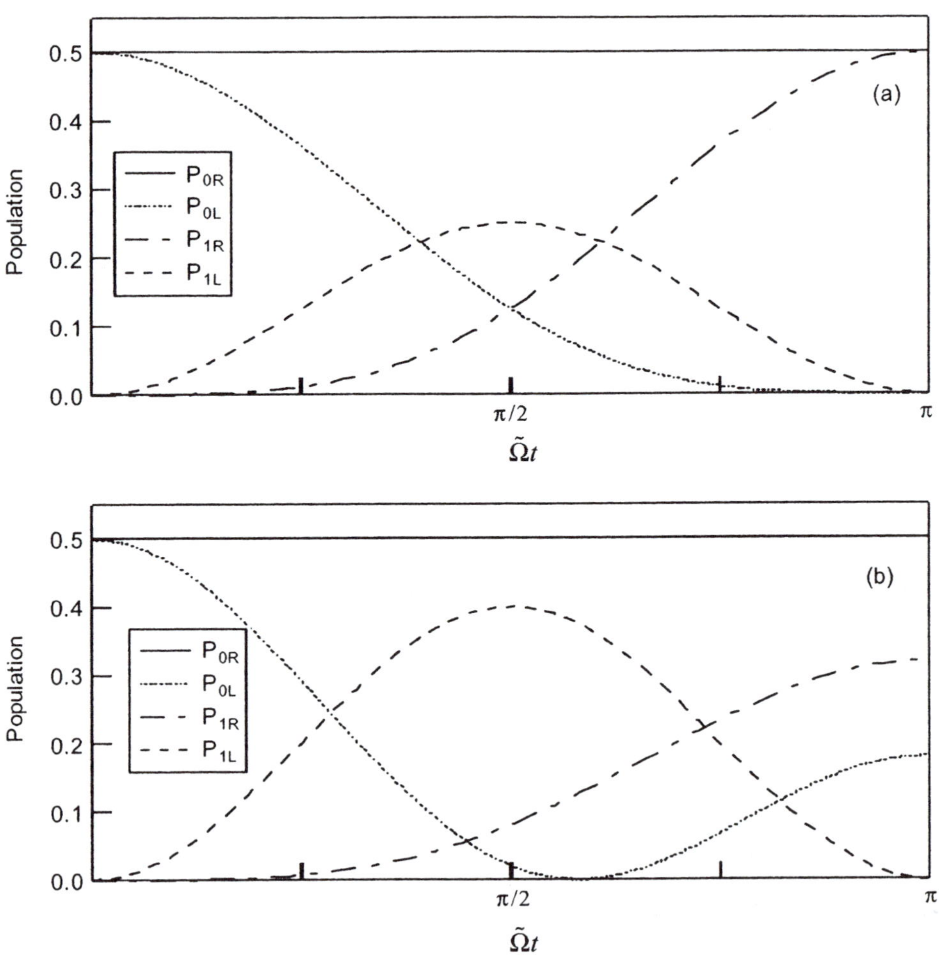

Figure 2. The time evolution of populations under an optimal field condition (a), in a strong field case (b), and in a weak field case (c). P_{0L} and P_{0R} denote the population of the L-form of H_2POSH in the ground state and that of the R-form, respectively. P_{1L} and P_{1R} denote the population of the L-form in the first excited torsional, vibrational state and that of the R-form, respectively.

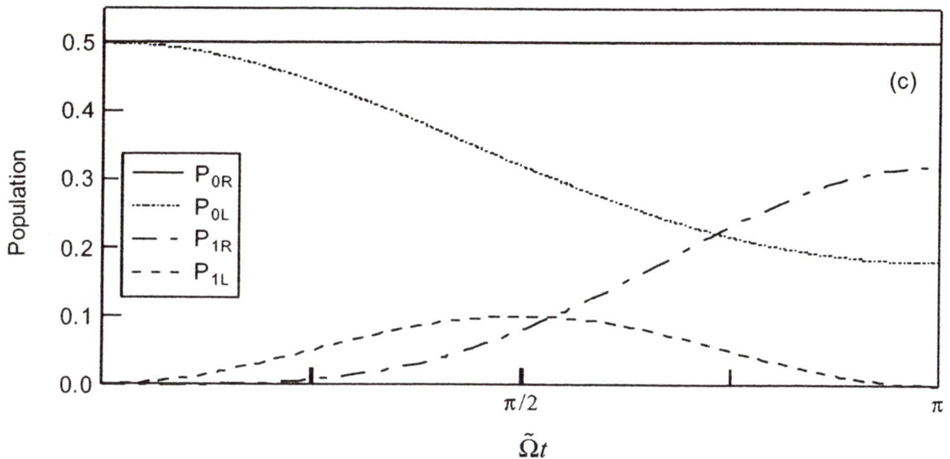

Figure 2. *Continued.*

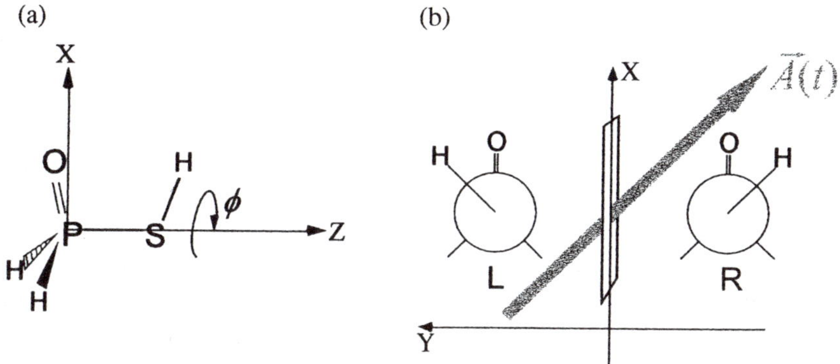

Figure 3. The reaction coordinate ϕ (a) and the polarization direction of the synthesized laser fields for dynamic chirality control of H_2POS (b). The pulse polarization is parallel to SH bond of the R- enantiomer of H_2POSH.

2c). However, the maximum yield of R-enantiomer production, $\left(P_R\right)_{\max}$ is about 80% at $\tilde{\Omega}t = \pi$. The rest of the population remains at the $0L$ state. This is because stimulated emission processes take place in such a strong field case. Existence of the stimulated emission processes can be examined by noting the population change of P_{0L}.

In Fig. 2c, population changes in a weak field case compared with the optimal field case are shown. Here, $|\Omega| = \delta/(4\hbar)$, i.e., $\hbar\tilde{\Omega} = \sqrt{5}\delta/2$ is adopted. The maximum yield of R-enantiomer production, $\left(P_R\right)_{\max}$ is about 80%; the rest of the population remains at the $0L$ state, similar to the case of the strong field as shown in Fig. 2b. However, the mechanisms of enantiomer selective preparation in the strong and weak field cases are different. In the weak field case, the rate-determining step is the excitation processes to the 1L state from the initial $0L$ state. This can be seen from the decrease of the $0L$ population while the 1R population increases.

Now, to demonstrate how enantiomers in pre-oriented racemates are selectively prepared by applying a linearly polarized laser whose polarization direction is properly determined, we use dynamic chirality control of H$_2$POSH as a model. The reaction path is along the torsional coordinate of the SH bond around the PS bond. H$_2$POSH is pre-oriented as shown in Fig. 3. The potential function that was calculated by using an ab initio MO method is characterized by a double-well potential similar to that shown in Fig. 1 (9). The L-form enantiomer is located at a torsional angle of $-60°$, and the R-form is located at $60°$. Its barrier height is about $500 cm^{-1}$. The four-state model described in the preceding section can be applied to the dynamic chirality control of H$_2$POSH: this model consists of torsional ground, first, second and third excited states. The condition of quasi-degenerate between the ground and first excited states is satisfied by enantiomer selection within a few hundred ps since the level splitting between them is only $0.053 cm^{-1}$ (9). This corresponds to a tunneling time of $630 ps$. The level splitting between the second and third excited states, 2δ, is, $1.6 cm^{-1}$ corresponding to a tunneling time of $21 ps$.

Dipole moment functions were also evaluated using an ab initio MO method (9). The magnitude of the matrix elements of the dipole moment is given as $|<1_+|\vec{\mu}|0_L>| = 0.19$ Debye. The direction of the photon polarization with respect to the X-axis was estimated to be $58°$. The value similar to that of the angle was obtained in controlling selective preparation of the same system from the racemate using a locally optimized control method in a previous paper (13). This direction is parallel to SH-bond of H$_2$POSH in the R-form as shown in Fig. 3. The amplitude of the laser used is $|\vec{A}| = 2.9 \times 10^7\ V/m$. In this parameter set, selective preparation time of enantiomers, $\pi/\tilde{\Omega}$ is $14 ps$.

In this study, we neglected rotational effects on selective preparation of enantiomers from their racemate. Taking into account the rotational degrees of freedom of enantiomers, we can understand from the results of a locally optimized control treatment that the control field consists of circularly polarized ones (12, 13).

In summary, we have developed a simplified theoretical treatment of enantiomer selective preparation from a racemate consisting of pre-oriented enantiomers. In the pre-oriented system, a linearly polarized laser field enables selective preparation of pure enantiomers from their racemate by choosing the polarization direction in a such a way that the laser field interacts with L-(R-) enantiomers but not with R-(L-) enantiomers. An analytical expression for time evolution of populations induced by a stationary laser field was presented within a four-state model for an enantiomer with one-dimensional, double-well potential. To demonstrate the effectiveness of the new treatment of the selective preparation of enantiomers, we considered dynamic chirality control of H_2POSH in their racemate. An extension of the present scenario to enantiomer control by pulsed lasers will appear elsewhere. In this paper, we dealt with laser control of pre-oriented enantiomers. Such a prerequisite to enantiomers as the initial condition can be realized by introducing laser fields for orientation of enantiomers (17, 18).

Acknowledgments

One of the authors (K. H.) acknowledges a support from Research Fellowship of the Japan Society for the promotion of Science (No. 6254). We would like to thank Professor J. Manz and Dr. L. Gonzalez for their stimulating discussions and fruitful comments. This work was partly supported by a German -Japan international joint research project, by Grants-in-Aid for Scientific Research (No. 10640480, 10044054 and 12640484), by Grant-in-Aid for Scientific Research on Priority Areas (No. 11166205) and by the Development of High-Density Optical Pulse Generation and Advanced Material Control Techniques.

References

1. Inoue, Y.; *Chem. Rev.* **1992,**92,741-770.
2. Avalos, M.; Babiano, R.; Cintas, P. ; Jimenez, J. ; Palacios, J. C. and Barron, L. D. *Chem. Rev.* **1998,** 98, 2391-2404.
3. Shapiro, M. ; Brumer, P. *J. Chem. Phys.* **1991,** 95, 8658-8661.
4. Cina, J. A. ; Harris, R. A. *J. Chem. Phys.* **1994,** 100, 2531-2536; Science **1995,** 267, 832-833.
5. Salam, A.; Meath, W. J. *J. Chem. Phys.* **1997,** 106, 7865-7868.
6. Shao, J. ; Hänggi, P. ; *J. Chem. Phys.* **1997,** 107, 9935-9941; Phys. Rev. A **1997,** 56, R4397-R4400.
7. Maierle, C. S. ; Harris, R. A. *J. Chem. Phys.* **1998** 109,3713-3720.

8. Shapiro, M. ; Frishman , E.; Brumer, P. *Phys. Rev. Lett.* **2000,** 84, 1669-2000.

9. Fujimura, Y. ; González, L.; Hoki, K. ; Manz, J.; Ohtsuki, Y. *Chem. Phys. Lett.* **1999,** 306,1-8; ibid, **1999**, 310, 578-579.

10. Fujimura, Y. ; González, L.; Hoki, K. ; Kröner, D. ; Manz, J.; Ohtsuki, Y. *Angew. Chem. Int. Ed.* **2000**, 39, 4586-4588.

11. González, L.; Hoki, K. ; Kröner, D. ; Leal, A. S.; Manz, J.; Ohtsuki, Y. *J. Chem. Phys.* **2000,** 113, 11134-11142.

12. Hoki, K.; Kröner, D. ; Manz, J. *Chem. Phys.* **2001**, 267, 59- 79.

13. Hoki, K. ; Ohtsuki, Y.; Fujimura, Y. *J. Chem. Phys.* **2001**, 114, 1575-1581.

14. Fujimura, Y.; González, L.; Hoki, K.; Manz, J; Ohtsuki, Y.; Umeda, H. *Advances in Multiphoton Processes and Spectroscopy;* vol. 14, World Scientific: Singapore, 2001; p 30-46.

15. Doslic, N.; Fujimura, Y.; González, L.; Hoki, K.; Kröner, D; Kühn, O.; Manz, J.; Ohtsuki, Y. Eds.; de Schryver, F.C.; de Feyter, S. ; Schweitzer, G. , *Femtochemistry*;Wiley-VCH, Weinheim, in press.

16. Kounuma, N.; Zijistra, R. W.; van Delden, R. A.; Harada, N.; Feringa, B. L. *Nature* **1999**, 401, 152-154.

17. Dion,C.M.; Bandrauk, A.D.; Akabek,O.; Keller, A.; Umeda,H.; Fujimura, Y. *Chem. Phys. Lett.* **1999**, 302, 215-223.

18. Hoki, K.; Fujimura, Y. *Chem. Phys.* **2001**, 267, 187-193.

19. Quack, M. *Femtosecond Chemistry*, Vol. 2 edited by Manz, J. ; Wöste, L. (VCH. Weinheim, 1995) p 781.

Chapter 4

Experimental and Theoretical Studies of the Channel Phase in the Coherent Control of Molecular Processes

Richard Billotto[1], Ani Khachatrian[1], Langchi Zhu[1],
Robert J. Gordon[1], Hélène Lefebvre-Brion[2], and
Tamar Seideman[3]

[1]Department of Chemistry, Mail Code 111, University of Illinois at Chicago,
845 West Taylor Street, Chicago, IL 60607–7061
[2]Laboratoire de Photophysique Moléculaire, Bât. 213, Université
Paris-Sud, 91405 Orsay Cedex, France
[3]Steacie Institute for Molecular Sciences, National Research Council
of Canada, Ottawa, Ontatio K1A 0R6, Canada

The source of the phase lag between different product channels in coherent control experiments is explored. It is shown theoretically that a structured continuum is required to obtain a non-zero channel phase. An isotope effect in the photoionization of HI and DI suggests a direct coupling between electronic and nuclear degrees of freedom. The occurrence of a channel phase is explained also in the context of multichannel quantum defect theory, resolving an earlier controversy.

Introduction

Interference effects are ubiquitous in nature. If two or more coherent paths connect an initial and a final state (*i.e.*, if a phase is associated with each path, and if the phase difference for each pair of paths is well defined), then the probability of reaching the final state varies sinusoidally with the relative phases of the paths. This property has been used to great advantage in controlling photo-induced reactions *(1)*. In the most commonly studied scenario *(2)*, the two paths consist of excitation by three photons of frequency ω_1 (wavelength λ_1) and one photon of frequency $\omega_3 = 3\omega_1$ (wavelength λ_3). As shown by Brumer and Shapiro *(3)*, the probability of producing product S is given by

$$P^S = P_1^S + P_3^S + 2 \mid P_{13}^S \mid \cos(\varphi + \delta_{13}^S),\tag{1}$$

where P_3^S (P_1^S) is the one- (three-) photon transition probability, P_{13}^S is the modulus of the interference term, $\varphi = \varphi_3 - 3\varphi_1$ is the relative phase of the two laser beams, and δ_{13}^S is termed the *channel phase (4)*. The quantities P_{13}^S and δ_{13}^S are defined by

$$\left| P_{13}^S \right| e^{i\delta_{13}^S} = e^{-i\varphi} \int d\hat{k} \langle g | D^{(1)} | ES\hat{k} \rangle \langle ES\hat{k} | D^{(3)} | g \rangle,\tag{2}$$

where $|g\rangle$ is the initial state, $|ES\hat{k}\rangle$ is the excited-state wave function for channel S at energy E and scattering angle $\hat{k}$, and $D^{(j)}$ is a j-photon electric dipole operator. The difference between two channel phases is an experimental observable known as the *phase lag (4)*,

$$\Delta\delta(A,B) = \delta_{13}^A - \delta_{13}^B.\tag{3}$$

If the channel phase for some particular process (e.g., δ_{13}^B) has a known value, then the channel phase for a process of interest (e.g., δ_{13}^A) may be determined from the phase lag between the two channels. The channel phase is of great interest both because it is a signature of control (i.e. if $\Delta\delta(A,B)$ is significantly different from zero, then the branching ratio into channels A and B can be controlled) and because it provides fundamental information about the continuum properties of the excited molecule *(4)*.

The first measurement of a phase lag was reported for the photoionization and photodissociation of hydrogen iodide *(5)*. At that time the physical origin of the channel phase was not understood and was the subject of some controversy *(6,7,8)*. One study *(6)*, using the formalism of multichannel quantum defect theory (MQDT), argued that the channel phase is identically zero. Subsequently, a general formalism of the channel phase was developed *(9,10)* that identified the conditions under which the channel phase does not vanish, and experimental data illustrating these results were published *(11,12,13)*. In the present paper we review the theoretical results and show that MQDT can be used to compute the channel phase. In addition, we report an isotope effect in the channel phase for photoionization and suggest a mechanism that is consistent with the general theory.

Experimental Results

An experimental example of an energy-dependent channel phase is shown in Figure 1 for the photodissociation and photoionization of HI,

$$HI + \omega_3, 3\omega_1 \rightarrow H + I^* \tag{4a}$$
$$\rightarrow HI^+ + e^-, \tag{4b}$$

measured in the vicinity of the $5d(\pi,\delta)$ Rydberg state *(14)*. Details of the experiment have been published elsewhere *(15)*. This state, which has an $X\ ^2\Pi_{1/2}$ ion core, decays by predissociation and spin-orbit autoionization. The circles are the phase lag between the two decay channels. The peak near 356.2 nm lies near the maximum in the one-photon absorption spectrum produced by the $5d(\pi,\delta)$ resonance *(16)*. The secondary maximum near 355.1 nm does not correspond to any resolved feature in the absorption spectrum and is possibly caused by the $(X\ ^2\Pi_{3/2}, v=1)\ 9d\pi \leftarrow X^1\Sigma^+$ transition. The diamonds in Figure 1 show the phase lag between the ionization of HI and H_2S. The latter was introduced as a marker to isolate the individual channel phases for HI. Because the ionization channel phase of H_2S^+ was shown to vanish, the diamonds give the channel phase for ionization of HI, and, similarly, the triangles give the channel phase for dissociation of HI. We see that the structure observed in $\Delta\delta(HI^+,I^*)$ is produced almost entirely by the channel phase for ionization, which rides on top of a zero baseline. In contrast, the channel phase for dissociation has a large nonzero value that varies only weakly with energy.

We have also measured the phase lag between the photoionization of DI and HI in the vicinity of the $5d(\pi,\delta)$ resonance. Preliminary measurements show that $\Delta\delta(DI^+,HI^+)$ is non-zero, having, for example, a value of $120.0 \pm 14.2°$ at λ_1 = 355.48 nm. Previously *(12)* we reported phase lag data near the $5s\sigma$ resonance, where $\Delta\delta(DI^+,HI^+)$ is essentially zero over the width of the resonance. The nonzero values of $\Delta\delta(DI^+,HI^+)$ near the $5d(\pi,\delta)$ resonance is a clear example

50

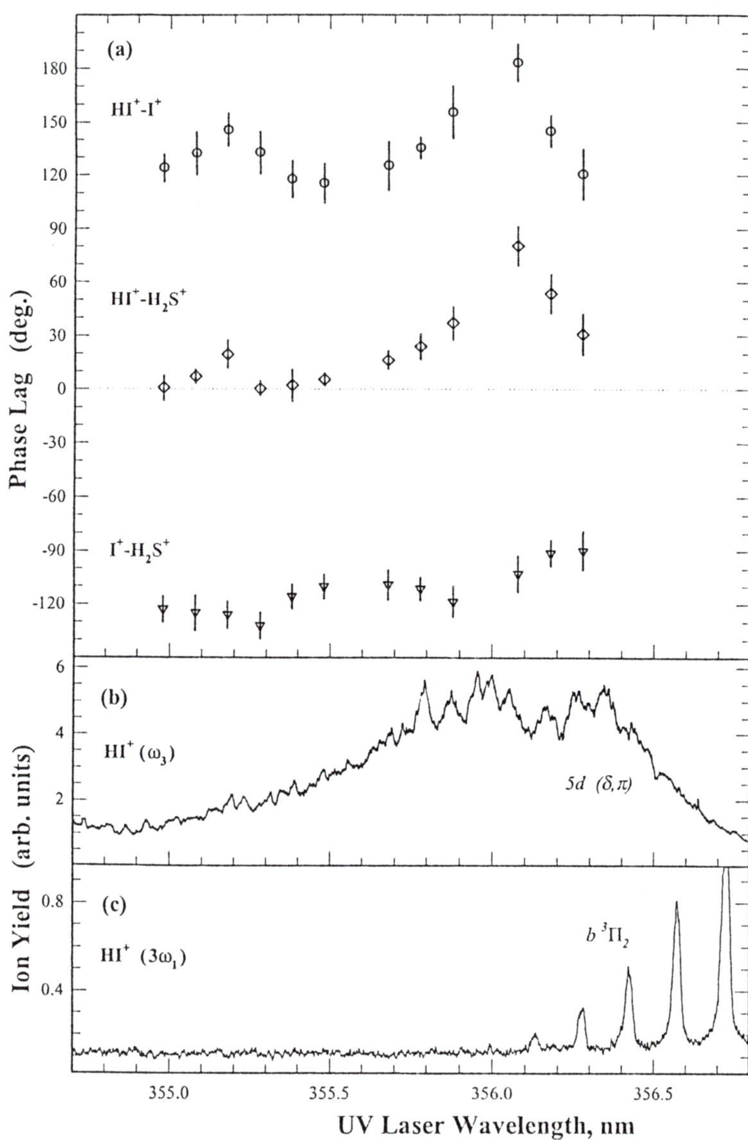

Figure 1. Phase lag spectrum for a mixture of HI and H$_2$S in the vicinity of the 5d(π,δ) resonance of HI (top panel). The various symbols show the phase lags between different pairs of ions. The bottom two panels are the one- and three-photon ionization spectra of HI.

of an isotope effect in the ionization channel. The physical origin of this effect is discussed below.

Mechanisms for Producing a Nonzero Channel Phase

Examination of Eq. (2) shows that the possible sources of nonzero δ_{13}^S are the three-photon dipole operator, $D^{(3)}$, and the complex scattering wave function, $| E S \hat{k} >$. Assuming for the moment that $D^{(3)}$ is real *(13,17)*, the problem is reduced to determining the circumstances under which the complex nature of the scattering states gives rise to a phase that survives integration over the scattering angles. It is shown in ref. 10 that for a direct transition to an uncoupled continuum the phases of the scattering state partial waves factor out of the matrix elements and cancel out. This case is depicted schematically in Figure 2a and is characteristic of a direct transition to a continuum wherein the potential induces only elastic scattering. An experimental example of this case is the zero baseline for the ionization channel phase of HI^+. (A similar effect was observed in the vicinity of the $5s\sigma$ resonance *(11)*). A necessary condition for a non-vanishing channel phase is that the continuum state reached by the photon(s) be coupled to some other state *(9)*. One possibility, depicted in Figure 2b, is coupling of the product continuum to some other continuum, which need not in itself produce observable products. In this case, substituting $\left| E S \hat{k} \right\rangle = c_1 \left| E S_1 \hat{k} \right\rangle + c_2 \left| E S_2 \hat{k} \right\rangle$ into the integrand of Eq. (2) produces a complex cross term with an argument that is only weakly energy dependent. An example of this case is the non-zero channel phase for dissociation of HI shown in Figure 1.

Another type of coupling that leads to a channel phase is that of a discrete state embedded in the continuum. If the continuum is elastic (Fig. 2c), the channel phase vanishes far off resonance and reaches a maximum in absolute value on resonance, whereas if the continuum is coupled (Fig. 2d), the channel phase is non-zero off-resonance and reaches a minimum in absolute value on resonance. An example of case c is the maxima in the channel phases for autoionization of the $5d(\pi,\delta)$ resonance HI. An example of case d is the predissociation of the $5s\sigma$ resonance shown in ref. *(11)*. Still more complex coupling cases arise from multiple resonances embedded in the continuum *(9)*.

The detailed properties of the phase lag spectrum depend on the number of interfering paths. For direct excitation of the continuum (cases a and b) there are just two paths, corresponding to one- and three-photon excitation, whereas for resonance-mediated excitation (cases c and d), there may be as many as four paths (one- and three-photon direct and one- and three-photon resonance-

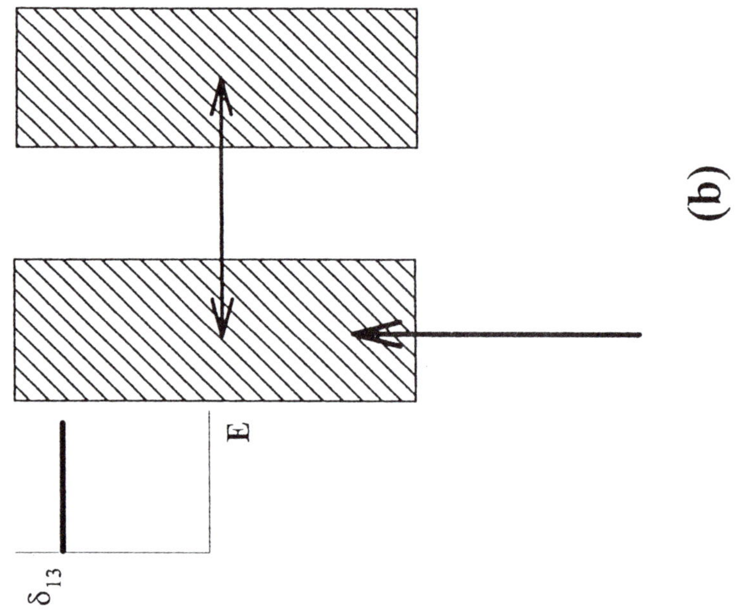

δ13
E
(b)

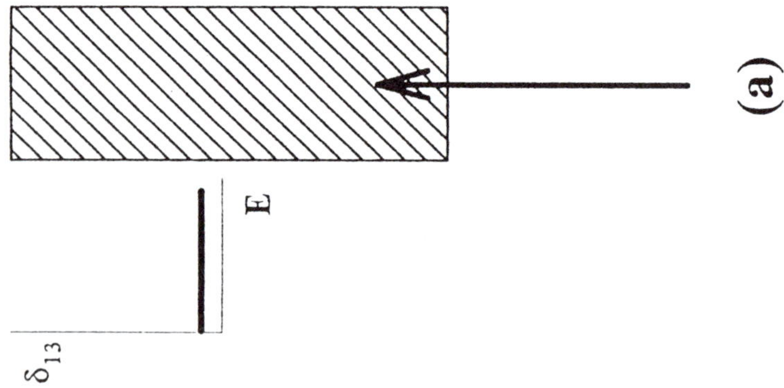

δ13
E
(a)

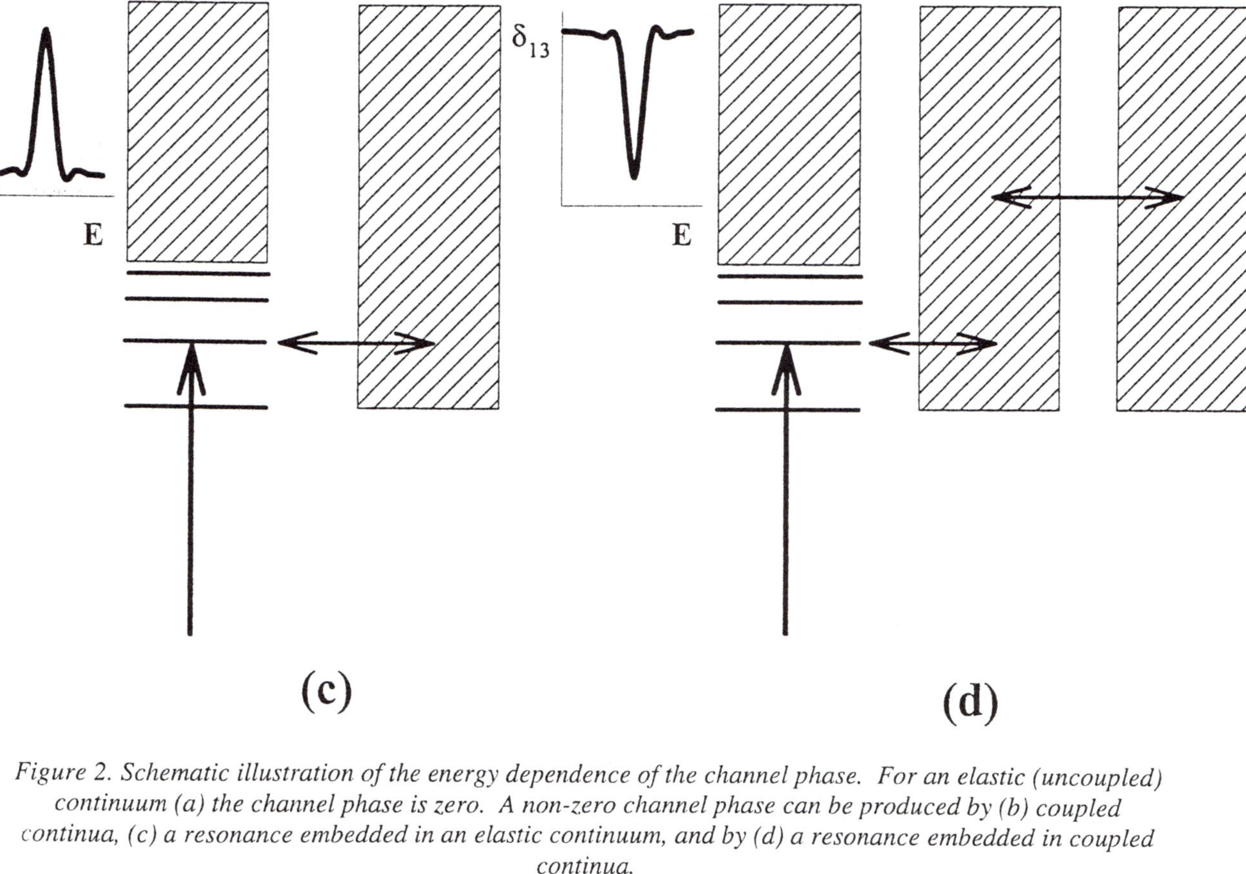

Figure 2. Schematic illustration of the energy dependence of the channel phase. For an elastic (uncoupled) continuum (a) the channel phase is zero. A non-zero channel phase can be produced by (b) coupled continua, (c) a resonance embedded in an elastic continuum, and by (d) a resonance embedded in coupled continua.

mediated excitation). In the limit that only the resonance-mediated paths contribute, Seideman *(9)* proved in general that for an isolated resonance the cross term is rigorously real; i.e.,

$$P_{13}^S = \int d\hat{k}\, \frac{\langle g|D^{(1)}|i\rangle \langle i|H_M|ES\hat{k}_1^-\rangle|^2 \langle i|D^{(3)}|g\rangle}{\left(E - E_i - \Delta_i\right)^2 + \Gamma_i^2/4}, \tag{5}$$

where H_M is the molecular Hamiltonian *(18)*, $|i\rangle$ is the resonance wave function, E_i is the unperturbed resonance energy, and Δ_i and Γ_i are, respectively, the energy shift and width of the resonance caused by the interaction with the continuum.

The properties of the phase lag spectrum of HI may be understood in light of these general results. The baseline for $\delta_{13}^{HI^+}$ in the vicinity of the 5d(π,δ) and 5sσ resonances is zero because there are no low-lying ionization continua coupled to the ionic state excited by the laser *(19)*. The nearly constant, non-zero channel phase for dissociation across the 5d(π,δ) resonance and the non-zero baseline for the dissociation channel phase in the vicinity of the 5sσ resonance may be attributed to the large density of coupled Born-Oppenheimer states near the ionization threshold, which produce an inelastic dissociation continuum. The source of the isotope effect is more speculative. A possible explanation of the non-zero channel phase for DI$^+$ near the 5d(π,δ) resonance is a coupling between electronic and nuclear degrees of freedom. (See Figure 3b.) The Kepler period *(20)* for the 5d(π,δ) Rydberg state is 16 fs, which is comparable to the vibrational periods of HI (14 fs) and DI (20 fs). The similarity in time scales for the two degrees of freedom suggests that coupling between them is possible. The higher vibrational frequency and greater tunneling probability for H vs. D may explain the much weaker coupling in HI. The absence of an isotope effect near the 5sσ resonance is still not understood.

MQDT Analysis of the Channel Phase

MQDT provides a framework for quantitative analysis of (auto)ionization and (pre)dissociation. In the present study it is necessary to use a unified version of MQDT that treats the competition between the two decay paths *(21,22)*. Previously, spin-orbit (SO) autoionization was treated by the standard MQDT *(23)*, which in the present study is incorporated into the unified theory.

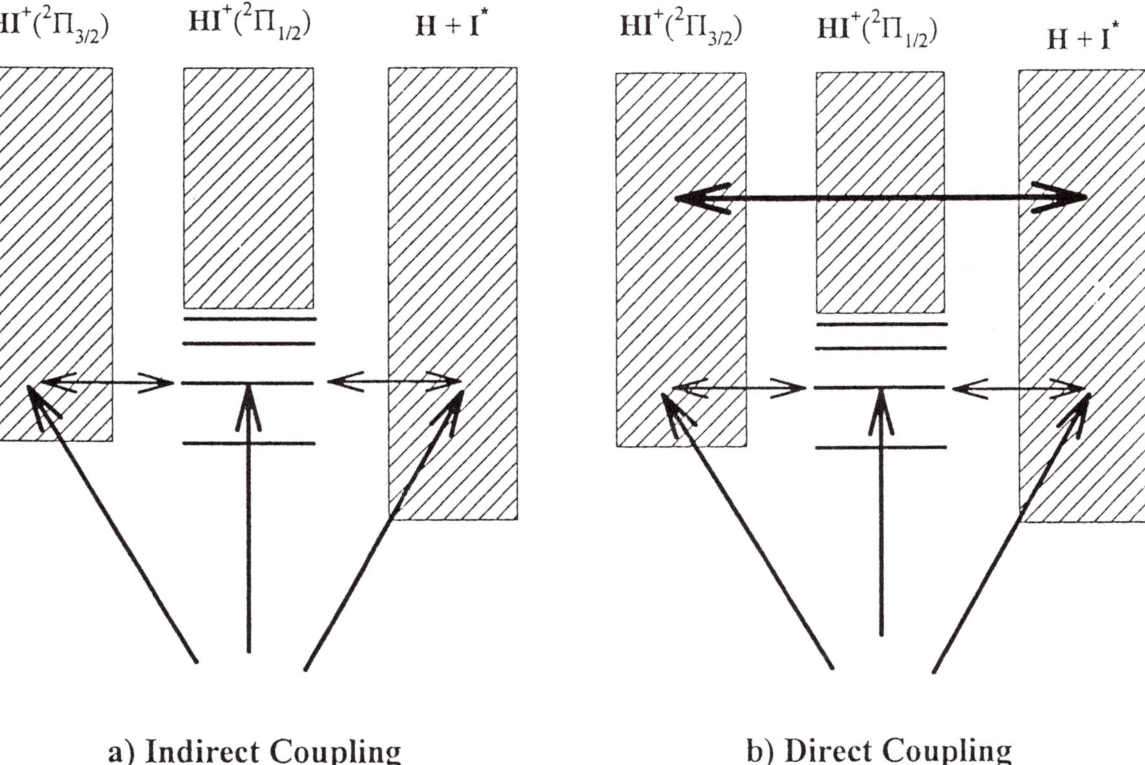

Figure 3. Possible coupling schemes for ionization and dissociation of HI. In scheme (a) the ionization and dissociation continua are coupled indirectly through a resonance, whereas in scheme (b) they are coupled directly.

An earlier MQDT calculation of the channel phase argued that this quantity should vanish identically *(6)*. Our recent study *(24)* traced this finding to an error in the transformation from the standing to the traveling wave boundary conditions in Ref. 6 *(vide infra)*, resolving the controversy that triggered much of the research on the channel phase problem and illustrating the sensitivity of δ_{13}^{s} to details of the scattering wave function that are not reflected in more conventional observables. Details of the formalism of unified MQDT applied to calculation of the channel phase appear elsewhere *(25)*, and only an overview is presented here

The basic concept in MQDT is the distinction between short-range and long-range interactions, which results in the subdivision of the configuration space into two regions. The inner "reaction zone" is characterized by strong interactions, whereas in the external zone the particles are subjected to long range local potentials such as the Coulomb potential, which lead to ionization or dissociation. In the unified theory, the ionization channel is defined as the set of all Rydberg states converging to an ionization limit as well as the continuum above this limit, whereas the dissociation channel consists of the set of vibrational bound levels of a given potential and the adjoining dissociation continuum. For the ionization channels, the effects of the inner region are represented by a quantum defect, μ, or a phase shift, $\pi\mu$, which is often treated as an adjustable parameter, to be determined from a fit to data. The couplings between all the channels in the reaction zone are similarly taken in general to be empirical parameters.

The adjustable input parameters are used in the external region, where the total wave function is expressed as a linear superposition of eigenfunctions of the long range potential. Boundary conditions applied at infinity yield a linear set of equations from which the superposition coefficients are determined. It is only at this stage that closed and open channels are distinguished.

In the present case there are three channels to consider: the ionization channel (i1) with the limit $X^2\Pi_{3/2}$, $v^+=0$ the ionization channel (i2) with the limit $X^2\Pi_{1/2}$, $v^+=0$ and the dissociation channel (d) containing the continuum of the dissociative state. Channel i2 consists of the experimentally observed resonances, including the $5d(\pi,\delta)$ state. The interaction between channels i1 and i2 has a spin-orbit character, whereas the interaction between channels i1 and d is assumed to have an electronic character.

The main results of the unified MQDT analysis are as follows: The eigenchannel wave functions, Ψ_α, are eigenfunctions of the reactance matrix and contain contributions from both the ionization and dissociation wave functions. The eigenfunctions of the scattering matrix outside of the reaction zone are linear combination of the eigenchannel functions, $\Psi_\rho = \sum_\alpha A_\alpha \Psi_\alpha$. The final open channel wave function is a linear combination of the S-matrix eigenfunctions, $F_{iv} = \sum_{\rho} T_{iv,\rho} \Psi_{\rho} e^{i\tau_{\rho}}$, where i and v.are electronic and vibrational indices, the coefficients $T_{iv,\rho}$ contain contributions from both ionization and dissociation, and τ_{ρ} are the eigenphases of the S-matrix. Last, the matrix element for the j-photon dipole operator between the ground and final open channel wave function is given by

$$D_{iv}^{(j)}(E) = 2^{-1/2} e^{i\tau_{\rho}} T_{iv,\rho} \left\{ A_1^{\rho} \left(D_1^{(j)} - D_2^{(j)} \right) + A_2^{\rho} \left(D_1^{(j)} + D_2^{(j)} \right) \right\}. \qquad (6)$$

In this equation, subscripts 1 and 2 refer, respectively, to the zero-order discrete and continuum states, and $D_{1(2)}^{(j)}$ is the corresponding primitive dipole matrix element.

In the conventional MQDT it is customary to use standing wave boundary conditions to construct the electronic radial wave functions in the reaction zone *(26)*. In this case $D_{1(2)}^{(j)}$ is real, and the argument of $D_{iv}^{(j)}$ is the eigenphase, τ_{ρ}. Because τ_{ρ} is independent of j, the channel phase is necessarily zero. When the proper traveling wave boundary conditions are imposed, however, the zero-order discrete and continuum states used to calculate $D_{1(2)}^{(j)}$ are complex, leading to a non-zero channel phase.

The channel phases calculated with the unified MDQT are compared with the data in Figure 4. The broad features of the data – a maximum in the ionization channel phase near the $5d(\pi,\delta)$ resonance, zero asymptotes for the ionization phase, and a broad, negative value for the dissociation channel phase – are reproduced by the calculation. The secondary maximum in the data caused by a weak transition is not included in the model, and the predicted minimum in the ionization channel phase is not observed experimentally. More recent unified MQDT calculations that include the $X^2\Pi_{3/2}$, $v^+=1$ channel display secondary structure that differ for the two isotopomers *(24)*.

A limitation of the present unified MQDT is that the two continua are assumed to be coupled only indirectly through the resonance (see Figure 3a). Coupling in either or both continua (possibly direct coupling of the ionization and dissociation continua, Figure 3b), which is accounted for in the more general scattering theory *(9,10)*, play an important role in determining the observed channel phase. Future MQDT calculations will account for direct coupling of the ionization and dissociation continua, overlapping resonances, and rotations.

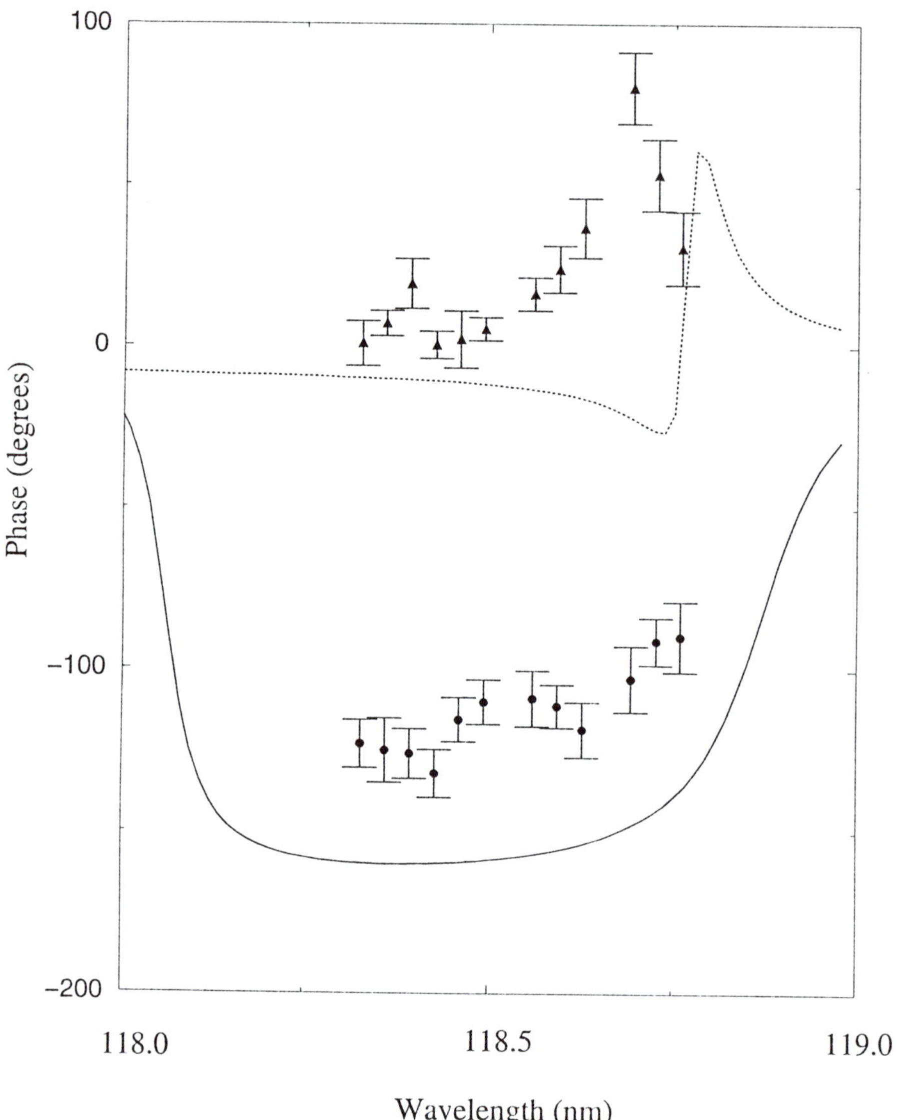

Figure 4. Channel phases calculated with MQDT for the ionization (dashed curve and triangles) and dissociation (solid curve and circles) of HI, as a function of VUV wavelength. The data points are the same as in Figure 1. Reprinted with permission from ref. 24. Copyright 2001 American Institute of Physics.

Conclusions

Experiments and theory are now capable of providing a detailed understanding of how continuum structure affects the channel phase that is observed in the coherent control of chemical branching ratios. This understanding provides new insights into the simultaneous ionization and dissociation dynamics of HI and identifies a useful tool in general coherent control experiments. Future experiments will attempt to generalize these results to polyatomic molecules, where bond-selective chemistry becomes possible.

Acknowledgment

Support of this research by the National Science Foundation is gratefully acknowledged.

References

1. Gordon, R. J.; Zhu, L, Photoionization and Photodetachment; Ng, C. Y., Ed,; World Scientific: Singapore, 2000; pp 47-90.
2. Shapiro, M.; Hepburn, J. W.; Brumer, P. *Chem. Phys. Lett.* **1988**, *149*, 451.
3. Brumer, P.; Shapiro, M. *Accts. Chem. Res.* **1989**, *22*, 407.
4. Gordon, R. J.; Zhu, L.; Seideman, T. *J. Phys. Chem. A* **2001**, *105*, 4387.
5. Zhu, L.; Kleiman, V. D.; Li, X.; Lu, S.; Trentelman, K.; Gordon, R. J. *Science* **1995**, *270*, 77.
6. Lefebvre-Brion, H. *J. Chem. Phys.* **1997**, *106*, 2544.
7. Lee, S. *J. Chem. Phys.* **1997**, *107*, 2734.
8. Lambropoulos, P.; Nakajima, T. *Phys. Rev. Lett.* **1999**, *82*, 2266.
9. Seideman, T. *J. Chem. Phys.* **1998**, *108*, 1915.
10. Seideman, T. *J. Chem. Phys.* **1999**, *111*, 9168.
11. Zhu, L.; Suto, K.; Fiss, J.; Wada, R.; Seideman, T.; Gordon, R. J. *Phys. Rev. Lett.* **1997**, *79*, 4108.
12. Fiss, J. A.; Zhu, L.; Gordon, R. J.; Seideman,T. *Phys. Rev. Lett.* **1999**, *82*, 65.
13. Fiss, J. A.; Khachatrian, A.; Truhins, K.; Zhu, L.; Gordon, R. J.; Seideman, T. *Phys. Rev. Lett.* **2000**, *85*, 2096.

14. Fiss, J. A. ; Zhu, L.; Gordon, R. J.; Seideman, T. *Disc. Faraday Soc.* **1999**, *113*, 61.

15. Lu, S.; Park, S. M.; Xie, Y.; Gordon, R. J. *J. Chem. Phys.* **1992**, *96*, 6613.

16. Hart, D. J.; Hepburn, J. W.; *Chem. Phys.* **1989**, *129*, 51.

17. A quasibound state located at energy ω_3 or $2\omega_3$ will give rise to complex $D^{(3)}$. In the absence of coupling in the continuum, the resulting channel phase equals the Breit-Wigner phase of the resonance. Details are provided in ref. 13.

18. H_M is the total molecular Hamiltonian, but only the interaction term (e.g., the spin-orbit operator) in it has a nonvanishing matrix element between $\langle g|$ and $|ES\hat{k}_1^-\rangle$. The subscript 1 distinguishes between the eigenstates $|ES\hat{k}\rangle$ of the complete Hamiltonian and those of the scattering portion of H, conventionally denoted PHP, from which the bound component has been projected out.

19. The $X^2\Pi_{1/2}$ continuum lies 5,359 cm^{-1} above the ground ionic state.

20. Gallagher, T. F. *Rydberg Atoms;* Cambridge: Cambridge, UK, 1994.

21. Giusti-Suzor, A,; Jungen, Ch. *J. Chem. Phys.* **1984**, *80*, 986.

22. Lefebvre-Brion, H.; Keller, F, *J. Chem. Phys.* **1989**, *90*, 7176.

23. Lefebvre-Brion, H.; Giusti-Suzor, A,; Raseev, G. *J. Chem. Phys.* **1985**, *83*, 1557.

24. Lefebvre-Brion,H., unpublished results.

25. Lefebvre-Brion,H.; Seideman, T,; Gordon, R. J. *J. Chem. Phys.* **2001**, *114*, 9402.

26. Giusti, A. J. Phys. B: Atom. Molec. Phys. **1980**, *13*, 3867.

Chapter 5

Four-Wave Mixing and Coherent Control

Vadim V. Lozovoy[1], Matthew Comstock[2], and Marcos Dantus[1,*]

Departments of [1]Chemistry and [2]Physics, Michigan State University,
East Lansing, MI 48824–1322
*Corresponding author: Dantus@msu.edu

Optimum control of a chemical reaction typically involves a number of nonlinear optical processes caused by a strong shaped laser pulse. Here we explore the role that three-pulse four-wave mixing techniques can play in unraveling the different nonlinear photophysical and photochemical processes that take place in the presence of the shaped laser pulse. Two types of experiments are included here. The first illustrates how electric field interactions (up to third order) manipulate the electronic, vibrational, and rotational coherence and population transfer processes in the sample molecules. The second set of experiments explores strong field effects using off-resonance four-wave mixing. Evidence of alignment and structural deformation is presented.

1. Introduction

The dream of controlling chemical reactions with lasers was conceived at about the same time when lasers were invented (*1*). Progress towards this goal has not been straightforward because a number of issues were not fully understood at the time, the most important being (a) the rate of energy redistribution, (b) the role of coherence and quantum mechanical interference,

(c) the influence of strong fields on molecules, and (d) nonlinear interactions between laser fields and molecules. The modern successes in the field are the result of progress in the understanding of these issues (*2-6*). The main lesson we have learned in the last two decades is that our accomplishments in coherent control depend on how well we understand the behavior of molecules in the presence of strong laser fields. In this chapter we discuss the role that four-wave mixing techniques can play in shedding light on the physical and chemical changes induced by laser-molecule interactions.

The fundamental concept behind optimal control is that "an electric field can optimize the yield of a given product in a chemical reaction" (*7, 8*). This paradigm shifts the focus to finding (and creating) the optimal field. The optimal field is capable of accessing all necessary pathways (intra and intermolecular) to maximize a particular measurement. The challenges in laser control are therefore the creation of optimal pulses, a technical challenge for achieving as much bandwidth, resolution, and intensity as possible; and understanding what a particular optimal pulse can teach us about the processes that bring about the desired chemical change. Although, by definition, the optimal pulse includes any number of collinear pulses that could be required to enhance a particular process, additional pulses can be used to understand how the optimal field achieves its goal or to probe the final outcome.

Analysis of optimal pulse shapes, when they are reduced to their essential components using appropriate pressure in the cost functional (*9*), shows that they usually consist of a discrete number of sub-pulses (*10*). Because three-pulse four-wave mixing (FWM) techniques provide a simple way of combining coherently three laser pulses, one is able to learn about the different ways that three electric fields can control the coherence and population transfer in a molecule. These aspects are demonstrated by a series of experiments involving resonant laser pulses. The sample chosen to illustrate these concepts is gas phase molecular iodine.

The main target of an optimum control experiment, for example the increased yield of a particular product and concomitant reduction of parasitic channels, is achieved through a combination of several processes such as multiphoton excitation, ionization, dissociation, predissociation, adiabatic passsage, etc. This undetermined combination of photochemical and photophysical processes is initiated by the strong optimum field; here we refer to them simply as nonlinear optical processes. When the same target can be achieved by a discrete combination of laser pulses one has the opportunity to learn about the nonlinear optical processes that are responsible for achieving the desired result (*4,9*). In order to take maximum advantage of constructive and destructive interference effects, the pulses must be phase-coherent with each other. The coherent combination of multiple pulses can be achieved with a series of phase locked pulses or by combining non-collinear pulses and detecting a phase-matched signal. As stated earlier, the maximum yield is achieved by the optimized pulse; however, the multiple pulse approach can shed light into the individual processes that are required to give the desired product.

The concepts in the previous paragraph can be illustrated with the schematic in Figure 1. The top-down aproach towards the main target is best illustrated by the experiments from the Gerber group and more recently from the Levis group (*11, 12*). In both cases a large number of photons (>8) is needed to produce the target ion mass. The precise pathway leading to the desired product is unknown in both cases (*11, 12*). The bottom-up approach (left) involves multiple collinear pulses and is best illustrated by the stimulated Raman adiabatic passage (STIRAP) experiments where a discrete number of pulses is used to achieve a maximum target yield that can be as large as 100% (*13*). These experiments, because of their relative simplicity, allow one to achieve maximum target yield and insight into the process that leads to product formation. Very few of the reported optimum field experiments achieve both maximum target yield and maximum insight into the mechanism (*14*). The third approach, bottom-up (right), achieves maximum insight but not maximum yield. Most pump-probe experiments illustrate this approach. The total yield, especially for experiments involving weak fields, is very low, but such experiments provide excellent insight into the molecular dymnamics processes including those occuring in the transition state (*15*). Three-pulse FWM experiments fall into this category of achieving maximum insight but not maximum yield. The crossing of the laser pulses creates a transient polarization grating in the sample where the coherence is maintained; however, outside the grating the molecules are not controlled thereby reducing the overall yield. The ease with which three laser pulses can be combined *coherently* makes this approach much more powerful than pump-probe methods to address some of the nonlinear optical processes occuring during a strong field optmium control experiment.

Nonlinear optical spectroscopic methods are concerned with the transient polarization of the sample induced by the electromagnetic field. The nonlinear polarization of the medium in response to a strong laser field can be expressed by the expression

$$P = P^{(1)} + P^{(2)} + P^{(3)} + \cdots \tag{1}$$

$$P = \chi^{(1)} E + \chi^{(2)} E E + \chi^{(3)} E E E + \cdots \tag{2}$$

where χ is the suseptibility and E the electric field. Higher order terms depend on multiple electric field interactions (*16, 17*). High order polarization can be achieved by a single intense laser pulse or by a series of less intense laser pulses that interact coherently. Both of these regimes are illustrated in this article. In the

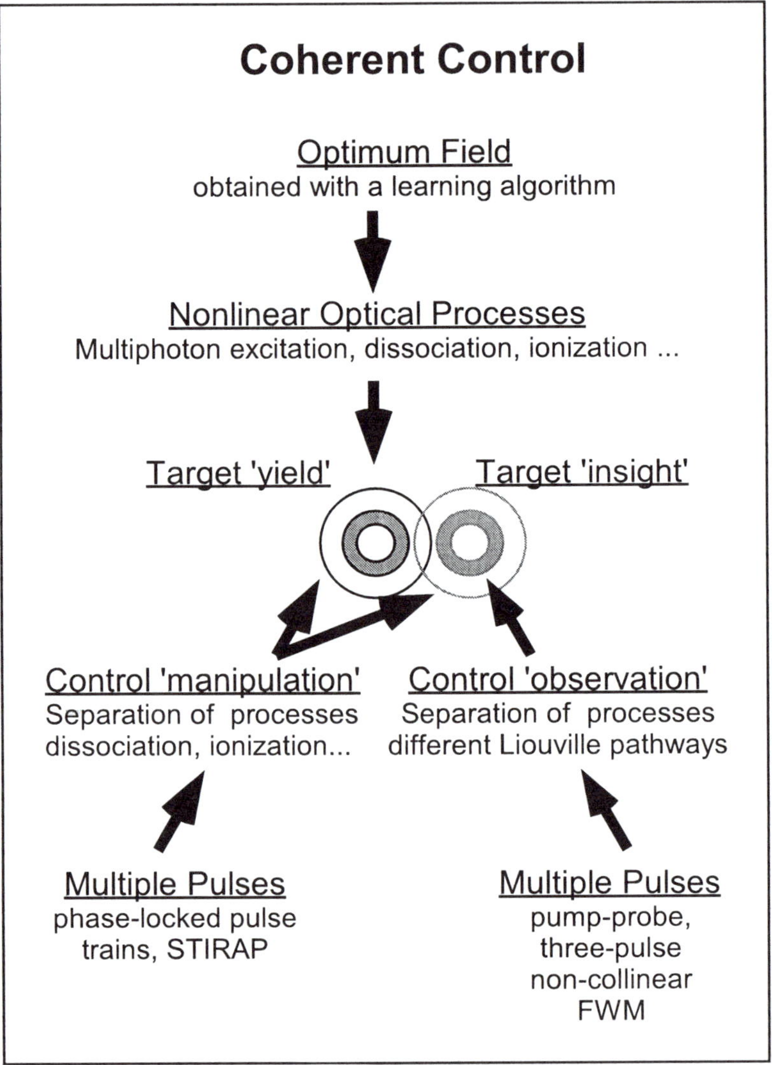

Fig. 1, Coherent control diagram indicating several approaches towards a specific target. The target can be either maximizing the yield of a product or gaining insight about a particular process (see text).

first set of experiments three weak fields are combined coherently in the sample, thereby inducing a third-order polariztion in the sample. Their time delay and wave vector determine the sample response (*18, 19*). We illustrate the process of coherence rephasing as well as the control of coherence and population transfer. The second set of experiments explores higher order processess initiated by lasers. For these experiments the sample is irraidated by two strong fields, and the structural changes that the molecules experience in the presence of the strong field are probed by a third weak pulse.

In order to distinguish between different FWM processes that contribute to the third order polarization, one can take advantage of the wave vector properties of the laser pulses. Each laser pulse can be represented by the expression

$$E_n(\mathbf{r}, t - t_n) = \varepsilon_n(t - t_n)\exp[-i(\omega_n t - \mathbf{k}_n \cdot \mathbf{r})] + \varepsilon_n^*(t - t_n)\exp[i(\omega_n t - \mathbf{k}_n \cdot \mathbf{r})] \quad (3)$$

where $\varepsilon_n(t - t_n)$ is the complex envelope of the nth pulse, ω_h ($\omega_h > 0$), the carrier frequency, $\mathbf{k}_n$, the wave vector, and $\mathbf{r}$, is the sample spatial coordinate. If each laser has a different wave vector, *e.g.*, they are not collinear, or they have different wavelengths, signals resulting from a particular combination of laser pulses can be isolated. This selection process is typically referred to as phase matching. Phase matching ensures conservation of energy and momentum for each nonlinear optical process. Here we will restrict the treatment to the case of three non-collinear and wavelength-degenerate pulses.

The phase matching condition restricts the sign of each electric field interaction, that is $\exp(i\omega t)$ or $\exp(-i\omega t)$, according to one of the terms in Equation 3. The system, after a single resonant electric field interaction, is left in a coherent state between the ground and the excited state. The sign of the electric field determines the time-dependence of the induced electronic coherence (*18*). Two electric field interactions are required to create a population in the ground or the excited state. This is consistent with the probability of excitation from a ground to an excited state, P_{eg}, which can be expressed as follows

$$P_{eg} \propto \left| \langle e | \mu E(t) | g \rangle \right|^2 = \langle e | \mu E(t) | g \rangle \langle g | \mu E(t)^* | e \rangle \quad (4)$$

where μ is the induced dipole moment and E is the electric field. It is also consistent with the linear dependence on laser intensity, $I \propto |E_0(t)|^2$.

In a non-collinear three-pulse FWM setup it is possible to control each electric field interaction. The timing, wavelength, and wave vector of the fields can be used to determine the final state of the system and the emission that is observed. The first two pulses determine if a ground or excited state population is formed and if rotational and vibrational coherences for that state are observed. The role of the first two electric fields is expressed in Equation 4 and more specifically in the Liouville pathways responsible for FWM (*18, 19*). The time evolution of a molecular system in the presence of three pulses can be best described using a density matrix approach (*18, 19*). For gas phase molecules, a description of wave packets in Hilbert space can be sufficient, so long as it is consistent with the time evolution of the density matrix. The time evolution of the density matrix does not need to be calculated according to the Liouville equation. There are a number of diagrammatic techniques that illustrate the different response functions. The full theoretical description has been presented elsewhere (*19*).

Strong field effects, typically involved in strong field optimal control experiments, are addressed using a different set of experiments. The principle of these experiments is to have the molecules interact with the strong field of the first two pulses, which are overlapped in time. The third pulse is used to probe the change in the molecular geometry tens of picoseconds after the strong field interaction. The goal of these experiments is to gain insight into the various nonlinear optical processes caused by strong field excitation.

2. Experimental methods

The laser system used for these experiments has been described elsewhere (*20, 21*), and here we give only a brief description. Femtosecond pulses were obtained with an amplified colliding-pulse mode-locked laser producing ~ 60 fs transform-limited pulses with 0.3 μJ of energy per pulse (*22*). The central wavelength of the pulses is 620 nm. This wavelength is important for resonant excitation of the B-X transition in molecular iodine (*23*) and for minimization of field ionization in off-resonance experiments (*24, 25*). The output is split into three arms, each one with an optical delay line. The three beams are recombined at the sample in an arrangement known as the forward box geometry (*26, 27*). The coherent coupling between the three electric fields is achieved spatially by the lasers, and it is not necessary to phase-lock the laser pulses (*21, 29*). Background free detection is achieved by placing the detector at the phase matching direction (*28*).

The different nonlinear phenomena are determined by the pulse ordering and phase matching direction. Each technique is known by a specific name. Here we show reverse transient grating RTG using $k_{RTG} = k_3 - (-k_1 + k_2)$, photon echo $k_{PE} = -k_1 + (k_2 + k_3)$, virtual echo $k_{VE} = k_1 - k_2 + k_3$, and stimulated pulse photon echo $k_{SPE} = -k_1 + k_2 + k_3$ (*19, 29*). Time-overlapped pulses are indicated with a parenthesis. Experiments are shown for two different samples, CS_2 and I_2. The I_2 sample was sealed in a quartz cell evacuated to 10^{-6} Torr and heated to 80 °C. Some data sets were obtained as a function of temperature to demonstrate the temperature dependence of inhomogeneous and homogeneous broadening. The CS_2 was kept at room temperature and a pressure of < 300 Torr.

3. Results

3.1 Three-pulse resonant interactions:

The experiments presented in this section address issues of coherence and population transfer. Coherence is essential for most forms of laser control because it maximizes the ability to use constructive and destructive interference among numerous quantum mechanical pathways (3-6). Coherence is quickly lost by collisions and inhomogeneities in the sample. Here we illustrate how certain pulse sequences can be used to extend the electronic coherence in the sample. This is of particular importance for experiments in a condensed phase where coherence can be very short lived.

As discussed earlier, in a three-pulse FWM measurement, the first laser pulse induces a polarization in the sample. If the laser is resonant with a spectroscopic transition, the polarization is long-lived; otherwise, for off-resonance excitation, the polarization is very short-lived. To measure the time dependence of the initial polarization (electronic decoherence time), a pulse sequence is chosen where the time-delay, τ, between the first pulse and the subsequent two pulses is scanned. Probing of the electronic coherence as a function of time is accomplished by the other two pulses that are usually overlapped in time. This measurement can be carried out with two different laser pulse arrangements known as photon echo (PE) and reverse transient grating (RTG) (29, 30).

In the photon echo process the first interaction is with a pulse that carries a negative wave vector (-**k**). All quantum mechanical states involved, as well as relaxation phenomena, evolve with a positive sign. The sign of the electric field is determined by the sign of the wave vector according to Equation 3. Subsequent interactions with two additional electric fields having a positive wave vector reverse the sign of the evolution, and the initial dephasing is reversed (18, 30, 21). In the RTG technique there is no rephasing or echo (18, 30, 21). Photon echo experiments cancel inhomogeneous broadening in the sample, and this leads to a much longer electronic coherence. Inhomogeneities in the sample are caused by the fact that not all the molecules are in the same initial quantum mechanical state or environment. At any specific temperature, the molecules have different velocities (Doppler broadening) and are in different rotational and vibrational states (hot bands). These differences lead to a loss in the coherence of the emitted RTG signal, but they are cancelled in a PE experiment (*30*).

Data for RTG and PE measurements are shown in Figure 2. The differences observed in the background, undulation, and apparent signal-to-noise ratio in these data result from the difference in the first pulse interaction. The RTG data shows a strong background, and a slow undulation that results from the inhomogeneous rotational population of the sample. After the first three ps, the RTG data shows a mixture of ground and excited state dynamics. The observation of ground state dynamics results from the initial thermal population of different vibrational modes. In the PE data, only excited state vibrational dynamics are observed.

The cancellation of inhomogeneous broadening in PE measurements has been recognized since the first photon echo measurement in 1964 (*31, 32*). This advantage has been exploited, for example, to measure the homogeneous lifetime of complex systems such as large organic molecules in solution (*33-35*). In Figure 3, we present RTG and PE measurements for molecular iodine taken with long time delays. The measurements are taken as a function of temperature to illustrate the different mechanisms for coherence relaxation. Notice that the RTG measurements appear not to be temperature dependent in this temperature range (see Fig. 3a). The reason for this observation is that the inhomogeneous contributions are overwhelming the homogeneous relaxation. In the PE measurements, we can see that the homogeneous relaxation times are much longer and are found to decrease with temperature. The cause for the decreased coherence lifetime is an increase in the number density and hence an increase in the collision frequency.

The coherent nature of FWM experiments provides the opportunity to harness the coherent properties of lasers for controlling intra- and inter-molecular degrees of freedom. We have been exploring FWM methods in our group to achieve coherent control of molecular dynamics (*22, 36, 37*). The data in Figure 4, obtained with molecular iodine, show ground and excited state dynamics. The time delay between the first two pulses was set to 614 fs (upper transient) or 460 fs (lower transient). The upper transient shows mostly ground state vibrations with a 160 fs period. The lower transient shows exclusively excited state vibrations with a period of 307 fs.

The transient in Figure 4a can be understood in terms of a coherent anti-Stokes Raman scattering or a pump-dump-pump process. The second pulse creates a coherent superposition of vibrational states in the ground state. This process is enhanced when the time between the first and second pulses matches the vibrational period of the excited state (*22, 36*). The third laser pulse probes the resulting ground state vibrational coherence. For the transient in Figure 4b the time delay between the first two pulses does not permit the transfer of the excited state superposition of states to the ground state. In this case, signal arises

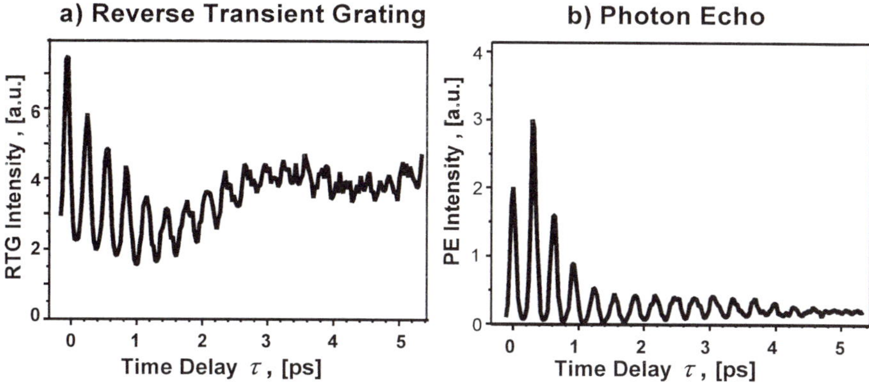

Fig. 2, Comparison between (a) reverse transient grating (RTG) and (b) photon echo (PE) signals for molecular iodine. The two scans were taken under identical condition except for the pulse sequence (see text). The RTG transient shows a large background, a slow modulation due to rotational dynamics and a mixture of ground and excited state vibrations. The PE transient shows only excited state vibrational motion.

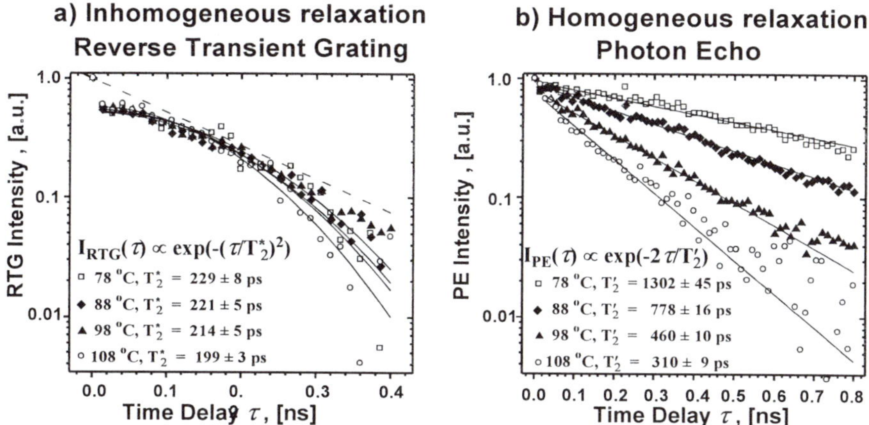

Fig. 3, (a) Reverse transient grating signal for molecular iodine obtained for long time delays. The data was taken as a function of sample temperature. Notice that the RTG data show very little temperature dependence because dephasing is overwhelmed by inhomogeneous broadening in this temperature range. (b) Photon echo signal obtained under the same conditions as the RTG data. Notice that the PE transients do show marked temperature dependence, and this reflects the homogeneous relaxation, which depends on sample density.

Spectrally Dispersed Femtosecond Three-Pulse Four-Wave Mixing

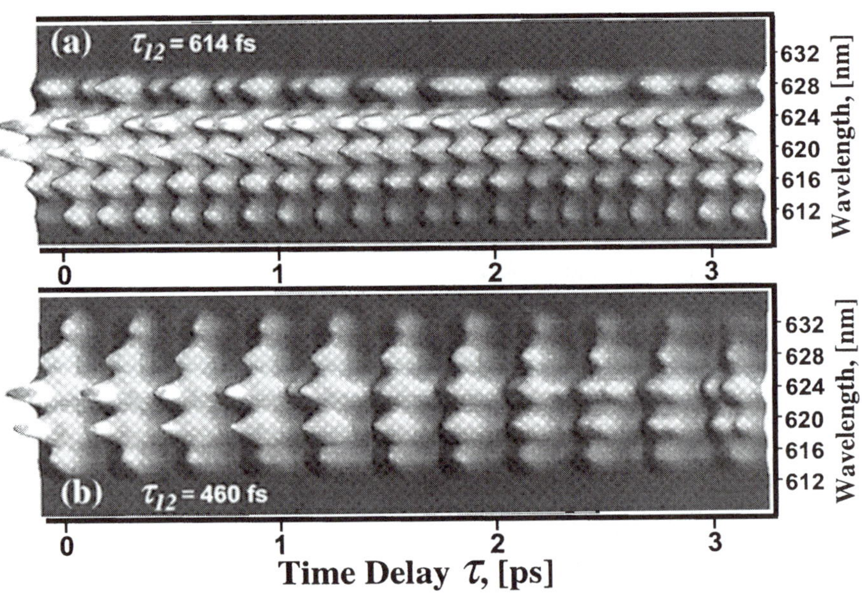

Fig. 4, Spectrally dispersed three-pulse FWM measurements on molecular iodine obtained for two different time delays between the first two pulses. (a) When the time delay between the first two pulses is 614 fs, ground state vibrational motion is observed. (b) When the time delay between the first two pulses is 460 fs, excited state vibrational motion is observed. The two transients were obtained as a function of time delay between the second and third pulses under otherwise identical conditions.

from excitation of a new excited state superposition by the second pulse creating a population in the excited state (*18, 19, 29*). Probing with the third pulse, in this case, results in the observation of excited state dynamics. The overall technique can be considered a pump-pump-dump process. These results show one of the several coherent manipulations that are possible using three-pulse FWM.

3.2 Off-resonance strong-field processes

When two ultrafast pulses cross the sample, a transient grating is formed. The grating depends on the polarizability of the medium and it is modulated by the molecular dynamics in the sample (*18, 21*). Impulsive excitation induces the formation of vibrational and rotational coherences in the ground state. This method has been used to interrogate gases, liquids, and solids (*38*). In our group, we have used the transient grating method to study ground state dynamics of molecules in the gas phase (*20, 21*). More recently, we have studied the effects of intense femtosecond laser beams on molecular systems (*21, 39*). The idea behind those measurements is that strong fields induce a number of processes such as alignment, bond softening, structural deformations, and field-induced ionization (*40, 41, 42*). By tracking the position and shape of rotational recurrences observed in the time evolution of the transient grating, one can detect small structural changes that the molecules experience in the presence of the strong laser fields. The strategy for our strong field measurements involves the creation of the grating using two intense laser pulses overlapped in time. The third pulse, delayed in time and attenuated, probes the rotational recurrences with high precision (10^{-5} cm^{-1}). For example, an increase of 0.001 Å in the CS bond length in CS_2 molecules translates into a 104 fs delay in the first rotational coherence; this change is at least one order of magnitude greater than the precision of our experiments.

The first full rotational recurrence of CS_2 occurs at 76.5 ps (*43, 20*). When the laser pulses are weak, this feature can be observed and an accurate rotational constant can be determined for the linear molecule (*20*). In Figure 5a, we show experimental data for the case when all laser pulses are weak. The simulation (line) of the data with no adjustable parameters fits the experimental results (dots in Figure 5a) well. At the center of the rotational recurrence, the polarizability of the sample achieves a maximum value; however, because the transient grating signal depends on the magnitude squared of the derivative of the transient polarization of the sample (*18, 20, 21*), the value at the center of the recurrence equals zero. The first feature corresponds to the initial increase in the polarizability of the medium caused by rotational rephasing and the second feature at longer times corresponds to a decrease in the polarizability caused by rotational dephasing. In principle, for linear molecules, the first and second

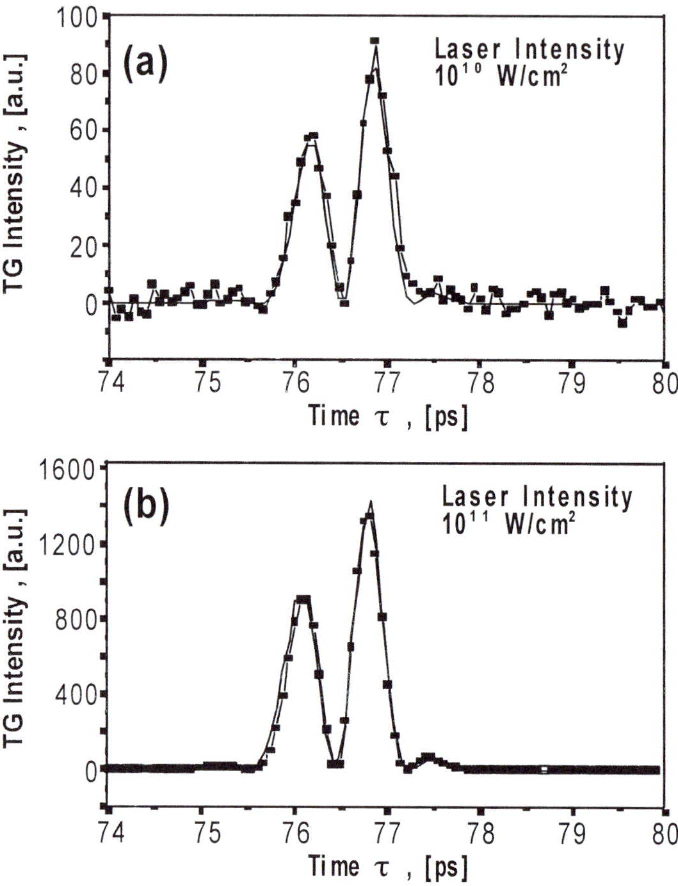

Fig. 5, Off-resonance FWM signals for CS$_2$ as a function of laser intensity. The experimental data show the first full rotational revival at 76.5 ps. The intensity of the first two pulses is indicated in the upper-right corner of each data set the intensity of the third pulse was always below 10^{10} W/cm^2. The experimental data is shown as dots, and a simulation of the data is shown with a thick continuous line. (a) Data obtained for low intensity.(b) Data obtained for higher laser intensity. (c) Data obtained with highest intensity. A portion of the data is also shown with five times magnification to highlight new features observed. The signal intensity is indicated to permit comparison among the different transients.

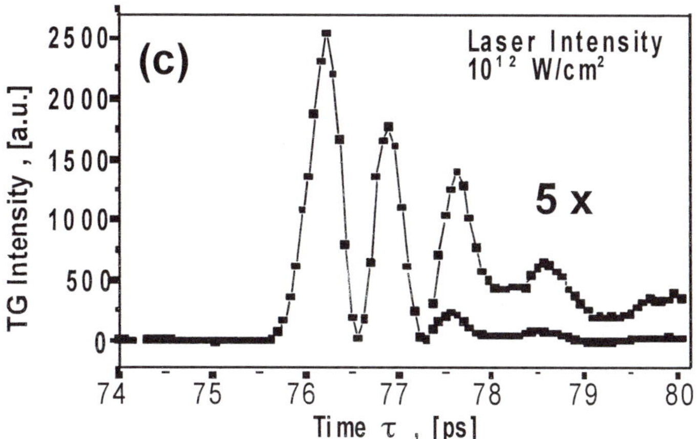

Figure 5. *Continued.*

peaks in a rotational recurrence have the same intensity; however, centrifugal distortion causes a slight increase in the molecular moment of inertia and hence a reduction of the rotational frequency. This effect is evidenced by a more intense second feature in Figures 5a and 5b. When the laser intensity of the first two pulses is increased by an order of magnitude, the full rotational recurrence observed increases in intensity. The data can still be fitted by a room temperature distribution of linear CS_2 molecules (see Figure 5b) with a minor (~ 100 fs) delay noticeable in the first rising edge. The 17x increase in the signal intensity occurs despite the fact the probe beam is kept at $\sim 10^{10}$ W/cm^2 for all measurements. The increase in the signal and slight time delay are consistent with impulsive molecular alignment (*44*).

When the intensity is increased above 10^{12} W/cm^2, the full rotational recurrence signal increases in intensity and changes shape (see Figure 5c). Most markedly, the first feature increases in intensity by a factor of 2.5 while the intensity of the second feature remains constant. The features observed at long time delays, > 77.5 ps, increase in intensity and new oscillations are observed. As mentioned earlier the two features in a rotational recurrence are expected to have the same intensity, or the second one is expected to be more intense. There is no mechanism allowing signal from a linear molecule to have a more prominent first feature, as observed in Figure 5c.

The simulations displayed in Figure 6 were obtained using the same formulas that fit the low intensity data, except for the following modifications caused by the high intensity fields. The simulations (lines) are superimposed on the high intensity experimental data (dots). It is possible that the molecules experience multiple Raman transitions during the first two laser pulses, changing the population of the various rotational levels. A simulation based on this assumption is shown in Figure 6a. Notice that the population in higher rotational levels results in an increase of the centrifugal distortion effects leading to a more prominent second feature. This simulation does not fit the observed experimental result. Figure 6b assumes that the intense pulses have stretched some of the molecules because of bond softening (*45*). The combination of two types of populations, including some with a 0.01 Å bond increase in the CS bond length, results in a simulation that has greater similarity with the experimental data. Figure 6c assumes that some of the molecules are bent by the strong electric fields; this would result from mixing with excited electronic states (*46*). Both singlet and triplet excited states are bent for this molecule (*47*). This simulation gives an excellent fit to the experimental data after 77 ps. Figure 6d is based on a combination of bending and stretching induced by the electric field. A bond angle of 160° and a 10% increase in the C-S bond length result in a simulation that is close to matching the experimental results. These parameters are consistent with the $^1B_2$ state of CS_2 (47). Optimization of the simulation using

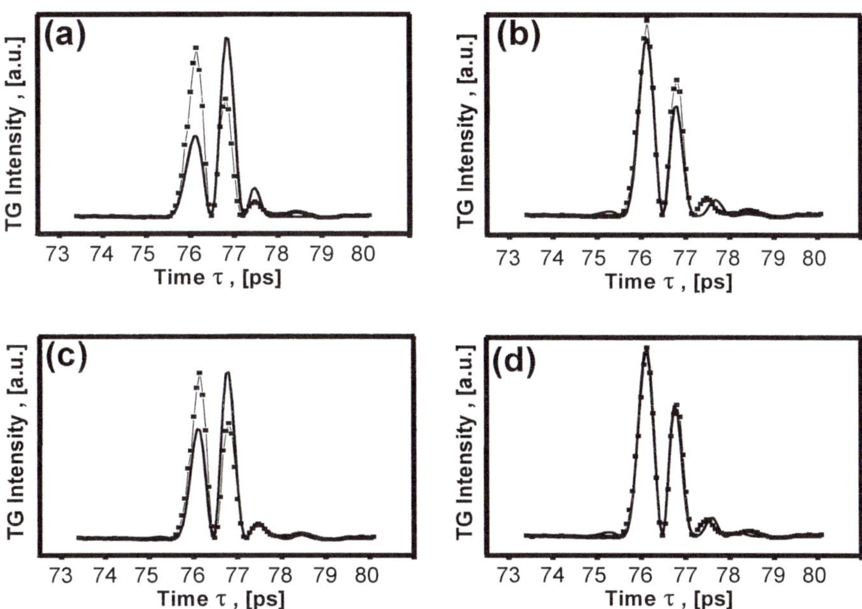

Fig. 6, Experimental FWM signal (dots) from CS_2 obtained at the highest laser intensity together with different simulation models (shown as thick continuous lines).(a) Simulation using a rotational distribution that resulted from 1000 Raman transitions. (b) Simulation assuming a 10% bond stretching caused by the strong field. (c) Simulation assuming molecules are bent by the strong field. (d) Simulation combining bending and stretching of some of the molecules in the sample volume.

nonlinear least squares fitting should result in a better fit of the experimental results.

4. Discussion

The goal of this article is to illustrate some of the fundamental processes that may take place during a strong-field optimal control experiment using three-pulse FWM. We have addressed two types of laser-molecule interactions: resonant third order nonlinear responses and strong-field molecular deformation caused by off-resonance excitation.

4.1 Third-order resonant nonlinear response

With these experiments, we have illustrated two types of nonlinear response: (a) Pulse sequences can manipulate the electronic coherence in the sample enhancing or canceling inhomogeneous broadening effects, and (b) pulse sequences can control the population transfer and vibrational coherence in the sample. The rationale behind these measurements is the fact that the electric field in many optimal control calculations can be reduced to a combination of discrete laser pulses (9). The decoherence times are much longer for photon echo pulse sequences. In liquids, the electronic coherence can be extended from a few femtoseconds to picoseconds using photon echo sequences (48). It is possible that the shaped pulse preserves or minimizes the electronic coherence during an optimal control experiment.

The two-dimensional information obtained in the VE measurements presented in Figure 4 can be used to elucidate the role of chirp in laser excitation (28, 49). These measurements are similar to other wave packet imaging methods (50) and provide an insight into the dynamics of wave packets launched by phase tailored pulses, such as wave packet following effects (51). The pulse sequences illustrated in Figure 4 take advantage of the initial thermal population of different vibrational levels in the ground state to control the microscopic (intramolecular) and macroscopic (intermolecular) coherence in the sample (52). The role of macroscopic coherence in laser control has typically been ignored because it involves interference among different molecules. We consider that the response of the sample, and hence the final yield of a photoinitiated process, depends on the laser induced polarization of the medium. This response is regulated by macro- and microscopic coherence contributions (19, 52). In gas phase samples, electronic coherence can be increased to the nanosecond time scale. While this may not be of direct interest to optimal control of a chemical reaction, the long coherence lifetimes of gas phase molecules provide the opportunity for designing arrangements where multiple coherent interactions are possible (*30*). We have explored different pulse sequences to control the different Liouville pathways that lead to signal formation (*29*). Such setups should be of interest for the coherent manipulation of large numbers of quantum mechanical states. Currently, Apkarian's and our group are exploring the coherent control afforded by FWM techniques and its possible applications to concepts of quantum computation (*53, 54*).

4.2 Off-resonance strong field interactions

The impressive work on optimum control (*11, 12*) involves a strong shaped field where a large number of photons are required to achieve the formation of specific ions. The precise photophysical and photochemical pathways that are followed to achieve the desired product have not been fully identified. The measurements we have presented here provide evidence for alignment, bond softening, and bond deformation processes caused by strong field excitation. While most of the evidence for these processes comes from measurements of ions and electrons generated in the presence of the strong field (*24, 25, 40-42, 44-46*), the results presented here probe the molecules after drifting field-free for 76 ps. Our measurements are sensitive primarily to neutral species and provide accurate structural information rather than detection of a specific ion mass (*11, 12*). We have carried out similar experiments on CS_2 with 800 nm excitation. In these cases, field-induced ionization (*24, 25*) takes place at 10^{13} W/cm^2 and prevents us from obtaining the kind of information we have observed at 620 nm.

5. Conclusion

In this article, we have illustrated a number of nonlinear optically induced processes that may play important roles during an optimum control experiment. These processes have been investigated using three-pulse FWM methods. Future experiments in our group will combine shaped laser pulses and FWM methods to continue to explore nonlinear spectroscopy and issues related to control of molecules and their reactions.

6. Acknowledgements

Support of this work from the National Science Foundation and the Department of Energy is gratefully acknowledged. Additional funding comes from a Packard Science and Engineering Fellowship, an Alfred P. Sloan Research Fellowship and a Camille Dreyfus Teacher-Scholar award.

References

1. Letokhov, V. S. *Science* **1973**, *180*, 451-458
2. Tannor, D. J., Rice, S. A. *J. Chem. Phys.* **1985**, *83*, 5013-5018
3. Brumer, P., Shapiro, M. *Acc. Chem. Res.* **1989**, *22*: 407-413
4. Shi, S., Rabitz, H. *J. Chem. Phys.* **1990**, *92*, 364-376
5. Shapiro, M., Brumer, P. *J. Chem. Phys.* **1993**, *98*, 201-205
6. Gordon, R. J., Rice, S. A. *Annu. Rev. Phys. Chem.* **1997**, *48*, 601-641
7. Tannor, D.J.; Kosloff R.; Rice S. A. *J. Chem. Phys.* **1986,** *85*, 5805-5820.
8. Neuhauser, D., Rabitz, H. *Acc. Chem. Res.* **1993**, *26*, 496-501.
9. Geremia JM, Zhu WS, Rabitz H. *J. Chem. Phys.* **2000**, *113*, 10841-10848
10. Ohtsuki, Y.; Nakagami, K.; Fujimura, Y.; Zhu, W. Rabitz, H. *J. Chem. Phys.* **2001,** *114,* 8867-8876.
11. Assion, A., Baumert, T., Bergt, M., Brixner, T., Kiefer, B., Seyfried, V., Strehle, M., Gerber, G. *Science* 1998, *282*, 919-922
12. Levis R.J., Menkir G.M., Rabitz H. *Science* **2001**, *292*, 709-713
13. Coulston G. W., Bergmann K. *J. Chem. Phys.* **1992**, *96*, 3467-3475
14. Meshulach D., Silberberg Y. *Phys. Rev. A* **1999**, *60*, 1287-1292
15. Dantus, M., Rosker, M. J., Zewail, A. H. *J. Chem. Phys.* **1987**, 87, 2395-2397
16. Shen, Y.R. *The principles of Nonlinear Optics*; Wiley: New York, 1984.
17. Boyd, R.W. Nonlinear Optics; Academic Press: San Diego, 1992.
18. Mukamel, S. *Principles of Nonlinear Optical Spectroscopy*; Oxford University Press: New York, 1995.
19. Grimberg, B.I.; Lozovoy, V. V.; Dantus, M.; Mukamel, S. *J. Chem. Phys. in press,* **2001.**
20. Brown, E. J.; Zhang, Q.; Dantus, M. *J. Chem. Phys.* **1999**, *110*, 5772-5788.
21. Dantus, M. *Annual Review Phys. Chem.* **2001**, *52*, 639-679.
22. Pastirk, I.; Brown, E. J.; Grimberg, B. I.; Lozovoy, V. V.; Dantus, M. *Faraday Discuss.* **1999**, *113*, 401-424.
23. Tellinghuisen, J. J. Quant. Spectrosc. Radiat. Transfer **1978**, *19*, 149-161.
24. Levis, R. J.; DeWitt, M. J. *J. Phys. Chem. A*, **1999**, *103*, 6493-6507
25. Hankin, S.M.; Villeneuve, D.M.; Corkum, P.B.; and Rayner, D.M. *Phys. Rev. Letters* **2000**, *84*, 5082-5088
26. Prior, Y. *Appl. Opt.* **1980**, *19*, 1741-143.
27. Shirley, J. A.; Hall, R. J.; Eckbreth, A. C. *Opt. Lett.* **1980**, *5*, 380-382.

28. Pastirk, I.; Lozovoy, V. V.; Grimberg, B. I.; Brown, E. J.; Dantus, M. *J. Phys. Chem. A.* **1999**, *103*, 10226-10236.
29. Lozovoy, V. V.; Pastirk, I.; Brown, E. J.; Grimberg, B. I.; Dantus, M. *Int. Rev. Phys.Chem.* **2000**, *19*, 531-552.
30. Pastirk, I.; Lozovoy, V. V.; Dantus, M. *Chem. Phys. Letters* **2001**, *333*, 76-82.
31. Kurnit, N. A.; Abella, I. D.; Hartmann, S. R. *Phys. Rev. Lett.* **1964,** *20,* 1087-1089.
32. Patel, C. K. N.; Slusher, R. E. *Phys. Rev. Lett.* **1968**, *20*, 1087-1089.
33. Fleming, G. R.; Cho, M. H. *Ann. Rev. Phys. Chem.* **1996**, *47*, 109-134.
34. Vohringer, P.; Arnett, D. C.; Yang, T. S.; Scherer, N. F. *Chem. Phys. Lett.* **1995**, *237*, 387-398.
35. de Boeij, W. P.; Pshenichnikov, M. S.; Wiersma, D. A. *Annu. Rev. Phys. Chem.* **1998**, *49*, 99-123.
36. Brown, E. J.; Pastirk, I.; Grimberg, B. I.; Lozovoy, V. V.; Dantus, M. *J. Chem. Phys.* **1999**, *111*, 3779-3782.
37. Lozovoy, V. V.; Grimberg, B. I.; Brown, E. J.; Pastirk, I.; Dantus, M. *J. Raman Spectrosc.* **2000**, *31*, 41-49.
38. Fayer, M. D. *Annu. Rev. Phys. Chem.* **1982,** *33,* 63-87
39. Dantus, M. *Faraday Discuss.* **1999**, *113*, 471.
40. Sanderson J.H.; Thomas R.V.; Bryan W.A.; Newell W.R.; Langley A.J.; Taday P.F.*J. Phys. B* **1998**, *31*, L599-L606
41. Graham P.; Ledingham K.W.D.; Singhal R.P.; McCanny T.; Hankin S.M.; Fang X.; Smith D.J.; Kosmidis C.; Tzallas P.; Langley A.J.; Taday P.F. *Phys. B* **1999**, *32*, 5557-5574, and references there in.
42. Banerjee, S. Mathur, D.; and Kumar, G.R. *Phys. Rev. A* **2001**, *63*, 045401-045405
43. Heritage, J. P., Gustafson, T. K., Lin, C. H. *Phys. Rev. Lett.* **1975**, *34*, 1299-1302
44. Dion, C. M., Keller, A., Atabek, O., Bandrauck, A. D. *Phys. Rev. A.* **1999**, *59*, 1382-1391
45. Zavriyev A, Bucksbaum P.H., Squier J, Saline F. *Phys. Rev. Letters* **1993**, *70*, 1077-1080
46. Safvan C.P., Bhardwaj V.R., Kumar G.R., Mathur D., Rajgara F.A. *J. Phys. B* **1996**, *29*, 3135-3149
47. Herzberg, G. Molecular spectra and Molecular Structure III. Electronic Spectra and Electronic Structure of Polyatomic Molecules; Van Nostrand Reinhold Company, New York, 1966.
48. Bardeen, C. J., Shank, C. V. *Chem. Phys. Lett.* **1993**, *203*, 535-539
49. Pastirk, I.; Brown, E. J.; Zhang, Q.; Dantus, M. *J. Chem. Phys.* **1998,** *108*, 4375-4378.

50. Walmsley I.A., Waxer L. *J. Phys. B- At. Mol. Opt.* **1998**, *31*, 1825-1863
51. Yakovlev, V., Bardeen, C. J., Che, J., Cao, J., Wilson, K. R. *J. Chem. Phys.* **1998**, *108*, 2309-2313
52. Lozovoy, V. V.; Grimberg, B. I.; Pastirk, I.; Dantus, M. *Chem. Phys.* **2001**, *267*, 99-114.
53. Zadoyan R., Kohen D., Lidar D. A., Apkarian V. A., *Chem. Phys.* **2001**, *266*, 323
54. Lozovoy V.V., and Dantus M. *Chem. Phys. Letters,* **2001,** submitted

Chapter 6

Control of Chemical Reactions by Using Chirped Laser Pulses

Junici Shimamura, Kenji Mishima, and Koichi Yamashita

Department of Chemical System Engineering, The University of Tokyo, 7–3–1 Hongo, Bunkyo, Tokyo 113–8656, Japan

Quantum wavepacket calculations have been performed to study ways of using chirped laser pulses to control (a) complete population transfer, (b) coherent Raman processes, and (c) vibrational cooling via photoassociation. It has been found that positively and negatively chirped pulses significantly influence the behavior of wavepackets for these elementary processes. Although some of the numerical results can be interpreted from the standpoint of intra-pulse pump-dump processes, we have discovered some novel characteristics of chirped pulses.

Introduction

The possibility of controlling elementary chemical processes in a special environment by specifically manipulating the quantum behavior of atoms and molecules, such as interference and tunneling, has been a subject of great interest in recent years. Various techniques have been developed for optical control of molecular dynamics (*1*). Chirped pulses provide a very simple, but very fundamental technique for optical control. The technique has been successfully applied to experimental studies on controlling molecular dynamics (*1*). During the last few years, we have studied theoretically the control of chemical reactions using chirped laser pulses and have published quantum wavepacket studies on several topics: (a) molecular nonadiabatic processes (*2*), (b) molecular photodissociation (*3–5*), and (c) vibrational dynamics in the presence of relaxation processes (*6*).

Chirped pulses have several characteristics (*7*): (a) the ability to control wavepacket localization, (b) the adiabatic rapid population transfer mechanism, and (c) an intra-pulse pump-dump process. The characteristic that we addressed in this study is the intra-pulse pump-dump process. The basic idea of intra-pulse pump-dump can be illustrated as follows for photo-excitation: Let us think about the excitation from the ground state to a repulsive excited state S_1. The motion of the non-equilibrium population on S_1 results in a dynamic Stokes shift, where the population moves from higher to lower energy. A negatively chirped pulse, where the ordering of the frequency components is from high to low, can follow this motion and thus absorption-stimulated emission or a pump-dump sequence is favored. A positively chirped pulse cannot enhance this process and pure absorption dominates. Thus, we expect that a high-power negatively chirped pulse should excite a lower population to S_1 than the corresponding positively chirped pulse, while a transform-limited pulse should fall somewhere in between.

In the present paper, we briefly describe our recent studies into the ways of controlling molecular vibrational distributions: (a) complete population transfer (*7*), (b) the coherent Raman process (*8*), and (c) vibrational cooling via photoassociation, by using chirped laser pulses. It has been found that positively and negatively chirped pulses significantly influence the behavior of wavepackets for these elementary processes. Although some of the numerical results can be interpreted from the standpoint of intra-pulse pump-dump processes, we have discovered some novel characteristics of chirped pulses.

Complete Population Transfer

In order to discover new mechanisms and the effects of short chirped laser pulses, we have investigated the vibrational population transfer from the electronic ground state S_0 to the excited state S_1 (*8*). Figure 1 shows the potential energy surfaces of our model system. The wavepacket dynamics of the vibrational populations in the electronic ground and excited states were studied by solving the time-dependent Schrödinger equation:

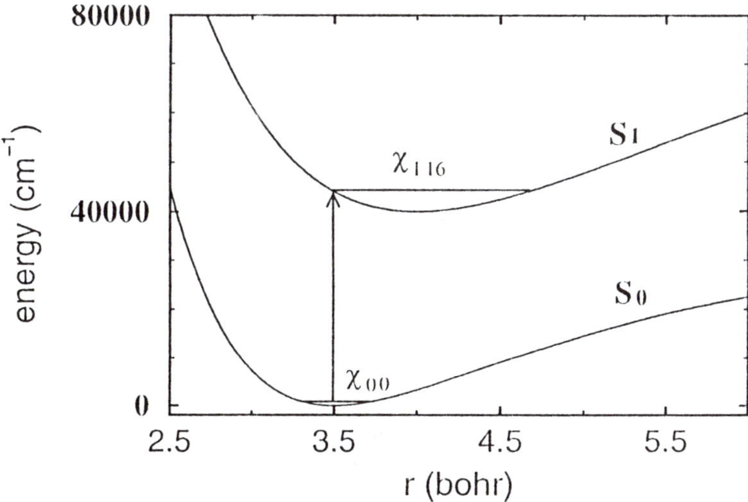

Fig. 1. Potential energy curves for rhe electronic ground and excited states of the model system. The vertical arrow indicates the dominant Franck-Condon transition from the vibrational ground state to the vibrationally excited state (Reproduced with permission from reference 8. Copyright AIP).

$$i\hbar\frac{\partial}{\partial t}\begin{pmatrix}\Psi_0\\\Psi_1\end{pmatrix}=\begin{pmatrix}H_{00} & H_{01}\\H_{10} & H_{11}\end{pmatrix}\begin{pmatrix}\Psi_0\\\Psi_1\end{pmatrix}$$

where the laser–molecule couplings are given by

$$H_{01}=H_{10}=-\varepsilon(t)\mu_{01}.$$

It was demonstrated that both positively and negatively chirped laser pulses with high intensities can achieve almost complete population transfer to the vibrational ground or low excited states in the electronically excited state. Figure 2 shows the populations of the vibrational states in the electronically excited state. Apparently, the transform-limited pulse leads to preferential populations of the states close to the target state $v = 16$. In contrast, both positive and negative chirps achieve preferential populations of lower vibrational states, corresponding to near vibrationally adiabatic transfer to the electronically excited state. Obviously, however, this effect is different, and in fact more efficient for negative than for positive chirps. These findings suggest a different mechanism for positively and negatively chirped pulses. In order to clarify this difference, we looked at the laser-driven wavepacket in detail, that is, the population, the average energy, the coordinate, and the momentum of the wavepacket in the excited state S_1. In the case of a positively chirped pulse, the transition was achieved in a very short time interval, from 150 to 250 fs. During this time interval, the transient frequency of the positive chirp is smaller than the Franck-Condon (FC) frequency, and therefore rather low vibrational states are preferentially excited. The corresponding coordinates were longer than the position of the FC transition. The momentum was positive and therefore the wavepacket ran towards larger values of r, because it was formed in the region of the repulsive wall of the excited state. Some pump-dump processes were also observed, but the pump process remained dominant. That is, there was suppression of multiple intra-pulse pump-dump processes. In contrast with the positively chirped laser pulse, the negatively chirped pulse transferred the population from 100 to 400 fs. This implies that the negatively chirped laser pulse remains nearly on resonance throughout most of the pulse duration, causing multiple series of pump, pump-dump-pump, etc., which transfer populations continuously from S_0 to S_1, and then from S_1 via S_0 back to S_1. That is, the compensation of the intra-pulse pump-dump process occurs. The process induced by the negatively chirped laser pulse was applied to design a complete $S_0 \rightarrow S_1$ population transfer to the lowest vibrational states of the first electronically excited state of 9-(carbazolyl)-anthracene, which could not be observed by means of traditional spectroscopy.

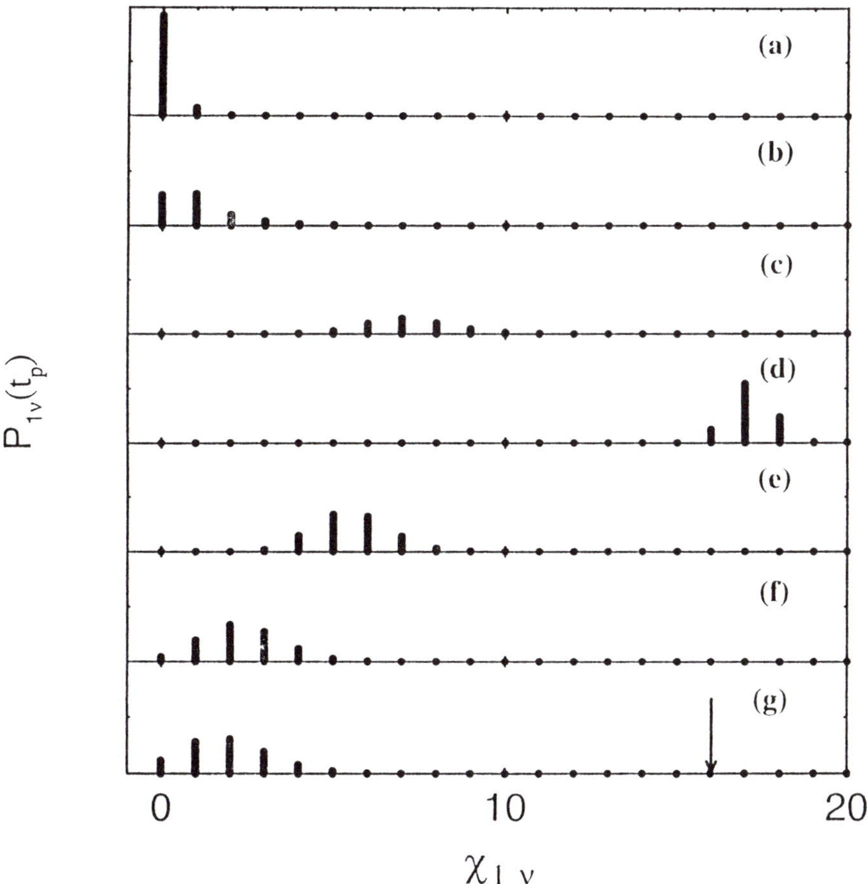

Fig.2. Populations of the vibrational states of the model system in the electronically excited state (pulse duration=500 fs, pulse intensity=26 TW/cm³, pulse frequency=43881 cm⁻¹). Panels (a)-(g) correspond to different chirps, from −140, -70, -35, 0, 35, 70 to 140 cm⁻¹fs⁻¹, respectively (Reproduced with permission from reference 8. Copyright AIP).

The Coherent Raman Process

Raman spectroscopy, which is a two-photon process involving incident and scattered photons, is receiving much attention as a powerful technique for extracting information about potential energy surfaces. Recently, we have derived some analytical formulas for describing spontaneous and stimulated Raman scattering of a molecular system in the presence of a weak pulse-mode chirped laser field, on the basis of second-order perturbation theory (9). We were concerned, in particular, with the effect of the chirping of the incident laser pulse on the Raman spectroscopy profile and the molecular vibrational distribution.

Consider the single molecule (O_2) placed in a weak radiation field. In the case of the $B^3\Sigma_u^-$ state that accommodates some bound vibrational states and many continuum states in the same frequency range as the excitation laser pulse, we have, in the interaction representation,

$$|\Psi(t)\rangle = |\{n_k\}\rangle \Big\{ a_i(t)|\psi_i^{gr}\rangle \exp(-iE_i^{gr}t/\hbar) + \sum_j b_j^{ex}(t)|\psi_j^{ex}\rangle \exp(-iE_j^{ex}t/\hbar - \gamma_{ex}t/2)$$

$$+ \sum_n \int dE\, b_n^E(t)|\psi_n^-(E)\rangle \exp(-iEt/\hbar - \Gamma_{ex}t/2)\Big\}$$

$$+ \sum_j c_j^{gr}(t)|\psi_j^{gr}\rangle \exp(-iE_j^{gr}t/\hbar)\Big\}$$

$$+ \sum_{L,l} |\{n_k + 1_L\}\rangle b_l^L(t)|\psi_l^L\rangle \exp(-iE_l^{gr}t/\hbar).$$

The stationary state $|\psi_i^{gr}\rangle$ is the initial state, and $|\psi_l^l\rangle$ and $|\psi_j^{gr}\rangle$ are the final states produced by spontaneous and stimulated Raman scatterings, respectively. The stationary states of the electronic state manifold $|\psi_j^{ex}\rangle$ and $|\psi_n^-(E)\rangle$ are intermediate bound and continuum scattering states, respectively. The constants γ_{ex} and Γ_{ex} are phenomenological spontaneous emission rates for the intermediate bound and continuum scattering states, respectively. It is a reasonable approximation to take γ_{ex} or Γ_{ex} to be equal for all levels, since all of the excited eigenstates have about the same energy and one does not expect γ_{ex} and Γ_{ex} to vary much from level to level. The ket vector $|\{n_k\}\rangle$ denotes the state of the multi-mode field of the incident light with respective modes and with quantum number n_{kl}. The state $|\{n_k + 1_L\}\rangle$ represents the state $|\{n_k\}\rangle$ plus the photon with mode L emitted spontaneously by the molecule. The spontaneous emission amplitude $b_l^L(t)$ and the stimulated amplitude $c_j^{gr}(t)$ can be derived by substituting the system total wave function into the time-dependent Schrödinger equation.

The light-matter interaction term $V(t)$ is the sum of the radiative interaction of the molecule in the dipole approximation $V_d(t)$ and the spontaneous emission term $V_s(t)$

$$V_d(t) = |\{n_k\}\rangle(-\hat{\mu}_E \cdot E_c(t))\langle\{n_k\}|$$

$$V_s(t) \approx i\sum_s \sqrt{\frac{\hbar\omega_s}{2\varepsilon_0 v}} |\{n_k + 1_s\}\rangle\hat{\mu}_s (\exp(-i\omega_s t) + c.c.)\langle\{n_k\}|$$

Using the instantaneous frequency of the carrier wave $\omega(t)$,

$$\omega(t) = \omega_0 + \alpha t$$

the classical external electric field $E_c(t)$ is defined by a polarization vector $\hat{\varepsilon}_c$, a pulse shape $S(t)$, and the instantaneous frequency of the carrier wave $\omega(t)$,

$$E_c(t) = \hat{\varepsilon}_c S(t)\left[\exp(i\int_0^t \omega(t')dt') + \exp(-i\int_0^t \omega(t')dt')\right]$$

$$S(t) = \frac{E_0}{2}\exp(-A|t|)$$

where the constant A determines the time length of the electric field, ω_0 is the center frequency of the laser pulse, and α is the linear chirp rate. Our idea here is that in the small chirp rate limit, $|\alpha t| \ll \omega_0$, the classical electric field $E_c(t)$ can be expanded to first order with regard to the chirp rate α.

$$E_c(t) = E_{c0}(t) + \delta E_c(t)$$

$$\approx \hat{\varepsilon}_c S(t)\left[(\exp(i\omega_0 t) + \exp(-i\omega_0 t)) + i\alpha t^2/2\,(\exp(i\omega_0 t) - \exp(-i\omega_0 t))\right]$$

After some tedious manipulations, the spontaneous emission amplitude $b_l^L(t)$, and the stimulated emission amplitude $c_j^{gr}(t)$, can be composed of the zeroth-order terms, $b_{l0}^{L}(t)$ and $c_{j,0}^{gr}(t)$, and the first-order perturbation terms, $\delta b_{l}^{L}(t)$ and $\delta c_{j}^{gr}(t)$, regarding the linear chirp rate α,

$$b_{l}^{L}(t) = b_{l0}^{L}(t) + \delta b_{l}^{L}(t),$$

$$c_{j}^{gr}(t) = c_{j,0}^{gr}(t) + \delta c_{j}^{gr}(t).$$

The zeroth-order term $b_{l0}^{L}(t)$ represents the resonance fluorescence part of the spontaneous emission amplitude, the spontaneous Raman process component, and the time-independent Kramers-Heisenberg-Dirac (KHD) term with damping factors. The first-order perturbation term $\delta b_{l}^{L}(t)$ includes the influence of the chirping on the spontaneous emission amplitude, which is also composed of two distinct time-dependent terms—resonance fluorescence and spontaneous Raman —as well as time-independent KHD terms.

In this paper, our interest is focused on the calculation of physical quantities: (1) the probability of observing a spontaneously emitted photon in the L mode at time t,

$$P_{L}(t) = \sum_{l} \left| b_{l}^{L}(t) \right|^{2}$$

and (2) the molecular vibrational distribution of the vibrational state $v = l$ of the ground electronic state due to spontaneous emission $Q_{l}^{sp}(t)$ and stimulated emission $Q_{l}^{st}(t)$:

$$Q_{l}^{sp}(t) = \frac{v}{\pi^{2} c^{3}} \int d\omega_{L}\, \omega_{L}^{2} \left| b_{l}^{L}(t) \right|^{2}$$

Since the time-dependent Raman and fluorescence terms of $b_{l}^{L}(t)$ vanish at the limit of $t \to \infty$, we can derive the Raman spectral intensity ratio $P_{rel}(t \to \infty) = P_{L}(t \to \infty) / P_{L0}(t \to \infty)$, with $P_{L}(t \to \infty)$ including chirp, and $P_{L0}(t \to \infty)$ without chirp, as follows:

$$P_{rel}(t \to \infty) = 1 + \frac{4\alpha^{2}}{A^{4}}$$

It can easily be seen that the Raman spectral intensity does not depend on the sign of the chirp rate, and that the intensity in the presence of the chirp increases relative to that of the transform-limited pulse, irrespective of the details of the molecular characteristics, e.g., the PES.

Figure 3 shows the molecular vibrational distributions $Q_l^{sp}(t)$ of the vibrational state $v = l$ of the ground electronic state due to the spontaneous emission. It should be noted that the final vibrational distributions are larger for the chirped pulse than for the transform-limited pulse, and that the difference resulting from changing the chirp rate sign is rather small. A more distinctive feature that can be seen is that the final vibrational distributions due to the chirped pulses are larger, compared with the transform-limited pulse, for the long duration pulse than for the short duration pulse. These numerical results agree well with the expectation based on $P_{rel}(t \to \infty)$.

The dependence on the sign of the chirp rate appears most prominently in stimulated emission. In Figure 4, it can be seen that the molecular vibrational distributions $Q_l^{st}(t \to \infty)$ are significantly different according to the chirp rate in the fundamental and the higher overtones. The distribution is widest when the incident laser pulse is negatively chirped, while it is narrowest when a positively chirped laser pulse is used. It may physically be assumed that the differences observed above among the positively, negatively, and transform-limited pulses are due to the intra-pulse pump-dump process. The excited state wavepacket photo-induced from the ground state slides down the curve because the wavepacket created in the excited electronic state is not an eigenstate of the vibrational manifold. The motion along the potential energy curve (PEC) has the effect of a dynamical Stokes shift: the transition energy $V_{ins}(t)$ between the excited state wavepacket and the ground state PEC decreases with time. The negatively chirped pulse whose instantaneous frequency is $\omega(t)$ follows the behavior of $V_{ins}(t)$, and the excited state wavepacket is dumped to the ground state effectively. However, positively chirped pulses do not follow the behavior of $V_{ins}(t)$ and the dumping effect does not work. It should be noted that the stimulated Raman process in the presence of the chirp is quite sensitive to the molecular characteristics, in contrast to the spontaneous Raman process.

In conclusion, the spontaneous emission probability is enhanced by chirping the incident laser pulse, irrespective of the sign of the chirp rate, although the spectral profile does not change much. The stimulated Raman emission probability shows a propensity readily attributable to the intra-pulse pump-dump process.

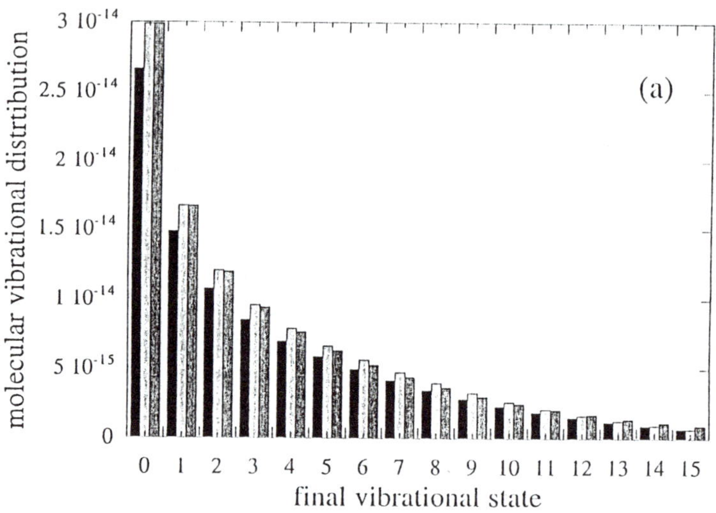

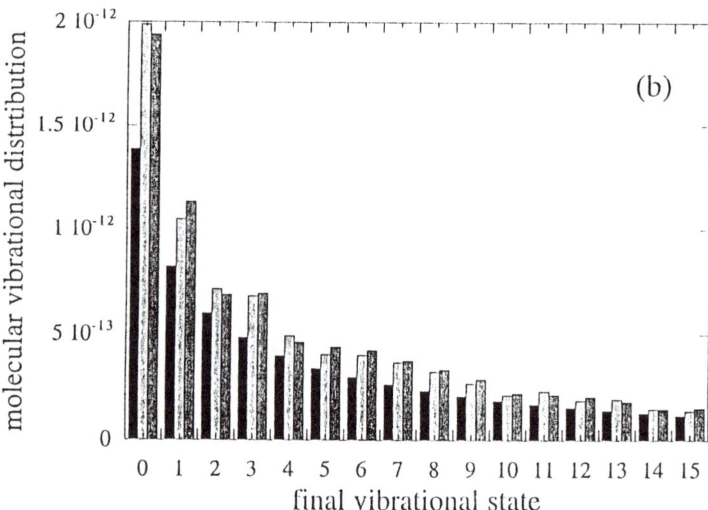

Fig.3. Final molecular vibrational distributions of the ground electronic state due to spontaneous emission $Q_l^{sp}(t \to \infty)$ in the resonance condition. The laser pulse intensity is $3.54 \times 10^6 W/cm^2$. Figure (a) is for the short ($A=2.27fs^{-1}$) and (b) for the long ($A=0.62fs^{-1}$) laser pulse. The light gray, black, and dark gray bars are for positively chirped, transform-limited, and negatively chirped laser pulses, respectively.

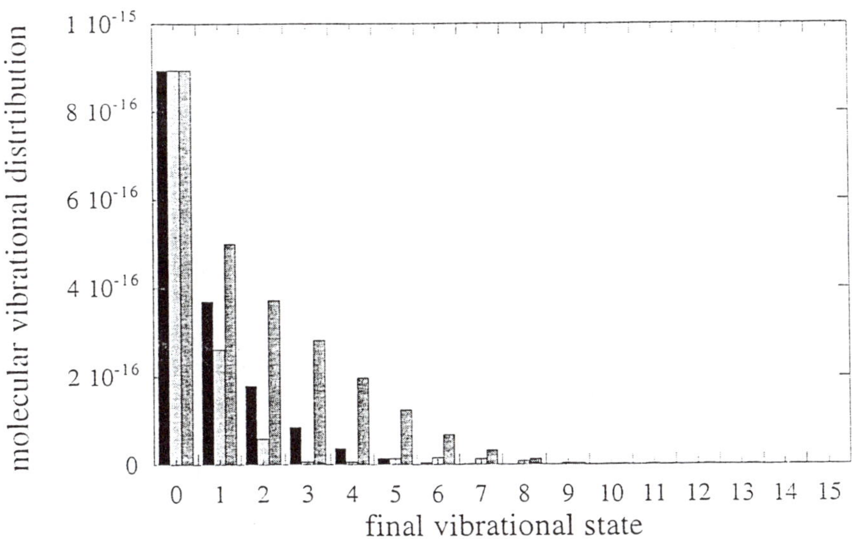

Fig. 4. *Final molecular vibrational distribution of the ground electronic state due to stimulated emission* $Q_i^{st}(t \to \infty)$ *in the resonance condition. The light gray, black, and dark gray bars are for positively chirped, transform-limited, and negatively chirped laser pulses, respectively.*

Vibrational Cooling via Photoassociation

In view of the desirability of making cold atoms, it is not surprising to wish to make cold molecules. However, strategies that work well for atoms fail for molecules, mainly because molecules contain many degrees of freedom compared with atoms, and therefore, while the cooling of atoms was demonstrated more than a decade ago, cooling of molecules is still a challenging task (*10*). The recent observation of ground state ultra-cold molecules obtained by spontaneous radiative decay of photoassociated molecules (*11*) is opening up a new research field. However, it should be noted here that although the experiment succeeded in producing translationally cold molecules via the photoassociation of cold atoms, the molecules were vibrationally hot (*11*). In this section, we propose a photoassociation reaction using chirped infrared (IR) laser pulses that can create a vibrationally cold molecule from two colliding cold atoms. It is expected that by using an IR laser the initial continuum state of the colliding pair will be transferred into the bound vibrational states of the colliding pair by a stimulated emission process. Our strategy is that efficient vibrational de-excitation can be achieved by a step-by-step descent of vibrational levels, and finally a vibrationally cold molecule can be created by sweeping the frequency of the IR laser, that is, by using a chirped IR laser pulse. This is just the reverse process of the control of vibrational excitation and dissociation of a diatomic molecule using chirped IR laser pulses that was successfully demonstrated by Chelkowski and Bandrauk (*12*).

Our model system is the photoassociation of $O(^3P) + H(1s)$. For this system, Korolkov et al. (*13*) investigated a photoassociation reaction controlled by an IR picosecond (transform-limited) laser pulse. The relative motion of colliding atoms is modeled by a Gaussian wavepacket. The interaction of a rotating molecule with the laser field is assumed to be linearly polarized along the projection on the molecular axis.

$$H_{\text{int}}(r,t) = -\mu(r)\cos\theta E(t)$$

The wavepacket of the system is represented by the expansion of the wave function in spherical harmonics $Y_{j,0}(\theta)$ (only $m_j = 0$ initial states are considered),

$$\Psi(r,\theta,t) = \sum_{j=0}^{j_{\text{max}}} \psi_j(r,t) Y_{j,0}(\theta)/r$$

The following differential equation is obtained by inserting the wavepacket of the system into the time-dependent Schrödinger equation and propagating it by the split operator method.

$$i\hbar\frac{\partial\psi_j}{\partial t} = -\frac{\hbar^2}{2m}\frac{\partial\psi_j}{\partial r} + V_j(r)\psi_j - \mu E(t)\left(P_{j,0}\psi_{j+1} + P_{j-1,0}\psi_{j-1}\right)$$

where

$$P_{j,0} = \left\langle Y_{j+1,0}\left|\cos\vartheta\right|Y_{j,0}\right\rangle = (j+1)\left[(2j+1)(2j-3)\right]^{-1/2}$$

$$V_j(r) = V(r) + \frac{\lambda^2 j(j+1)}{2mr^2}$$

In order to produce an efficient step-by-step vibrational de-excitation of the colliding pair through photoassociation, it is necessary to optimize the rate of the frequency sweep, that is the chirp rate. Our model is that the colliding pair first transit to a vibrational level of around $v = 15$ due to the dominant process of a free-bound transition by stimulated emission and then transfer to the lower vibrational levels, step by step. The OH potential function used in this study supports 22 bound states. We consider the system as a sequence of two level systems and adjust the frequency sweep. Based on the theory of two level systems, it is well known that the transition amplitudes are functions of the pulse area, given by

$$\sigma(t) = (p/\hbar)\int_{-\infty}^{t}\varepsilon(t')dt'$$

where p is the transition dipole moment, and $\varepsilon(t)$ is the electric field envelope. Since complete population inversion occurs when the pulse area equals π, one can expect that a pulse with frequency $\omega_{n,n+1}(t) = (E_{n+1} - E_n)/\hbar$ should completely transfer the population from the vibrational state of $v = n + 1$ to the state of $v = n$, that is,

$$(p/\hbar)\int_{t_{n+1}}^{t_n}\varepsilon(t')dt' = \pi$$

The population of the vibrational states can be calculated by

$$P_{v,j}(t) = \left|\left\langle\phi_{v,j}\left|\Psi\right\rangle\right.\right|^2$$

where $\phi_{v,j}$ is the eigenfunction of vibrational quantum number v, and rotational quantum number j.

Figure 5 shows the vibrational distribution of the OH molecule created by the photoassociation of colliding O and H atoms for $J = 0$. Clearly, a vibrationally de-excited OH molecule is created. The intensity of the pulse is 1.0×10^{13} W/cm^2 and the pulse duration is 3.0 ps. The optimized frequency modulation is positively chirped and quadratically dependent on time. On the other hand, in the case of $J \neq 0$, our calculations show a considerable sensitivity of the vibrational distribution to the initial rotational quantum number. Figure 6 shows the vibrational distributions for $J = 1$–3 in the case of the initial rotational quantum number being 5. The vibrational de-excitation is much less effective, and this may be due to vibration-rotation energy transfer. One might think about using another pulse to de-excite the molecular rotations. However, Tannor and Bartana (*14*) discussed the control of molecular rotations and concluded that the rotational cooling mechanism should be completely different from that of vibrational cooling. They suggest that rotational cooling can be achieved only through the inclusion of spontaneous emission, which is difficult to control by external fields. One possible way to avoid rotational cooling may be to perform the photoassociation reaction under the control of the orientation of the colliding atoms.

Acknowledgements

The present research is partially supported by a grant-in-aid for Scientific Research on the priority Area "Molecular Physical Chemistry" from the Ministry of Education, Science, Culture and Sports of Japan. KY thanks Drs. J. Manz, H. Naundorf and Y. Zhao for their fruitful collaborations.

References

1. Rice, S. A.; Zhao, M. *Optical Control of Molecular Dynamics*; Wiley Inter-Science: New York, 2000.
2. Mishima, K.; Yamashita, K. *J. Chem. Phys.* **1998**, *109*, 1801.
3. Mishima, K.; Yamashita, K. *Int. J. Quantum. Chem.* **1999**, *72*, 525.
4. Mishima, K.; Yamashita, K. *J. Mol. Struct. (THEOCHEM)*, **1999**, *461-462*, 483.
5. Mishima, K.; Yamashita, K. *J. Chem. Phys.* **1999**, *110*, 7756.

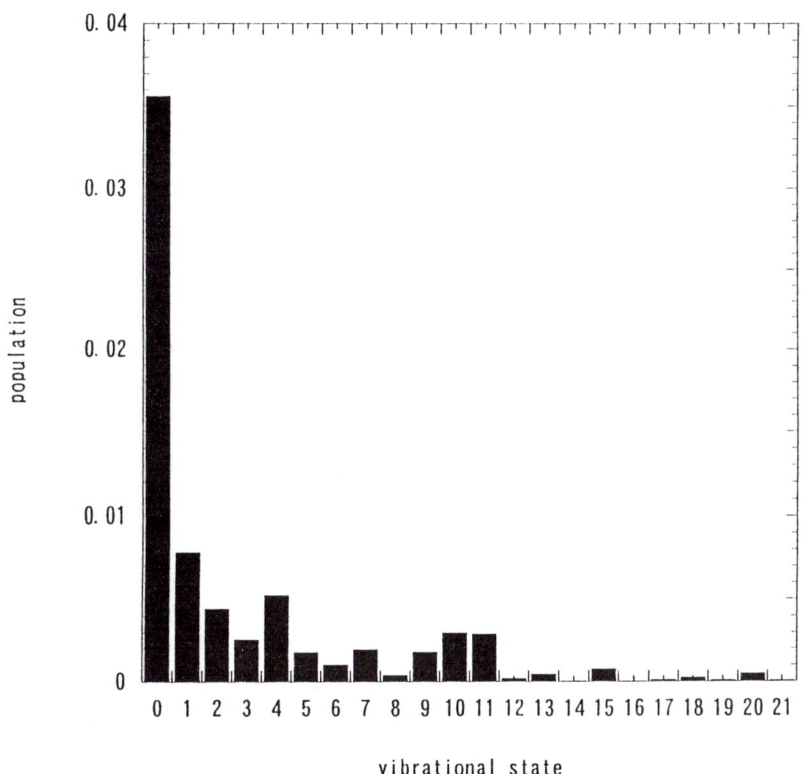

Fig.5. Vibrational distributions of the OH molecule created by photoassociation of colliding O and H atoms for J=0.

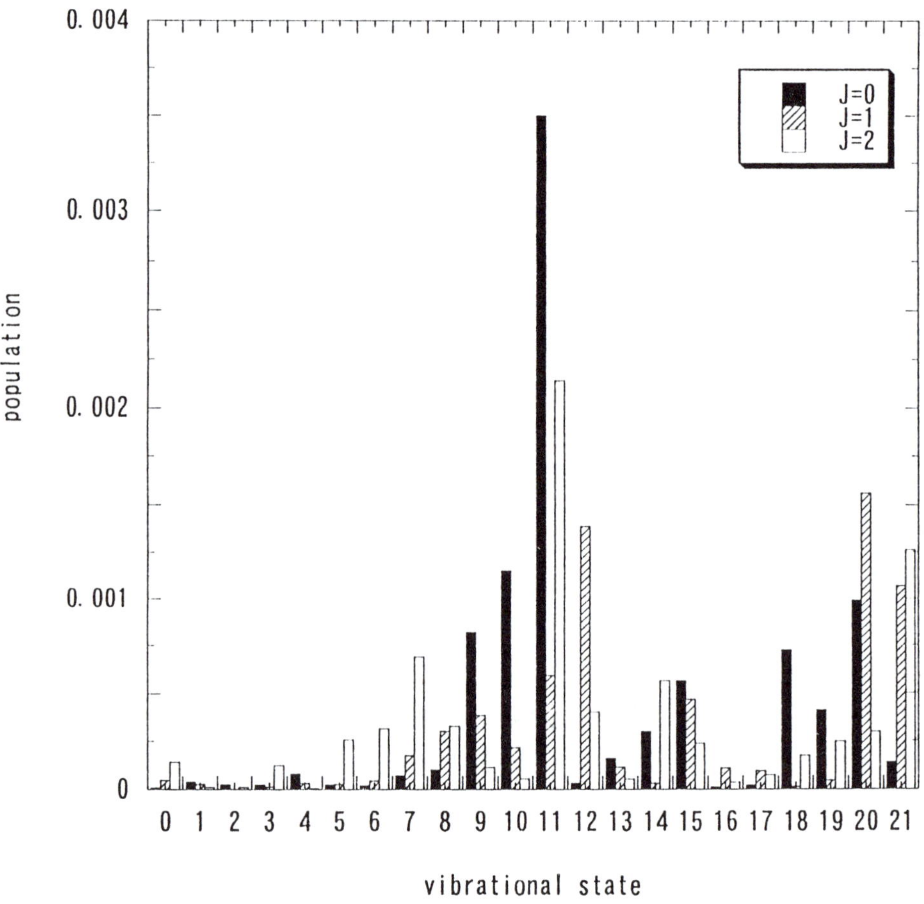

Fig.6. Vibrational distributions for J=1-3 in the case of the initial rotational quantum number being 5.

6. Mishima, K.; Hayashi, H.; Lin, J. T.; Yamashita, K.; Lin, S. H. *Chem. Phys. Lett.* **1999**, *309*, 279.

7. Cao, J; Bardeen, C. J.; Wilson, K. R. *J. Chem. Phys.* **2000**, *113*, 1898.

8. Manz, J; Naundorf, H; Yamashita, K; Zhao, Y. *J. Chem. Phys.* **2000**, *113*, 8969.

9. Mishima, K.; Yamashita, K. *J. Phys. Chem. A,* **2001**, *105*, 2867.

10. Bartana, A.; Kosloff, R.; Tannor, D. J. *J. Chem. Phys.* **1997**, *106*, 1435.

11. Fioretti, A.; Comparat, D.; Crubellier, A.; Dulieu, O.; Masnou-Seeuws, F.; Pillet, P. *Phys. Rev. Lett.* **1998**, *80*, 4402.

12. Chelkowski, S; Bandrauk, A.D. *J. Chem. Phys.* **1993**, *99*, 4279.

13. Korolkov, M. V.; Manz, J.; Paramonov, G. K.; Schmidt, B. *Chem. Phys. Lett.* **1996**, *260*, 604.

14. Tannor, D. J.; Bartana, A. *J. Phys. Chem. A* **1999**, *103*, 10359.

Chapter 7

Selective Excitation among Closely Lying Multi levels

Kuninobu Nagaya[1], Yoshiaki Teranishi[2], and Hiroki Nakamura[1,3]

[1]Department of Functional Molecular Science, School of Mathematical and Physical Science, The Graduate University for Advanced Studies Myodaiji, Okazaki 444–8585, Japan
[2]RIKEN, Hirosawa 2–1, Wako 351–01, Japan
[3]Department of Theoretical Studies, Institute for Molecular Science, Myodaiji, Okazaki 444–8585, Japan

A new idea is proposed to accomplish selective and complete excitation to any specified state among closely lying multi-levels. The basic idea is to control nonadiabatic transitions among dressed states by sweeping the laser frequency periodically. Both three- and four-level models are treated by the semiclassical theory of nonadiabatic transition and conditions of complete excitation are formulated. Numerical demonstrations are presented in comparison with the π-pulse and adiabatic rapid passage.

Recently, control of molecular processes or chemical reactions by laser fields has attracted much attention and has become a hot topic of science thanks to a remarkable progress of laser technology.[1,2] Several ideas have been proposed so far such as coherent control,[3] pump-dump method,[4] pulse-shape driven control,[5-8] and adiabatic rapid passage with use of the chirped pulse.[9-13]

By introducing the Floquet (or dressed) state formalism,[14] various processes in periodic external fields can be regarded as a sequence of nonadiabatic transitions among adiabatic Floquet states. Two types of control schemes from the viewpoint of nonadiabatic transitions have been proposed recently.[15-17] One is to control molecular photodissociation in a stationary laser field with use of the complete reflection phenomenon in time-independent nonadiabatic transition.[15] The other is to control population transfer by using nonstationary laser fields with explicit use of the time-dependent nonadiabatic transitions.[16,17]

In this chapter a new scheme of selective and complete excitation among closely lying multi-levels is proposed based on the above mentioned second method. Based on the basic two-state theory,[16,17] three (one ground and two excited)- and four (one ground and three excited)-level models are formulated and conditions of selective and complete excitation are derived in simple analytical forms. A brief discussion is also presented for a general multi-level case. Numerical examples are provided for the three- and four-level cases and advantages of the present scheme are demonstrated in comparison with the π-pulse and adiabatic rapid passage (ARP). For instance, the present method can complete the transition in a much shorter time compared to the latter methods, and would be useful for systems with fast relaxation.

Selective Excitation by Periodic Chirping

As mentioned above, dynamic processes in nonstationary laser fields can be treated as a sequence of time-dependent nonadiabatic transitions among adiabatic Floquet states within the Floquet formalism, in which the laser frequency and/or laser intensity can be regarded as adiabatic parameters. In the simplest case of an isolated two-level model with the laser frequency ω as an adiabatic parameter (see Figure 1), the control condition can be formulated analytically for any specified initial and final states with use of the semiclassical theory of time-dependent nonadiabatic transition.[16,17] The control parameters are the sweep rate, the sweep amplitude of laser frequency (ω_1, ω_2), and the number of sweeps. In the following we will consider generalization of this idea to multi-level systems.

Three-level (1+2) case

First, let us consider the three level system composed of $|1>$, $|2>$ and $|3>$, in which the two higher levels are close to each other (see Figure 2(a)), i.e. $\omega_{12} \gg \omega_{23}$, where $\omega_{ij} = (E_j - E_i)/\hbar$ and E_i is the energy of $|i>$. When the laser field of the frequency ω near resonant to the middle level $|2>$, i.e. $\omega \approx \omega_{12}$, is applied to this system, the Floquet diagram is given by Figure 2(b). Under this frequency condition a transition between $|2>$ and $|3>$ is negligible and the Floquet Hamiltonian is expressed as

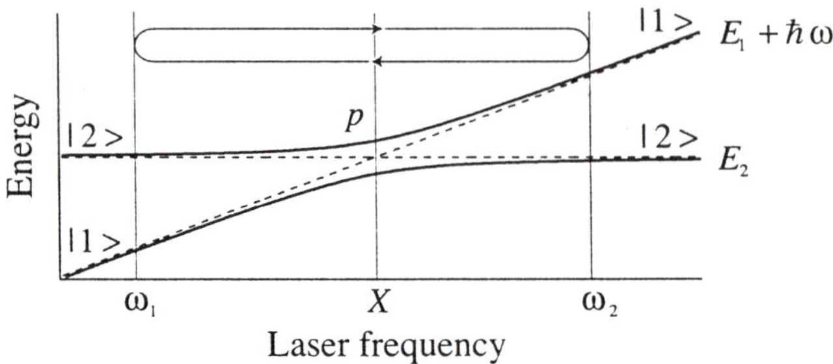

Figure 1. Schematic Floquet picture of a two-level model as a function of laser frequency ω. Dotted lines:diabatic Floquet states, solid lines:adiabatic Floquet states.

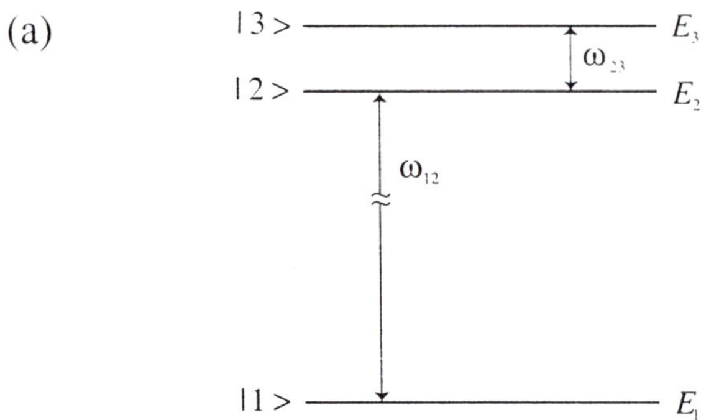

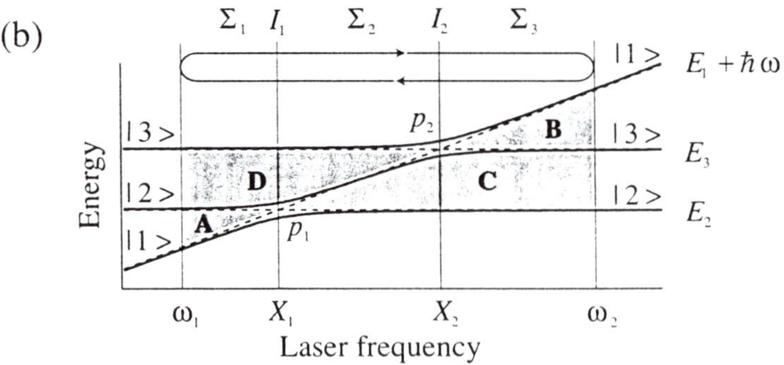

Figure 2. (a)Schematic level structure of a three-level model. (b)Floquet diagram of the three-level model shown in Fig.2(a) as a function of laser frequency.

$$H_{\text{Floquet}} = \begin{bmatrix} E_1 + \hbar\omega(t) & -\mu_{12}\varepsilon(t)/2 & -\mu_{13}\varepsilon(t)/2 \\ -\mu_{12}\varepsilon(t)/2 & E_2 & 0 \\ -\mu_{13}\varepsilon(t)/2 & 0 & E_3 \end{bmatrix}, \tag{1}$$

where μ_{ij} is the transition dipole moment between $|i>$ and $|j>$ and $\varepsilon(t)$ is an envelope function of the laser field. The Floquet states cross at two laser frequencies X_1 and X_2. When the laser frequency is swept from ω_1 to ω_2, the excitation process can be divided into two parts: adiabatic propagation along the adiabatic Floquet states and nonadiabatic transitions at the crossings. In the semiclassical treatment[16-18] the transition matrix $T_{\omega1\to\omega2}$ in the adiabatic state representation is explicitly given by

$$T_{\omega1\to\omega2} = \Sigma_3 I_2 \Sigma_2 I_1 \Sigma_1, \tag{2}$$

where Σ_j represents the adiabatic propagation from X_{j-1} to X_j, and I_j represents the nonadiabatic transition (I-matrix) at the crossing X_j, i.e.,

$$\Sigma_j = \begin{bmatrix} e^{-i\sigma_j^{(1)}} & 0 & 0 \\ 0 & e^{-i\sigma_j^{(2)}} & 0 \\ 0 & 0 & e^{-i\sigma_j^{(3)}} \end{bmatrix}, \tag{3}$$

$$\sigma_j^{(k)} = \frac{1}{\hbar}\int_{X_{j-1}}^{X_j} E_k^{(a)}(\omega)d\omega\frac{dt}{d\omega}, \tag{4}$$

$$I_1 = \begin{bmatrix} \sqrt{1-p_1}\,e^{i\varphi_1} & \sqrt{p_1}\,e^{i\psi_1} & 0 \\ -\sqrt{p_1}\,e^{-i\psi_1} & \sqrt{1-p_1}\,e^{-i\varphi_1} & 0 \\ 0 & 0 & 1 \end{bmatrix}, \tag{5}$$

and

$$I_2 = \begin{bmatrix} 1 & 0 & 0 \\ 0 & \sqrt{1-p_2}\,e^{i\varphi_2} & \sqrt{p_2}\,e^{i\psi_2} \\ 0 & -\sqrt{p_2}\,e^{-i\psi_2} & \sqrt{1-p_2}\,e^{-i\varphi_2} \end{bmatrix}. \tag{6}$$

Here p_j denotes the nonadiabatic transition probability for one passage of the avoided crossing X_j, φ_j and ψ_j are the dynamical phases due to the nonadiabatic transition at the avoided crossing X_j, $E_k^{(a)}(\omega)$ is the k-th adiabatic Floquet state, $X_0 = \omega_1$ and $X_3 = \omega_2$. $T_{\omega1\to\omega2}$ can be simplified as

$$T_{\omega1\to\omega2} = \begin{bmatrix} \sqrt{1-p_1} & \sqrt{p_1}\,e^{-iA} & 0 \\ -\sqrt{p_1(1-p_2)}\,e^{-iC} & \sqrt{(1-p_1)(1-p_2)}\,e^{-i(A+C)} & \sqrt{p_2}\,e^{-i(A+C+D)} \\ \sqrt{p_1 p_2}\,e^{-i(B+C)} & -\sqrt{(1-p_1)p_2}\,e^{-i(A+B+C)} & \sqrt{1-p_2}\,e^{-i(A+B+C+D)} \end{bmatrix}, \tag{7}$$

where

$$A = \varphi_1 - \psi_1 + \Delta\sigma_1^{(2,1)}, \tag{8.a}$$
$$B = \varphi_2 + \psi_2 + \Delta\sigma_3^{(3,2)}, \tag{8.b}$$
$$C = \varphi_1 + \psi_1 - \varphi_2 + \Delta\sigma_2^{(2,1)} + \Delta\sigma_3^{(2,1)}, \tag{8.c}$$

$$D = -\varphi_1 + \varphi_2 - \psi_2 + \Delta\sigma_1^{(3,2)} + \Delta\sigma_2^{(3,2)}, \qquad (8.d)$$

and

$$\Delta\sigma_j^{(k,l)} \equiv \sigma_j^{(k)} - \sigma_j^{(l)}. \qquad (9)$$

These phases roughly correspond to the areas shown in Figure 2(b). As is shown below, one period of sweeping $(\omega_1 \to \omega_2 \to \omega_1)$ is enough to excite the state $|1>$ to $|2>$ completely. The overall transition matrix $T^{(1)}$ can be expressed as

$$T^{(1)} = T_{\omega2\to\omega1} \cdot T_{\omega1\to\omega2} = \left(T_{\omega1\to\omega2}\right)^t T_{\omega1\to\omega2}. \qquad (10)$$

The transition probability from $|1>$ to $|2>$ for one period sweep is explicitly given by

$$P_{12}^{(1)} = \left|\left(T^{(1)}\right)_{21}\right|^2 = p_1(1-p_1)\left|e^{2iC} - 1 + p_2\left(1 - e^{-2iB}\right)\right|^2. \qquad (11)$$

By taking account of $0 \le p_1(1-p_1) \le 1/4$ and $|e^{2iC} - 1 + p_2(1 - e^{-2iB})|^2 \le 4$, the condition of the complete excitation from $|1>$ to $|2>$ for one period of sweeping (i.e. $P_{12}^{(1)} = 1$) is expressed as

$$p_1 = \frac{1}{2}, \quad B = m\pi \quad \text{and} \quad C = \left(n + \frac{1}{2}\right)\pi \quad (m,n:\text{integer}). \qquad (12)$$

The physical meaning of $B = m\pi$ is that no bifurcation into the diabatic state $|3>$ occurs at X_2 on the second half of the sweep $(\omega_2 \to \omega_1)$ whatever the value of p_2 is. The conditions of p_1 and C guarantee that the interference between $|1>$ and $|2>$ at X_1 on the way back leads to the complete excitation to $|2>$. It is not possible to make complete excitation from $|1>$ to $|3>$ by one period of sweeping, because some portion is always bifurcated into $|2>$ at X_1 on the first half period $(\omega_1 \to \omega_2)$. However, if we start from ω_2, then one period of sweeping is good enough and the condition is given by

$$p_2 = \frac{1}{2}, \quad A = m\pi \quad \text{and} \quad D = \left(n + \frac{1}{2}\right)\pi \quad (m,n:\text{integer}). \qquad (13)$$

Four-level (1+3) case

Next, let us consider the four-level system such as the one shown in Figure 3(a), in which three higher levels $|2>$, $|3>$ and $|4>$ are close to each other, i.e. $\omega_{12} \gg \omega_{23}, \omega_{34}$. When the laser field of the frequency ω near resonant to the second level $|2>$, i.e. $\omega \approx \omega_{12}$, is applied to this system, the Floquet diagram such as Figure 3(b) can be generated. When the laser frequency is swept from ω_1 to ω_2, the transition matrix in the adiabatic state representation can be expressed as

$$T_{\omega1\to\omega2} = \Sigma_4 I_3 \Sigma_3 I_2 \Sigma_2 I_1 \Sigma_1, \qquad (14)$$

where

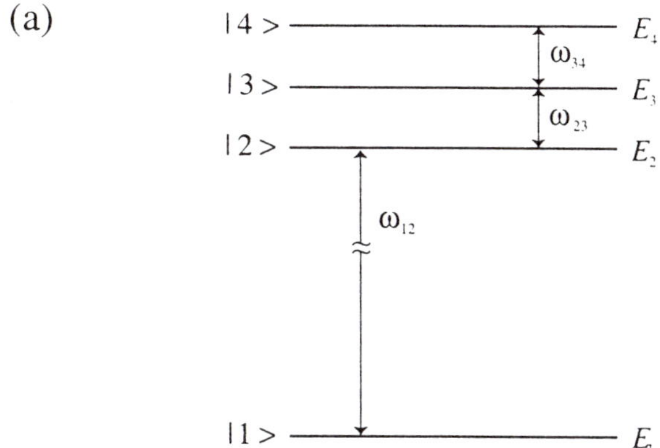

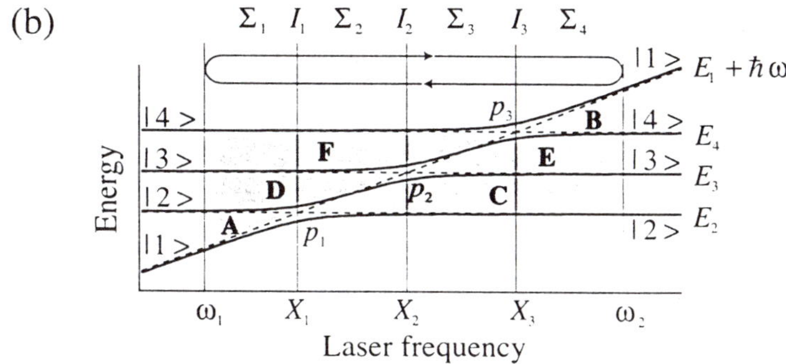

Figure 3. (a)Schematic level structure of a four-level model. (b)Floquet diagram of the four-level model shown in Fig.3(a) as a function of laser frequency.

$$\Sigma_j = \begin{bmatrix} e^{-i\sigma_j^{(1)}} & 0 & 0 & 0 \\ 0 & e^{-i\sigma_j^{(2)}} & 0 & 0 \\ 0 & 0 & e^{-i\sigma_j^{(3)}} & 0 \\ 0 & 0 & 0 & e^{-i\sigma_j^{(4)}} \end{bmatrix}, \tag{15}$$

$$I_1 = \begin{bmatrix} \sqrt{1-p_1}\,e^{i\varphi_1} & \sqrt{p_1}\,e^{i\psi_1} & 0 & 0 \\ -\sqrt{p_1}\,e^{-i\psi_1} & \sqrt{1-p_1}\,e^{-i\varphi_1} & 0 & 0 \\ 0 & 0 & 1 & 0 \\ 0 & 0 & 0 & 1 \end{bmatrix}, \tag{16}$$

$$I_2 = \begin{bmatrix} 1 & 0 & 0 & 0 \\ 0 & \sqrt{1-p_2}\,e^{i\varphi_2} & \sqrt{p_2}\,e^{i\psi_2} & 0 \\ 0 & -\sqrt{p_2}\,e^{-i\psi_2} & \sqrt{1-p_2}\,e^{-i\varphi_2} & 0 \\ 0 & 0 & 0 & 1 \end{bmatrix}, \tag{17}$$

and

$$I_3 = \begin{bmatrix} 1 & 0 & 0 & 0 \\ 0 & 1 & 0 & 0 \\ 0 & 0 & \sqrt{1-p_3}\,e^{i\varphi_3} & \sqrt{p_3}\,e^{i\psi_3} \\ 0 & 0 & -\sqrt{p_3}\,e^{-i\psi_3} & \sqrt{1-p_3}\,e^{i\varphi_3} \end{bmatrix}. \tag{18}$$

Here the definitions of p_j, φ_j, ψ_j and $\sigma_j^{(k)}$ are the same as before, and $X_0 = \omega_1$ and $X_4 = \omega_2$. This transition matrix $T_{\omega_1 \to \omega_2}$ can be simplified as

$$T_{\omega_1 \to \omega_2} = \begin{bmatrix} \sqrt{1-p_1} & \sqrt{p_1}\,e^{-iA} \\ -\sqrt{p_1(1-p_2)}\,e^{-iC} & \sqrt{(1-p_1)(1-p_2)}\,e^{-i(A+C)} \\ \sqrt{p_1 p_2(1-p_3)}\,e^{-i(C+E)} & -\sqrt{(1-p_1)p_2(1-p_3)}\,e^{-i(A+C+E)} \\ -\sqrt{p_1 p_2 p_3}\,e^{-i(B+C+E)} & \sqrt{(1-p_1)p_2 p_3}\,e^{-i(A+B+C+E)} \end{bmatrix}$$
$$\begin{matrix} 0 & 0 \\ \sqrt{p_2}\,e^{-i(A+C+D)} & 0 \\ \sqrt{(1-p_2)(1-p_3)}\,e^{-i(A+C+D+E)} & \sqrt{p_3}\,e^{-i(A+C+D+E+F)} \\ -\sqrt{(1-p_2)p_3}\,e^{-i(A+B+C+D+E)} & \sqrt{1-p_3}\,e^{-i(A+B+C+D+E+F)} \end{matrix}, \tag{19}$$

where

$$A = \varphi_1 - \psi_1 + \Delta\sigma_1^{(2,1)}, \tag{20.a}$$
$$B = \varphi_3 + \psi_3 + \Delta\sigma_4^{(4,3)}, \tag{20.b}$$
$$C = \varphi_1 + \psi_1 - \varphi_2 + \Delta\sigma_2^{(2,1)} + \Delta\sigma_3^{(2,1)} + \Delta\sigma_4^{(2,1)}, \tag{20.c}$$
$$D = -\varphi_1 + \varphi_2 - \psi_2 + \Delta\sigma_1^{(3,2)}, \tag{20.d}$$
$$E = \varphi_2 + \psi_2 - \varphi_3 + \Delta\sigma_3^{(3,2)} + \Delta\sigma_4^{(3,2)}, \tag{20.e}$$

and

$$F = -\varphi_2 + \varphi_3 - \psi_3 + \Delta\sigma_1^{(4,3)} + \Delta\sigma_2^{(4,3)} + \Delta\sigma_3^{(4,3)}. \tag{20.f}$$

These phases roughly correspond to the areas shown in Figure 3(b). First, let us consider the condition of the complete excitation from $|1\rangle$ to $|2\rangle$. It turns out that one period of sweeping is again good enough. Substituting Eq.(19) into Eq.(10), we obtain the $1 \to j$ ($j = 2-4$) matrix elements of $T^{(1)}$ as follows:

$$\left(T^{(1)}\right)_{21} = \sqrt{p_1(1-p_1)}\,e^{-iA}X, \tag{21.a}$$

$$\left(T^{(1)}\right)_{31} = -\sqrt{p_1 p_2(1-p_2)}\,e^{-i(A+2C+D)}Y, \tag{21.b}$$

and

$$\left(T^{(1)}\right)_{41} = \sqrt{p_1 p_2 p_3(1-p_3)}\,e^{-i(A+2C+D+2E+F)}Z, \tag{21.c}$$

where

$$X = 1 - e^{-2iC} + p_2 e^{-2iC}Y, \tag{22.a}$$

$$Y = 1 - e^{-2iE} + p_3 e^{-2iE}Z, \tag{22.b}$$

and

$$Z = 1 - e^{-2iB}. \tag{22.c}$$

From the conditions, $|(T^{(1)})_{21}| = 1$, $(T^{(1)})_{31} = 0$ and $(T^{(1)})_{41} = 0$, we finally obtain the following condition of the complete excitation from $|1\rangle$ to $|2\rangle$:

$$p_1 = \frac{1}{2},\ B = l\pi,\ E = m\pi\ \text{ and }\ C = \left(n+\frac{1}{2}\right)\pi\quad (l, m, n : \text{integer}). \tag{23}$$

Next, let us consider complete excitation from $|1\rangle$ to $|3\rangle$. It turns out that in this case at least one and half period of sweeping is necessary. For the sweep $\omega_1 \to \omega_2 \to \omega_1 \to \omega_2$, the overall transition matrix $T^{(1+1/2)}$ is given by $T^{(1+1/2)} = T_{\omega_1\to\omega_2}\cdot T^{(1)}$ and its $1 \to j$ ($j = 1-4$) matrix elements are explicitly expressed as

$$\left(T^{(1+1/2)}\right)_{11} = \sqrt{1-p_1}\left(1 - p_1 X + p_1 e^{-2iA}X\right), \tag{24.a}$$

$$\left(T^{(1+1/2)}\right)_{21} = \sqrt{p_1(1-p_2)}\,e^{-iC}\left\{-(1-p_1 X) + (1-p_1)e^{-2iA}X - p_2 Y e^{-2i(A+C+D)}\right\}, \tag{24.b}$$

$$\left(T^{(1+1/2)}\right)_{31} = \sqrt{p_1 p_2(1-p_3)}\,e^{-i(C+E)}\left\{1 - p_1 X - (1-p_1)e^{-2iA}X\right.$$
$$\left. -(1-p_2)Y e^{-2i(A+C+D)} + p_3 Z e^{-2i(A+C+D+E+F)}\right\}, \tag{24.c}$$

and

$$\left(T^{(1+1/2)}\right)_{41} = \sqrt{p_1 p_2 p_3}\,e^{-i(B+C+E)}\left\{-(1-p_1 X) + (1-p_1)e^{-2iA}X\right.$$
$$\left. +(1-p_2)Y e^{-2i(A+C+D)} + (1-p_3)Z e^{-2i(A+C+D+E+F)}\right\}. \tag{24.d}$$

From the conditions, $|(T^{(1+1/2)})_{21}| = 1$, $(T^{(1+1/2)})_{11} = (T^{(1+1/2)})_{31} = (T^{(1+1/2)})_{41} = 0$, and $(T^{(1)})_{41} = 0$, the following conditions are obtained:

$$\begin{cases} 4p_1(1-p_2)\sin^2 E = 1 \\ B = k\pi \\ D = l\pi \\ A + C = m\pi \\ C + E = n\pi \end{cases} \quad \text{or} \quad \begin{cases} 4p_1(1-p_2)\sin^2 E = 1 \\ B = k\pi \\ A + C + D + E = (l+1/2)\pi \\ C + D = (m+1/2)\pi \end{cases}, \tag{25}$$

where k, l, m, n are integers. Here, for simplicity, we assume the linear chirping with the same rate and the same transition dipole moments for the transitions $|1>\rightarrow|2>$ and $|1>\rightarrow|3>$, i.e. $\mu_{12} = \mu_{13}$. Using the condition $p_1 = p_2$ in Eq.(25), we can obtain the following conditions:

$$p_1 = p_2 = \frac{1}{2}, A = \left(k + \frac{1}{2}\right)\pi, B = l\pi, C + D = \left(m + \frac{1}{2}\right)\pi \text{ and } E = \left(n + \frac{1}{2}\right)\pi. \quad (26)$$

Finally, as can be easily known, the complete excitation from $|1>$ to $|4>$ can be realized by one period of chirping, if we start from ω_2: $\omega_2 \rightarrow \omega_1 \rightarrow \omega_2$, and the following condition is obtained:

$$p_3 = \frac{1}{2}, A = l\pi, D = m\pi \text{ and } E = \left(n + \frac{1}{2}\right)\pi \quad (l, m, n : \text{integer}). \quad (27)$$

Numerical Examples

Selective excitations in both three- and four-level systems are considered numerically. The present periodic chirping scheme is compared with the ARP (adiabatic rapid passage) and π-pulse.

Three-level model

The parameters of the system considered are as follows: ω_{12}=500 [cm^{-1}], ω_{23}=10 [cm^{-1}], $\mu_{12}= \mu_{13}= 1.0$ [a.u.]. The complete excitation from $|1>$ to $|2>$ by one period of sweeping is shown in Figure 4(a). The laser frequency is swept quadratically as a function of time,[18] $\omega(t) = -a(t - t_0)^2 + b - E_1$, where $a = 8(V_{13})^3 \alpha / \hbar^2$, $b = E_3 - 2V_{13}\beta$, $V_{13} = -\mu_{13}E_0 / 2$ and E_0 is the amplitude at the pulse peak. The pulse shape is a combination of hyperbolic-tangent functions: $\varepsilon(t) = E_0[1 + \tanh(\beta_e(t - t_{0e}))]/2$ for $t \le t_0$ and $E_0[1 - \tanh(\beta_e(t - t_{1e}))]/2$ for $t > t_0$. The parameters α, β and the peak intensity are determined to be 0.6005, 1.58142 and 0.1 [GW/cm^2] from the semiclassical condition Eq.(12). The other parameters of laser field are t_0= 3.5 [ps], t_{0e}= 1.7775 [ps], $t_{1e} = 2t_0 - t_{0e}$, and β_e= 6.515 [ps^{-1}]. As can be seen from Figure 4(a), the transition time is about 3 [ps]. This time is very close to the time $\Delta T = 2\pi / \Delta E \approx 3.3$ [ps] determined from the uncertainty principle, where ΔE is the energy spacing between $|2>$ and $|3>$ and is 10 [cm^{-1}] in the present case. This indicates that the present result is the possible shortest time of transition. The complete excitation from $|1>$ to $|3>$ is shown in Figure 4(b). As mentioned above, the laser frequency is swept in the way opposite to Figure 4(a). In Figure 5 the excitation from $|1>$ to $|2>$ by ARP is depicted. The laser frequency is linearly chirped, $\omega(t) = \omega_{12} + c(t - 20\text{ps})$. The chirp rate c and the laser intensity are determined to be 8.816 $[\text{cm}^{-1}/\text{ps}]$ and 0.1 $[\text{GW/cm}^2]$ so that the excitation probability is equal to 0.99. Selective excitation is possible, but it takes a quite long time (about 20 [ps]) and it is impossible to make a complete excitation. The excitations from $|1>$ to $|2>$ by two types of π-pulse are shown in Figure 6(a) and 6(b). The pulse shape is hyperbolic-secant: $\varepsilon(t) = E_0\text{sech}[\beta_e(t-t_0)]$. In Figure 6(a) [6(b)], $t_0 = 10$ [25] [ps], peak intensity = 0.05 [0.003] $[\text{GW/cm}^2]$, and $\beta_e = 1.56022$ [0.38218] $[\text{ps}^{-1}]$. In Figure 6(a) the laser parameters are determined so that the transition time is comparable to the case of one period chirping (~3 [ps]). The excitation to $|3>$ also occurs and the selective and complete excitation is not possible, because the time duration of the pulse is short, and thus the frequency band-width broadens and the level $|3>$ is covered. In Figure 6(b), a longer pulse is used to achieve the selective and complete excitation; then naturally a longer time (~14 [ps]) is required.

Four-level model

The parameters used are as follows: $\omega_{12}=500$ $[\text{cm}^{-1}]$, $\omega_{23}=10$ $[\text{cm}^{-1}]$, $\omega_{34}=10$ $[\text{cm}^{-1}]$, $\mu_{12}=\mu_{13}=\mu_{14}=1.0$ [a.u.]. Here we consider the complete excitation to $|3>$. The result of the one and half period of sweeping is shown in Figure 7. The laser frequency is chirped linearly as follows:

$$\omega(t) = \begin{cases} \omega_{14} + c(t-t_0) & (t \leq t_1) \\ \omega_{14} - c(t-t_2) & (t_1 < t \leq t_3), \\ \omega_{14} + c(t-t_4) & (t > t_3) \end{cases}$$

where $t_1 = t_0 + \Delta t_1$, $t_2 = t_0 + 2\Delta t_1$, $t_3 = t_2 + \omega_{24}/c + \Delta t_2$, $t_4 = 2t_3 - t_2$, $c = 346.6$ $[\text{cm}^{-1}/\text{ps}]$, $\Delta t_1 = 726.3$ [fs], $\Delta t_2 = 90.08$ [fs] and the peak intensity is 0.5917 $[\text{GW/cm}^2]$. These values are estimated from the semiclassical condition Eq.(26). The pulse shape is again a combination of hyperbolic-tangent functions defined above with $t_{0e} = 981.2$ [fs] and $\beta_e = 4.257$ $[\text{ps}^{-1}]$. The transition time is about 3 [ps] and this is close to the time limit determined by the uncertainty principle. In Figure 8(a) [8(b)] the excitation by π-pulse with a short [long] pulse duration is shown. The pulse shape is a hyperbolic-secant function. In Figure 8(a) [8(b)] $t_0 = 10$ [25] [ps], the peak intensity is 0.05 [0.005] $[\text{GW/cm}^2]$ and $\beta_e = 1.56022$ [0.493386] $[\text{ps}^{-1}]$. In the case of Figure 8(a) the pulse duration is short and thus the band-width is broad, and the transitions to $|2>$ and $|4>$ cannot be avoided. In Figure 8(b) the pulse duration is long, and the selective and complete excitation to $|3>$ is possible. Needless to say, the ARP takes a much longer time, because two ARP processes, $|1>$ to $|2>$ and $|2>$ to $|3>$, are necessary to make a selective excitation from $|1>$ to $|3>$.

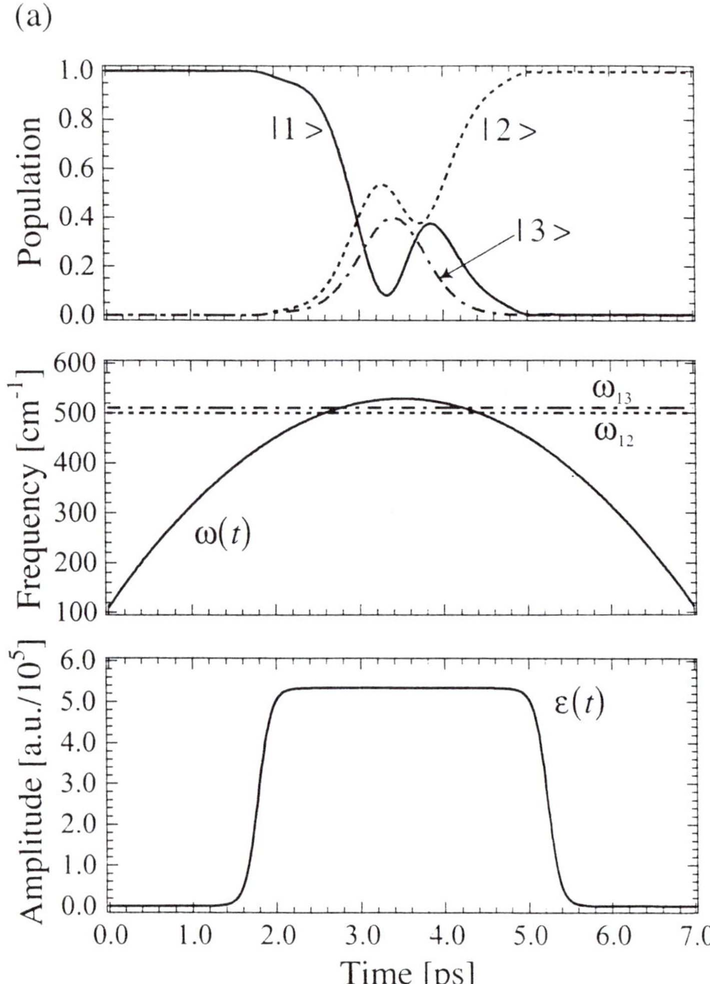

Figure 4. (a)Complete excitation from |1 > to |2 > by one period of frequency chirping in the case of three-level model. Time variations of the population (upper part), laser frequency (middle part) and laser envelope (bottom part) are shown. (b)Complete excitation from |1 > to |3 > by one period of frequency chirping.

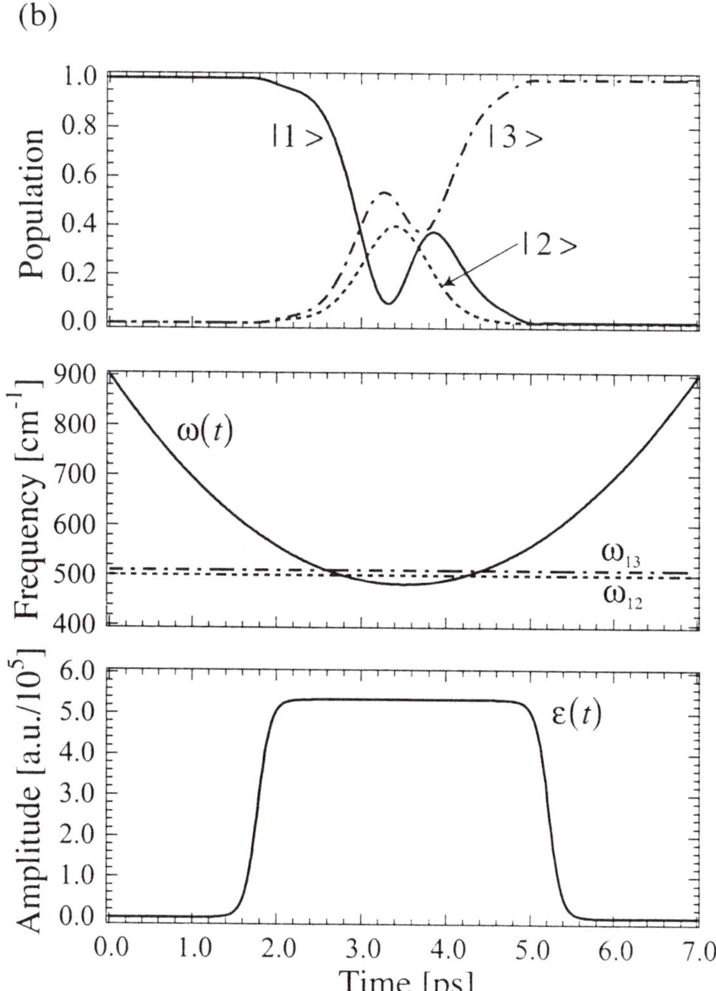

Figure 4. *Continued.*

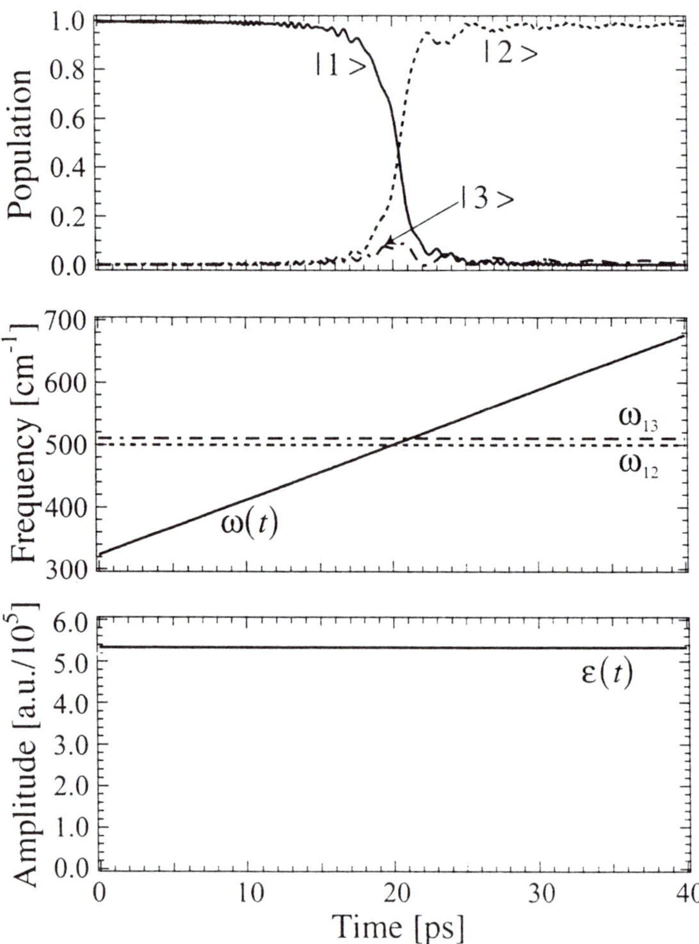

Figure 5.Selective excitation from $|1>$ to $|2>$ by ARP in the case of three-level model. Time variations of the population (upper part), laser frequency (middle part) and laser envelope (bottom part) are shown.

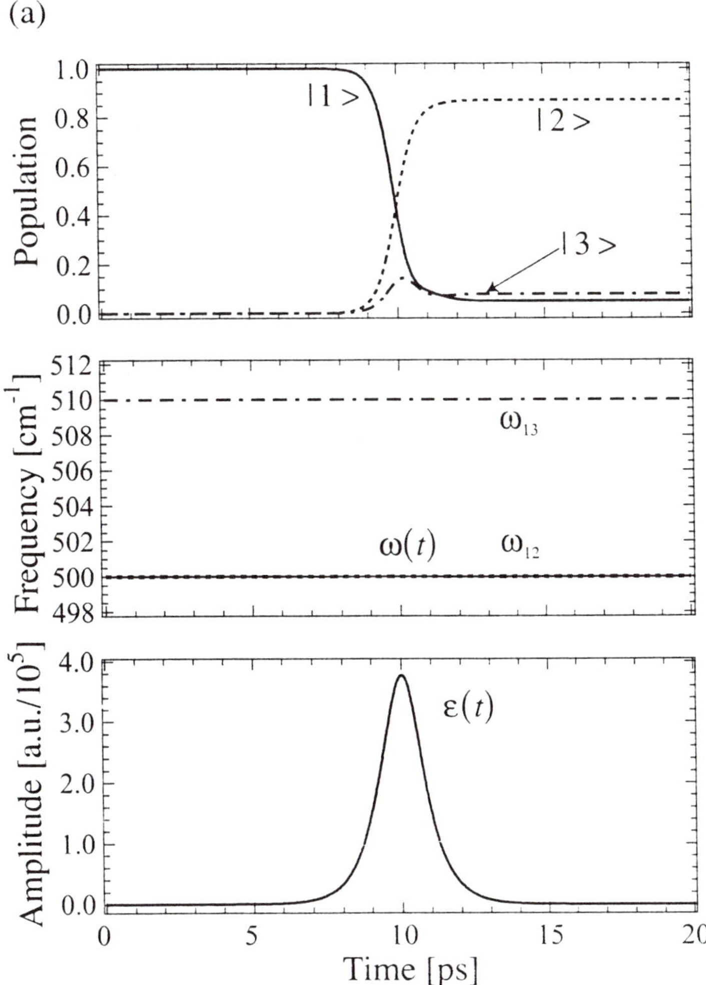

Figure 6. Excitation from |1> to |2> by π-pulse with (a)a short time duration and (b)a long time duration in the case of three-level model. Time variations of the population (upper part), laser frequency (middle part) and laser envelope (bottom part) are shown.

Continued on next page.

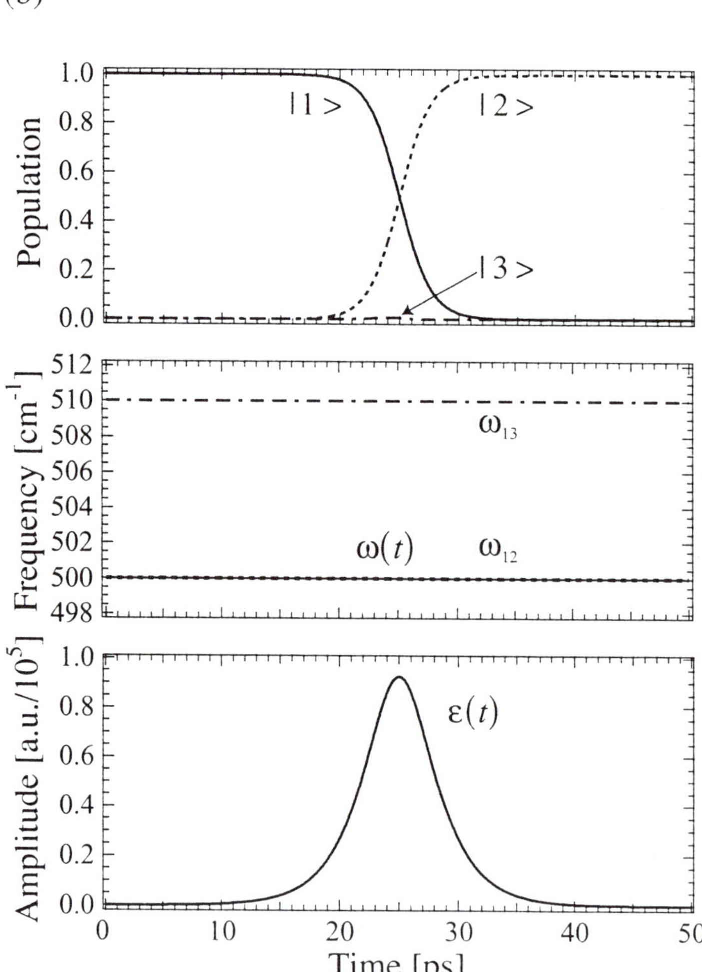

Figure 6. *Continued.*

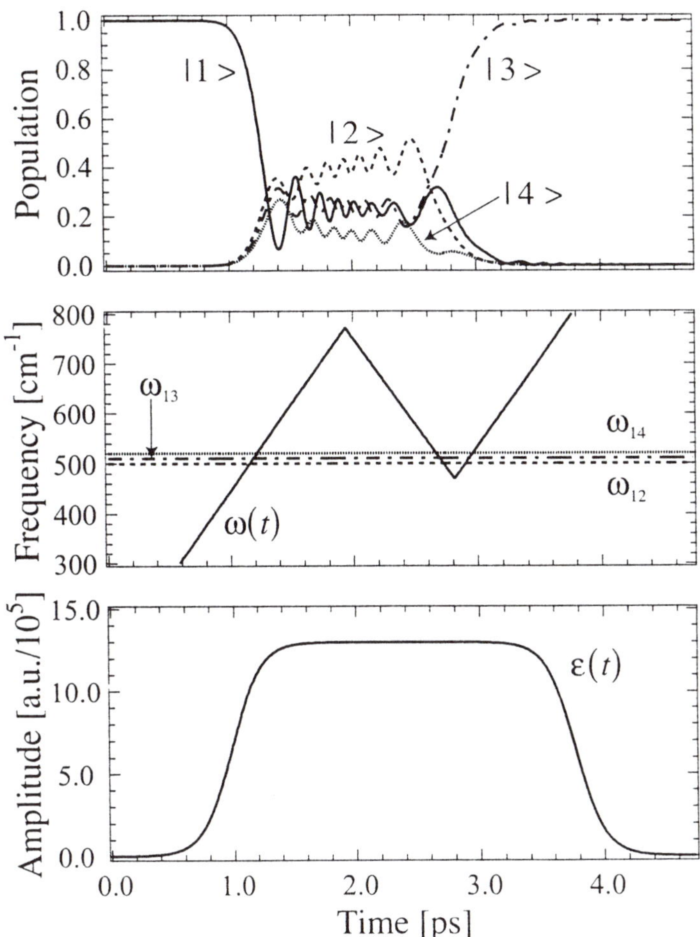

Figure 7. Complete excitation from $|1>$ to $|3>$ by one+half period of frequency chirping in the case of four-level model. Time variations of the population (upper part), laser frequency (middle part) and laser envelope (bottom part) are shown.

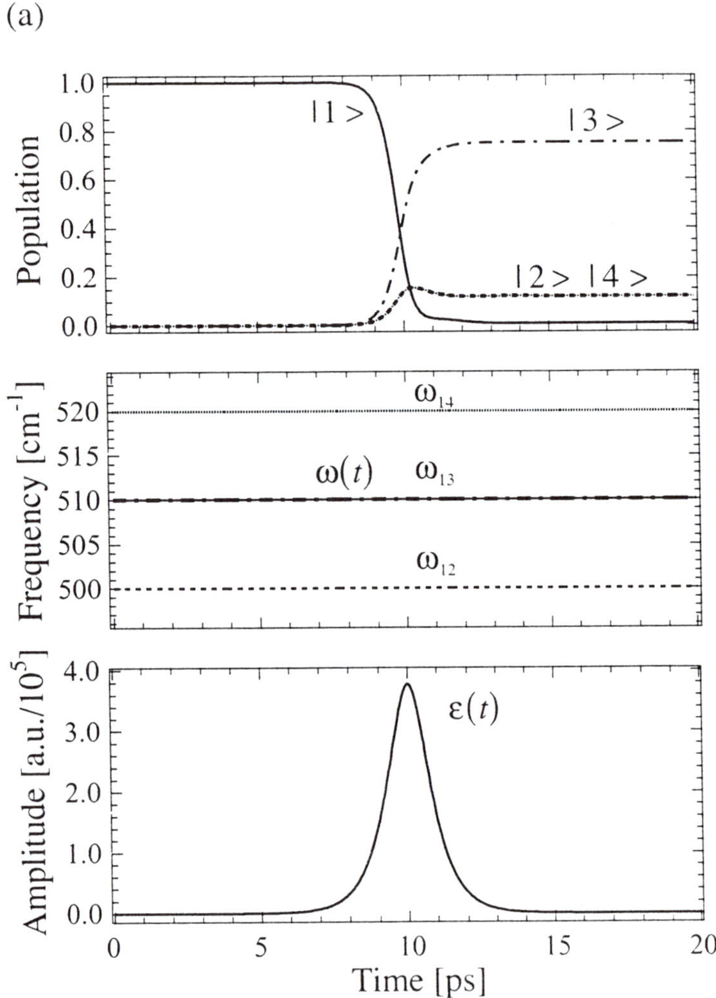

Figure 8. Excitation from |1 > *to* |3 > *by* π*-pulse with (a)a short time duration and (b)a long time duration in the case of four-level model. Time variations of the population (upper part), laser frequency (middle part) and laser envelope (bottom part) are shown.*

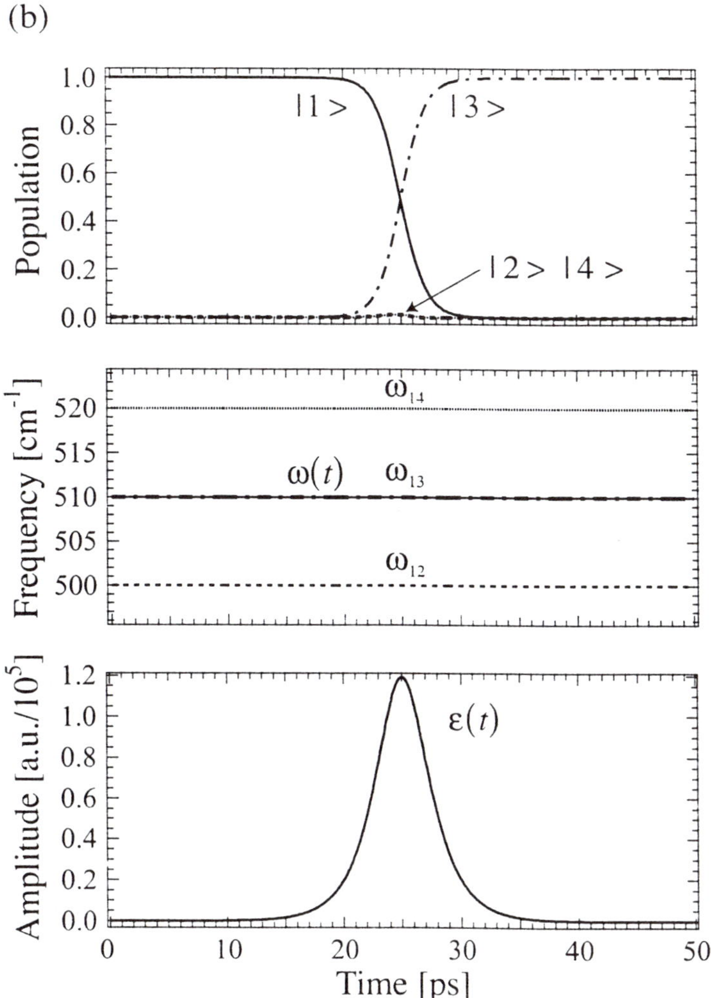

Figure 8. *Continued.*

Concluding Remarks

The selective and complete excitation among closely lying multi-levels by the periodic chirping is proposed and is numerically demonstrated by using the three- and four-level model systems. The advantage of the present scheme is that the transition time can be greatly shortened compared to the ARP and π-pulse and thus the present scheme would probably be quite useful and effective for the systems with fast relaxation.

Qualitative arguments about the transition time in comparison with the π-pulse can be made as follows: In the case of π-pulse it is necessary for the band-width $\Delta\omega$ to be narrow enough compared with the level separation ΔE, i.e. $\Delta\omega \ll \Delta E$. This corresponds to a condition that the excitation process can be treated as a two level problem. Since in the case of π-pulse the transition time ΔT_{π} is roughly equal to the pulse duration, the uncertainty principle between the pulse duration and the band-width ($\Delta\omega \cdot \Delta T_{\pi} \geq 2\pi$) leads to

$$\Delta T_{\pi} \geq \frac{2\pi\delta}{\Delta E}, \tag{28}$$

where $\Delta\omega = \Delta E / \delta$ and $\delta \gg 1$. Thus, we have

$$\Delta T_{\text{chirp}} \approx \frac{2\pi}{\Delta E} \ll \Delta T_{\pi}, \tag{29}$$

where ΔT_{chirp} is the transition time in the case of periodic chirping.

Although the numerical applications here were made to the particular case of $\Delta E = 10$ cm^{-1}, the control scheme proposed here can be scaled as follows when the level spacing ΔE is changed to $\alpha\Delta E$:

$$\Delta T_{\text{chirp}} \to \Delta T_{\text{chirp}} / \alpha \quad \text{and} \quad I\left(\text{laser intensity}\right) \to \alpha^2 I.$$

Furthermore, this idea of selective and complete excitation can in principle be applied to more general multi-level systems. Although it would probably be necessary to sweep the frequency more than one period, the transition time may be shortened to the limit determined from the uncertainty principle.

The periodic chirping within one laser pulse may not be realized easily in experiments. But the same effects can be realized by applying a train of linearly chirped pulses, in which the chirp rate and the pulse shape should be adjusted properly for each pulse. This will be discussed elsewhere. The present scheme would also be useful for controlling population transfer in ESR and NMR. For these various types of controls, a variety of semiclassical theories of time-dependent nonadiabatic transitions such as that derived from the Zhu-Nakamura theory and the exponential model are very useful.[19,20] In the present treatment no effect of relaxation has been considered. In order to do that Eq.(1) and the control scheme should be modified by incorporating an imaginary potential, for instance; but this is beyond the present scope and will be discussed in future.

Acknowledgments

This work was supported by the Grant-in-Aid for Scientific Research on Priority Area "Molecular Physical Chemistry" and by Research Grant No. 10440179 from the Ministry of Education, Science, Culture, and Sports of Japan.

References

1. R. J. Gordon and S. A. Rice, *Annu. Rev. Phys. Chem.* **1997**, *48*, 601.
2. A. D. Bandrauk, *Molecules in Laser Fields* (Marcel Dekker, New York, 1994).
3. P. Brumer and M. Shapiro, *Annu. Rev. Phys. Chem.* **1992**, *43*, 257.
4. D. J. Tannor and S. A. Rice, *J. Chem. Phys.* **1985**, *83*, 5013.
5. R. Kosloff, S. A. Rice, P. Gaspard, S. Tersigni, and D. J. Tannor, *Chem. Phys.* **1989**, *139*, 201.
6. D. Neuhauser and H. Rabitz, *Acc. Chem. Res.* **1993**, *26*, 496.
7. B. Kohler, J. L. Krause, F. Raksi, K. R. Wilson, V. V. Yakovlev, and R. M. Whitnell, *Acc. Chem. Res.* **1995**, *28*, 133.
8. M. Sugawara and Y. Fujimura, J. Chem. Phys. **1994**, *100*, 5646.
9. S. Chelkowski and A. D. Bandrauk, *J. Chem. Phys.* **1993**, *99*, 4279.
10. M. M. T. Loy, *Phys. Rev. Lett.* **1974**, *32*, 814.
11. J. S. Melinger, S. R. Gandhi, and W. S. Warren, *J. Chem. Phys.* **1994**, *101*, 6439.
12. S. Guerin, *Phys. Rev. A* **1997**, *56*, 1458.
13. K. Mishima and K. Yamashita, *J. Chem. Phys.* **1998**, *109*, 1801.
14. S. I. Chu, *Advances in Multiphoton Processes and Spectroscopy*, Vol. 2 (World Scientific, Singapore, 1986).
15. K. Nagaya, Y. Teranishi and H. Nakamura, *J. Chem. Phys.* **2000**, *113*, 6197.
16. Y. Teranishi and H. Nakamura, *Phys. Rev. Lett.* **1998**, *81*, 2032.
17. Y. Teranishi and H. Nakamura, *J. Chem. Phys.* **1999**, *111*, 1415.
18. Y. Teranishi and H. Nakamura, *J. Chem. Phys.* **1997**, *107*, 1904.
19. H. Nakamura, *Dynamics of Molecules and Chemical Reactions*, edited by R. E. Wyatt and J. Z. H. Zhang (Marcel Dekker, New York, 1996), p.473.
20. C. Zhu, Y. Teranishi, and H. Nakamura, *Adv. Chem. Phys.* **2001**, *117*, 127.

Chapter 8

Laser Manipulation of Differential Autoionization Yields: Pump–Dump Control Through Coupled Channels

R. R. Jones, S. N. Pisharody, and R. van Leeuwen

Department of Physics, University of Virginia, Charlottesville, VA 22904

Atoms with two optically accessible electrons provide pristine systems for testing the limitations of laser control scenarios. We present the results of two experiments in calcium in which ultra-short laser pulses are used to *pump* population from the atomic ground state into a non-stationary, singly excited wavepacket. After the system has evolved, a second short pulse *dumps* the bound population into a number of degenerate continua by resonantly driving the system into a rapidly decaying autoionizing configuration. In one experiment, autoionization from a radially localized Rydberg wavepacket is found to be non-exponential in time, occurring in a series of discrete steps. This phenomenon is explained using a simple intuitive picture of the time-dependent interaction of the Rydberg wavepacket with the other electron. In a second experiment, control of the angular distribution of electrons ejected during the autoionization of angularly localized wavepackets is examined. Only limited control is observed due to significant coupling between continuum channels as the system fragments.

Introduction

Current interest in the use of coherent radiation to alter chemical processes stems from a number of theoretical works dating from the mid 1980's (*1-3*). Since that time, the notion of coherent- or quantum-control has stimulated an enormous amount of theoretical activity (*1-5*). Experimental efforts, while substantial, have struggled to keep pace due to technological limitations and the sheer complexity of laboratory realizations of control scenarios (*4,5*).

In one control method, known as the "pump-dump" scheme (*3,6*), laser radiation (the "pump") creates a coherent wavepacket in a mixed configuration that is optically accessible from the ground state. With the passing of time, this non-stationary state evolves into a different configuration that may not be accessible from the ground state, but is optically connected to a desired final-state. A second radiation pulse then "dumps" the wavepacket into the target channel. Due to the difficulties associated with manipulating all of the relevant degrees of freedom associated with bi-molecular reactions, most gas-phase control studies have focussed on unimolecular fragmentation or ionization where the target states lie in the continuum, e.g. $AB + h\nu \rightarrow (AB)^+ + e^-$; $A + B^+ + e^-$; $A^* + B;\dots$ In these situations, the "dump" pulse deposits the system into a configuration that asymptotically approaches the desired final-state. However, immediately following the dump pulse, the system is still "close-coupled", i.e. the molecular constituents are still in close proximity. Although it is possible to identify real systems where the interaction between dump channels is negligible due to the topology of the potential energy surface and orthogonality of the respective reaction coordinates, in general, coupling between channels cannot be neglected as the system propagates towards one or more asymptotic targets.

Recently, we have investigated, experimentally, the role of channel mixing in a pump-dump control scenario where the propagation time of the system from the initial close-coupled configuration to the asymptotic target channels is not negligible for all possible final states. In our experiments and subsequent analysis, we consider a relatively simple, model system – a doubly excited, alkaline-earth Rydberg atom. Doubly excited two-electron systems have enough degrees of freedom to be interesting for proof-of-principle quantum control experiments and have a number of advantages over molecules in this regard. We note some of these advantages in detail below. Before doing so, however, it is useful to consider the analogy between dielectronic and molecular systems.

In molecules, energy and angular momentum in various degrees of freedom, e.g. vibrational, rotational, and electronic, are not locally conserved due to couplings between different modes. As a result, the eigenmodes at a given energy are superpositions of different rotational, vibrational, and electronic configurations. When viewed in the time-domain, the intramolecular couplings lead to complex dynamics in the form of energy and angular momentum flow, between modes, at fixed total energy and angular momentum.

As shown schematically in Figure 1, a similar, albeit simpler situation exists for doubly-excited Rydberg atoms (*7*). Consider the case where one electron is in

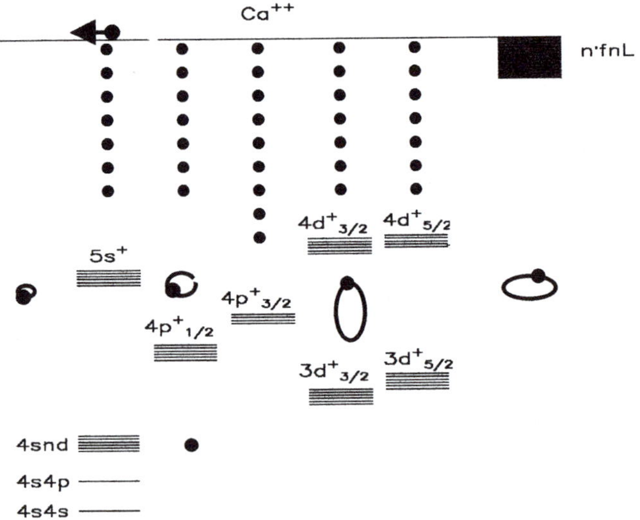

Figure 1: Schematic diagram of the energy level structure of Ca, a typical two-electron system. Classical orbit cartoons illustrate, in the independent electron approximation, a few of the bound and continuum modes that appear at different energies. At low energies only singly-excited bound configurations are present. Conversely, in the two-electron approximation, no bound orbitals exist above the second ionization limit. Between the first and second ionization limits, bound and continuum channels co-exist and are coupled by the electron-electron interaction. For example, at an energy just above the Ca⁺ 4p threshold, representative configurations include the 4sεL and 4p₁/₂,₃/₂ εL' continua and the 4d₃/₂nL'' and 4d₅/₂nL'' bound states.

a highly excited Rydberg orbital and a second, "core," electron is in a localized state that is tightly bound to the atomic nucleus. The "modes" in this system correspond to dielectronic configurations with different, discrete core-electron energies, core-electron angular momenta, or Rydberg angular momenta. The eigenstates of the atom, at fixed total energy and angular momentum, are linear combinations of different core- and Rydberg-electron configurations. In the time-domain, the electron-electron Coulomb repulsion provides the coupling between modes, and in analogy to the molecular case, is responsible for energy and angular momentum flow between dielectronic configurations.

As stated above, there are a number of reasons to study two-electron atoms rather than molecules for proof-of-principle coherent control experiments (*8-10*). First, the number of available configurations and the coupling strength between these is readily tunable in the atom. The number of interacting two-electron channels increases with energy in the doubly-excited system and the average coupling between these modes is a function of the overlap between the core and Rydberg electron wavefunctions. This overlap decreases with increasing angular momentum and/or energy of the Rydberg electron. Second, *for relatively low core-elecron energy*, the level structure, wave-function composition, and autoionization decay rates of alkaline-earth Rydberg atoms are well understood from experimental frequency domain spectroscopy and theoretical simulations that take advantage of *ab initio* K-matrix scattering parameters and multi-channel quantum defect theory (MQDT) (*11,12*). Third, because laser excitation of atoms can proceed from a pure rather than rotationally mixed initial state, and because singly excited Rydberg atoms have simple energy level structure, manipulation and observation of Rydberg electron dynamics is more easily achieved in atoms (*13*), as opposed to molecules.

We have been exploiting these three factors, to manipulate and view dynamics in multi-configurational systems. Our ultimate goal is to explore the use of wavepacket methods to probe and control time-dependent configuration between the near infinite number of coupled "double-Rydberg" modes that exist just below the second ionization limit of the atom. The experiments described here are relevant to both our long-term goal of controlling quantum correlation between electrons, and to the experimental demonstration of laser control of differential yields in unimolecular reactions. In the following sections, we describe two experiments in calcium. In the first, the time-dependence of electron ejection from autoionizing radial wavepackets is measured. The results of this experiment agree with the expectations of an intuitive model based on the dynamics of analogous, singly-excited wavepackets. In the second experiment, control over the initial orientation of an angularly localized Rydberg wavepacket is exerted in an attempt to manipulate the angular distribution of electrons ejected via autoionization (*14*). We show that while some modification of the angular distributions is indeed possible, it is less than one might expect from the simplest intuitive model. MQDT simulations based on *ab initio* K-matrices indicate that the limited control that is observed can be attributed to unexpectedly slow propagation of the system from the bound "close-coupled" configuration into certain ejection angles in the autoionization continuum.

Experimental Procedure

The apparatus and general procedure for the two experiments are similar, and have been presented elsewhere (*14,15*). We now describe the basic apparatus and approach that is common to both experiments. The experiments take place in a vacuum chamber with a base pressure of $<10^{-6}$ Torr. An effusive, Ca beam with a density of roughly 10^8 atoms/cm^3 passes through a laser /atom interaction region that is defined by two, grounded, parallel aluminum capacitor plates. One or more "pump" laser pulses are used to promote a fraction of the Ca atoms from their electronic ground-state to singly-excited Rydberg wavepackets. The electronic wavepackets evolve for some time, with an "outer" electron moving as a Rydberg wavepacket and an "inner" electron remaining in the ionic ground state. A short "probe" pulse then initiates a sudden isolated-core excitation (ICE) of the inner electron (*16*), promoting it to an excited state of the ion. During ICE, negligible direct photoionization of the Rydberg electron occurs since the resonant, core-electron transition moment is orders of magnitude larger. The Rydberg wavepacket eventually scatters from the excited core-electron, exchanging energy and angular momentum with it, and autoionizes into a number of degenerate ionic continua. A fraction of the electrons ejected from the atom travel through a small hole in the upper capacitor plate and strike a micro-channel plate detector (MCP). The branching ratio for electron emission into different ionic continua is determined by measuring the time-of-flight distribution of electrons that reach the MCP. The angular distribution of the ejected electrons can be measured by monitoring the MCP signal as the common direction of linear polarization of all the laser beams is rotated relative to the vertical axis. The effect, on the differential electron yields, of the initial configuration of the wavepacket at the instant of the ICE, is measured by monitoring the electron yield a function of the delay between the laser(s) that produce the wavepacket, and the ICE laser.

Stair-Step Autoionization of a Radial Wavepacket

Singly-excited radial Rydberg wavepackets are common subjects for investigation due to their relatively simple one-dimensional motion and ease of excitation (*13*). A radial wavepacket is launched when a short laser pulse photo-excites a tightly-bound electron to an energy just below the ionization threshold of an atom (or molecule). If the coherent bandwidth of the laser pulse is greater than the spacing between adjacent Rydberg states at the nominal excitation energy, then several Rydberg states with different principal-, but identical angular-quantum numbers, are populated. This coherent superposition is known as a radial wavepacket. Immediately following the short-pulse excitation, the electron is near the nucleus, and the radial extent of its wavefunction is significantly less than the extent of the Coulomb binding-potential at the Rydberg energy. The localized packet moves toward the outer turning point in the radial potential well, reflects, and travels back toward the nucleus. The round-trip time for the periodic radial beating is equal to the classical Kepler period , $\tau_K = 2\pi N^3$, of an electron whose energy equals the average energy of the wavepacket, $E_N = -1/2N^2$. At long times, the non-uniform inter-state energy spacing leads to the collapse of the wavepacket followed by integer and fractional revivals of the localized spatial structure.

Our first experiment is designed to investigate the evolution of a radial wavepacket in a rapidly autoionizing, doubly excited configuration. Specifically, we measure the time-dependent autoionization probability of the wavepacket to determine if one can extend the simple picture of a radially beating Rydberg electron from the singly-excited to the doubly-excited regime. If so, then we expect to observe significant autoionization only when the Rydberg electron is near the nucleus and the excited core-electron. We also intend to confirm our expectations, based on previous time- and frequency-domain spectroscopies of autoionization lifetimes in this specific 4pnd Ca Rydberg system (15), that the lifetime of these low angular momentum autoionizing wavepackets is approximately one Kepler period.

In the experiment, a dye-laser pulse (423 nm, 5 nsec) excites Ca atoms from the ground state to an intermediate 4s4p $^1P_1$ level. A second laser pulse (392 nm, 1 psec) then launches a 4snd $^1D_2$ radial wavepacket. The second laser-pulse is produced by spectrally filtering and then frequency doubling the 787($\pm$7)nm, output of an amplified Ti:Sapphire laser. The average energy of the radial wavepacket and number of constituent eigenstates can be continuously varied by changing the width and central frequency of the spectral filtering aperture. After the wavepacket has evolved for a time, T_1, a third laser pulse (ICE1, 393 nm, 200 fsec) that is also produced by frequency doubling the Ti:Sapphire output, drives the 4s-4p ionic resonance in the 4snd Rydberg atoms, producing atoms in the doubly-excited 4pnd configuration (see Figure 1). These atoms rapidly autoionize, producing 4s$^+$ Ca ions and 3.1 eV electrons, or 3d$_{3/2,5/2}^+$ Ca ions and 1.5 eV electrons.

The *time-dependence* of the autoionization is probed by performing a second ICE, after a delay T_2 relative to the first ICE, that further excites atoms remaining in the 4pnd configuration to 4dnd states (see Figure 1). The 300 fsec, 318 nm ICE2 pulse is produced by mixing the 200 fsec, 530 nm output of an optical parametric amplifier (OPA) with the 787 nm Ti:Sapphire fundamental. The 4dnd wavepacket also autoionizes, producing 4s$^+$ Ca ions and 7.0 eV electrons, 3d$_{3/2,5/2}^+$ Ca ions and 5.4 eV electrons, or 4p$^+$ Ca ions and 3.9 eV electrons. By monitoring the number of electrons produced with energies > 3.5 eV as a function of delay, T_2, we obtain the time-dependent probability for finding the electron in the 4pnd autoionizing wavepacket configuration. In addition, by varying the delay T_1, between the creation of the 4snd wavepacket and the first ICE, we can measure the time-dependence of autoionization as a function of the position of the wavepacket at the instant of the ICE.

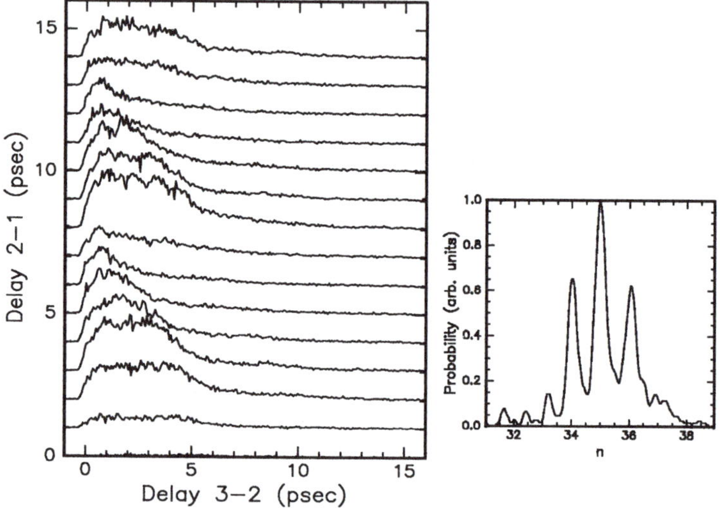

Figure 2: Time-dependent probability for finding an electron in the 4p35d wavepacket as a function of time (Delay 3-2) and as a function of the initial radial position of wavepacket at the instant of the 4s35d-4p35d ICE (Delay 2-1). The inset shows the 4snd eigenstate composition of the wavepacket.

In Figure 2, the time-dependent decay of a 4pnd wavepacket, centered at an average principal quantum number N = 35 with Kepler period τ_K=6.5 psec, is shown as a function of T_1 (denoted Delay 2-1 in the figure). Successive scans are taken at increasing values of T_1 at a constant interval. The eigenstate probability distribution for the wavepacket as measured using state-selective field ionization, is shown in the inset. The decay of the 4pnd wavepacket is obviously non-exponential, and is better characterized as stair-step in nature. The probability of finding the wavepacket in the 4pnd configuration remains relatively constant, for a period that depends on T_1, and then drops rapidly. An intriguing feature of the data in Figure 2 is the steady decrease in the duration of the stair-step with increasing delay T_1, with sudden periodic jumps in the step duration from a minimum to maximum value.

Figure 3 shows similar data for a wavepacket centered at N=40 (τ_K=9.7 psec). Close inspection of this figure shows that the 4pnd population does not drop to zero after the first stair-step, but instead, falls to a constant level for approximately one Kepler period before undergoing a second rapid decay. The insets in Figure 3 are snapshots of the theoretical time-dependent wavepacket probability distribution in the XZ plane (shown as gray-scale density plots with dark grey indicating regions of high probability and light grey indicating low probability). With the help of the insets, the data can be intuitively understood as follows. When $T_1 \approx 0$, the 4snd wavepacket has just left the nucleus at the instant the core electron is excited, and does not return for approximately one Kepler period. Upon its return, it scatters from the core electron, autoionizes with high probability, and the 4pnd population drops abruptly. The wavepacket returns to the core one Kepler period later and again suffers significant autoionization. Autoionization is nearly complete after one core-scattering event and essentially zero population remains after the second core collision. Alternatively, if $T_1 = \tau_K/2$, the wavepacket is at its outer radial turning-point at the instant of ICE1. Therefore, the wavepacket requires only half a Kepler period to return to the core. This reduced return time is reflected in the duration of the primary stair-step in the decay. However, the duration of second step is a full Kepler-period as expected. The sudden change in the duration of the primary step for ICE1 delays slightly greater than $T_1 = \tau_K$, can be understood intuitively as well. When T_1 is a little less than one Kepler period, the wavepacket crashes into the excited core immediately after its excitation, producing a 'step' of minimum duration. For T_1 slightly greater than one Kepler period, the situation is essentially identical to the $T_1 = 0$ case, and probability survives for one Kepler period.

The results of the radial wavepacket experiment indicate that we can reasonably draw several conclusions about wavepacket evolution in low-angular momentum Ca 4pnL autoionizing states. First, autoionization of Rydberg wavepackets seems to proceed identically to what one would expect by considering the evolution of the singly excited wavepacket. The doubly-excited packet has a time-varying autoionization rate that depends only on the probability for finding the electronic wavepacket near the nucleus. Second, although it is not immediate, autoionization from core-penetrating low-angular momentum states is rapid - if the electron passes near the nucleus, then the probability of autoionization is greater than 80% per passage. In the following section, we describe an experiment in which we attempt to use this behavioral insight to implement intuitive wavepacket control over differential autoionization yields in the same Ca system.

Manipulation of Differential Autoionization Yields via Angular Wavepacket Control

We now describe an experiment in which we attempt to control the angular distribution of autoionized electrons by varying the orientation of a 4snL angular wavepacket at the instant of a rapid ICE to a 4pnL configuration. Complete details on the experimental approach can be found elsewhere (14). Briefly, the 150 fs, 1.6 μm output of an OPA excites Ca atoms from an intermediate 4s5p $^1P_1$ level to a coherent superposition of two Rydberg states, $|4sNs\ ^1S_0\rangle$ and $|4sNd$

$^1D_0>$, with effective quantum number N = 9 or 10. Because of the difference in the quantum defects of the s and d series, the energies of the two states are not identical, and the wavepacket evolves with a period, τ_{ang}=1.06 psec or 0.70 psec for the N=9 and 10 packets, respectively. This angular period is significantly greater than the Kepler period of radial motion at these energies (< 150 fsec).

Figure 4 shows three (theoretical) density-plot snapshots of the evolution of the wavepacket as observed in the XZ-plane. Immediately after excitation, the distribution is "hour-glass" shaped, then becomes isotropic before assuming a "donut" form after a time, $\frac{1}{2}\tau_{ang}$. The wavepacket returns to its initial configuration after one angular period. It is important to note that since there are only two states in the wavepacket, there is a single evolutionary time-scale, and the motion is perfectly periodic. Furthermore, since the radial eigenfunctions of the two constituent states are nearly identical, the wavepacket experiences negligible radial motion.

Given the results of the radial wavepacket experiment and the properties of the angular wavepacket just described, we might expect that we could control the angular distribution of electrons ejected from the atom by inducing a sudden ICE at a particular phase of packet's angular evolution. This expectation is based on two fundamental assumptions. First, the autoionization lifetime of the low angular momentum 4pnL states is comparable to one Kepler period of the classical motion at that energy. Thus, since the angular motion takes place on a longer time scale ($\tau_{ang} >> \tau_K$), it appears that the wavepacket will be essentially frozen during the autoionization process. In other words, the orientation of the angular wavepacket at the instant of the ICE determines the angular distribution of the ejected electrons. Second, the previous experiment and intuitive picture of autoionization suggest that the autoionization rate depends only on the distance between the Rydberg electron and the core. Since there is negligible variation in the radial probability distribution of the angular wavepacket, we expect the autoionization *rate* in this experiment to be time-independent as well.

To test our expectations, we measure the energy and angularly resolved electron emission from these autoionizing angular wavepackets as a function of T_1, the delay between the laser pulses that launch the 4sNL wavepacket and perform the 4pNL ICE, respectively. The ICE is driven by a 150 fsec, 393 nm pulse that is created by frequency doubling the amplified output of a 787 nm, amplified Ti:Sapphire laser. As in the radial wavepacket experiment, decay of the 4pNL wavepacket results in the production of $4s^+$ Ca ions and 3.1 eV electrons or $3d_{3/2,5/2}^+$ Ca ions and 1.5 eV electrons. The extraction aperture in the upper capacitor plates allows detection of only those electrons that are emitted in the vertical direction. The angular distribution of the electrons is measured by recording the number of detected electrons in the 3.1eV (fast) and 1.5 eV (slow) channels as the common direction of linear polarization of the three laser pulses is rotated. In contrast to the radial wavepacket experiment, we measure the time-integrated angular- and energy-resolved electron yield from a single ICE. No information on the time of electron ejection is obtained.

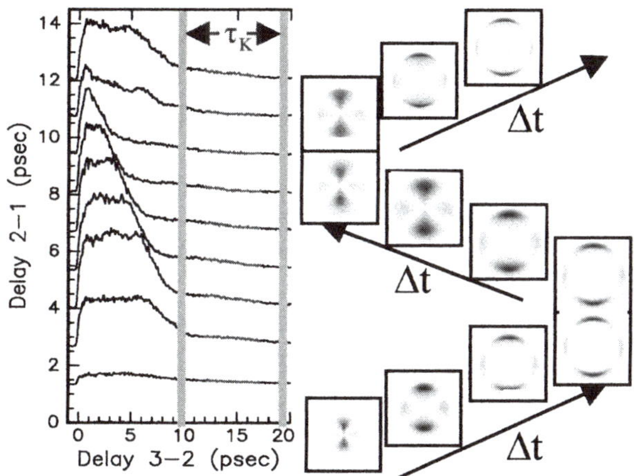

Figure 3: Analogous to Figure 2, but for a 4s40d wavepacket. The insets show the theoretical wavepacket probability distribution in the xz-plane, as a grey scale density plot, as time evolves. Note the abrupt change in the wavepacket lifetime if the ICE occurs just prior to, or just after a delay of 1 Kepler period relative to the wavepacket's launch.

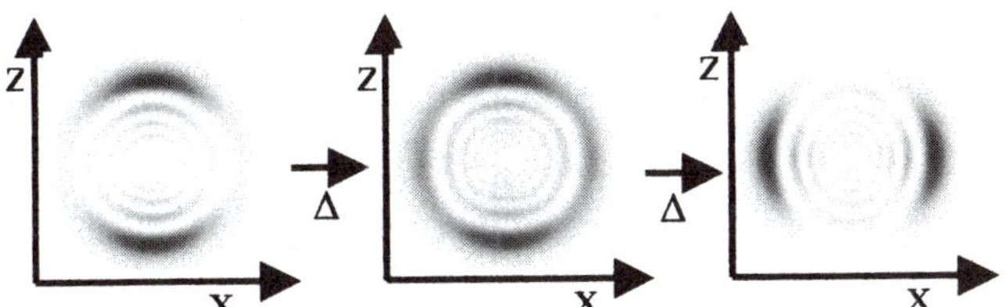

Figure 4: Density plot snapshots during the evolution of the angular wavepacket taken at delays of 0, $\frac{1}{4}\tau_{ang}$, and $\frac{1}{2}\tau_{ang}$, after its excitation.

128

Figure 5 shows (A) the number of electrons ejected along the laser polarization direction (0°), and (B) the number of ions produced as a function of the delay T_1 (i.e. nitial wavepacket orientation). Both the total ionization yield and the branching ratio for electron ejection into the two different energy channels are independent of T_1. However, the number of electrons ejected along the laser polarization axis shows a clear sinusoidal modulation with a period equal to τ_{ang}. We note that the independence of the ionic signal and electron branching-ratio to the wavepacket orientation indicates that control over the angular distributions is not at the expense of total yield. This desirable feature has not been realized in some previous differential control scenarios in atoms (*8*).

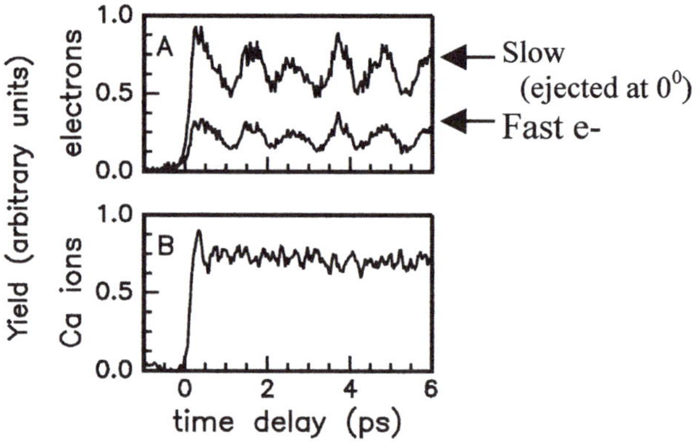

Figure 5: Autoionization yield vs. relative ICE/wavepacket-creation delay.
Adapted with permission from Reference (14).

Figure 6 shows the angular distribution of fast and slow electrons for two different values of T_1 that differ by $\tfrac{1}{2}\tau_{ang}$. The insets above the two sets of curves are density plots of the wavepacket configuration at the instant of the ICE. The data in Figure 6, shown as closed and open circles, shows significant electron ejection along the laser polarization (Z) axis with much less ejection in the XY-plane. Interestingly, it is more likely to have emission along the Z-axis if the wavepacket is initially oriented in the form of a donut in the XY-plane. The smooth curves drawn through the data are the results of a theoretical simulation based on MQDT aided by *ab initio* K-matrices (*11,12*). The only parameters that are varied to achieve the good agreement shown are: 1) the precise center of the laser spectrum (the chosen value differed from our measured value by a fraction of the laser bandwidth), and 2) the precise relative amplitude of the s and d components in the wavepacket (the s:d amplitude was chosen to be 2:1 in agreement with our expectations from photoexcitation cross-sections). Details of the calculation may be found elsewhere (*14*).

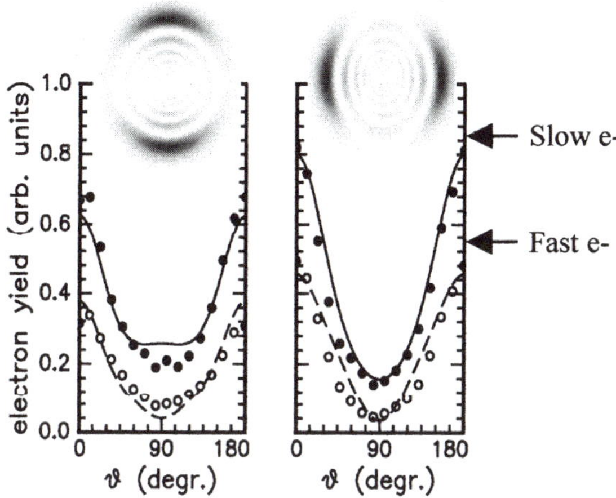

Figure 6: Angular distribution of fast and slow electrons for initial wavepacket configurations having an "hour-glass" distribution (left plot) or "donut" distribution (right plot). The open and closed circles are the data and the smooth curves are the result of the MQDT simulation described in the text. Adapted with permission from Reference (14).

Although control over the angular distributions has been demonstrated, the lack of significant delay-dependent change in the coarse features of the distributions in Fig. 7 is puzzling, especially considering the significant difference in the initial wavepacket configurations. The good agreement between theory and experiment further indicates that the similarity between the distributions is not due to some unidentified artifact of the measurement process. The distributions are clearly anisotropic, indicating a preferred direction for electron ejection. This is not unexpected. What is surprising, however, is that the preferred emission angle does not change with T_1. As the initial orientation of the wave packet changes, the most probable direction at which the low angular momentum Rydberg electron approaches the ionic core changes as well. Nevertheless, the dielectronic scattering event that causes autoionization preferentially ejects electrons in a direction that, apparently, is independent of the initial direction of approach of the Rydberg electron.

The MQDT formalism provides an explanation for the limited sensitivity of the angular emission to the initial wave-packet orientation. Calculations of the electron ejection probability as a function of excitation energy, for fixed emission angle and delay, indicate that the *rate* of decay of the autoionizing wave packet has an unexpected, orientation dependence. Figure 7 shows the theoretical autoionization yield as a function of excitation energy for three different fast electron emission directions and three different initial wave-packet orientations. The curves in Fig. 7 are the product of the energy-dependent ICE cross section and a 150 cm^{-1} (FWHM) Gaussian ICE laser spectrum. Broad resonance features indicate rapid autoionization into the specific angular channel while narrow structures correspond to slow decay rates. Apparently, electron ejection at $\theta=0°,180°$ occurs with high probability on a fast (200 fs) time scale. Therefore, when the initial wave packet is oriented such that electron emission at $\theta=0°,180°$ is most favorable, the frozen-wave-packet model discussed above is approximately valid. However, based on the width of the calculated resonance profiles, emission at $\theta=90°$ proceeds more slowly. Consequently, when the initial wave packet is oriented such that emission at 90° is preferred, very little autoionization actually occurs until the Rydberg electron has precessed into a configuration that decays through the rapid ejection of electrons at $\theta=0°,180°$. As a result, electron emission along the laser polarization axis is favored, regardless of the initial wave-packet orientation.

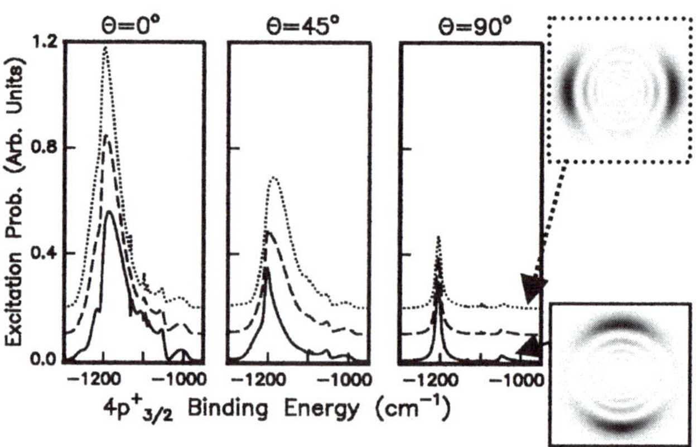

FIG. 7. Theoretical fast electron yield as a function of excitation energy for three different emission directions (0°, 45°, and 90°) and three different initial wave-packet orientations as described in the text. The curves in each plot are offset vertically for clarity and give (from bottom to top) the electron yields at $T_1=0$, $\frac{1}{4}\tau_{ang}$, and $\frac{1}{2}\tau_{ang}$, respectively. The density plot insets are snapshots of the wavepacket probability distribution for the lower and upper curves. Adapted with permission from Reference (14).

The apparent frustration of control over the preferential direction for electron emission can be attributed to configuration interaction, or mode-coupling, during the relatively slow propagation of the doubly-excited system through a subset of the "dump" channels. Similar mode-coupling and variation in the propagation-time of systems, toward asymptotic fragment channels, should be expected in more complex systems as well. This effect can, in general, limit the effectiveness of pump-dump control scenarios.

In summary, we have used Rydberg wavepacket manipulation techniques to investigate the pump-dump wavepacket control in "toy" two-electron systems. We find that even in relatively simple systems with well-understood wavepacket dynamics, that non-intuitive configuration couplings can limit the effectiveness of wavepacket control. This work has been supported by the Packard Foundation and the U.S.Department of Energy, Office of Basic Energy Sciences, Chemical Sciences, Geosciences and Bio-sciences Division. RVL was supported through a NSF IGERT grant.

References

1. Shapiro, M.; Hepburn, J.W.; Brumer P. *Chem. Phys. Lett.* **1988**, *149*, 451.
2. Judson, R.S.; Rabitz, H. *Phys. Rev. Lett.* **1993**, *68*, 1500.
3. Tannor, D.J.;Kosloff, R.; Rice, S.A.; *J. Chem. Phys.* **1986,** *85*, 5805.
4. Warren, W.S.; Rabitz, H.; Dahleh, M.; *Science* **1993**, *259*, 1581 and references therein.
5. Rabitz, H.; de Vivie-Riedle, R.; Motzkus, M.; Kompa, K. *Science* **2000**, *288*, 824 and references therein.
6. Shapiro, M.; Brumer, P.; *Chem. Phys. Lett.* **1993**, *208*, 193.
7. Gallagher, T.F. *Rydberg Atoms*; Cambridge University Press: Cambridge, UK; 1994 and references therein.
8. Wang, F.; Chen, C.; Elliott, D.S. *Phys. Rev. Lett.* **1996**, *77* 2416.
9. Nakajima, T.; Zhang, J.; Lambropolous, P. *Phys. Rev. Lett.* **1997**, *79*, 3367.
10. van Leeuwen, R.; Bajema, M.L.; Jones, R.R. *Phys. Rev. Lett.* **1999**, *82*, 2852.
11. Cooke, W.E.; Cromer, C.L. *Phys. Rev. A* **1985**, *32*, 2725 and references therein.
12. Greene, C.H.; Aymar, M. *Phys. Rev. A* **1991**, *44*, 1773 and references therein.
13. Jones, R.R.; Noordam, L.D. In *Advances in Atomic, Molecular, and Optical Physics*; Bederson, B.; Walther, H., Ed.; 38; Academic Press: San Diego, CA, 1998, pp. 1-38, and references therein.
14. van Leeuwen, R.; Vijayalakshmi, K.; Jones, R.R. *Phys. Rev. A* **2001**, *63*, 033403.
15. Thoma, J.E.; Jones, R.R.; *Phys. Rev. Lett.* **1999**, *83*, 516.
16. Cooke, W.E.; Gallagher, T.F.; Edelstein, S.A.; Hill, R.M. *Phys. Rev. Lett.* **1978**, *40*, 178.

Chapter 9

Optimal Quantum Control in Dissipative Environments: General Formalism and Perturbative Limits

Seogjoo Jang and Jianshu Cao

Department of Chemistry, Massachusetts Institute of Technology, Cambridge, MA 02139

Optimal quantum control in dissipative environments is formulated combining the optimal control theory from the Wilson group with the projection operator technique. In the weak response limit, a set of formally exact equations defines the optimal weak field for arbitrary target states and arbitrary system-bath interactions. Perturbative approximations are then made to derive general equations for weakly dissipative non-Markovian systems, which are amenable to analytical and/or numerical solutions. Potential applications of these results are discussed.

Introduction

Quantum processes are counter-intuitive in many respects. The notable features are uncertainty, interference, and tunneling, which seem to make the quantum system more difficult to control. With the advance of laser techniques, however, active utilization of such novel features for control has become possible through the manipulation of matter-radiation interactions on the relevant microscopic time scale. Impressive advances have been made for isolated gas phase systems, and recent reviews [1-4] provide a general perspective.

Most natural or practically important processes occur in condensed phases. The underlying microscopic events are governed by quantum principles, although the complexity of the system in general disguises such details. Therefore, the lessons learned from the studies of gas phase systems can in principle be extrapolated to condensed phase systems. Such possibilities have been demonstrated by recent experiments [4-9] and new theoretical works [10-14] suggest various novel ideas.

Two fundamental issues arise in condensed phase quantum control. One is the complexity of the Hamiltonian, and the other is how to deal with dissipation and decoherence [11,15]. The former aspect causes difficulties in finding robust schemes insensitive to small details of the underlying Hamiltonian. Recent advances in learning algorithms [4] provide ways to bypass these difficulties and to obtain information of the underlying Hamiltonian. The latter issue of decoherence and dissipation is under active investigation and some successful schemes have been devised in various limiting situations [11-14]. These studies indicate that a key factor is for the variation of the optical field to be in time scales comparable with those of the bath relaxation dynamics, thus invoking subtle interplay between the two processes. Therefore, the non-Markovian nature of the bath dynamics should be considered in devising a successful control scheme in condensed phase.

Quantum control in dissipative environments can be considered as an extension of time dependent dissipative quantum dynamics [11,16], where general formalisms have been introduced. However, extension of these formalisms to controlling molecular systems has not been well-established yet. Semi-group approaches are available [17,18], but these implicitly assume Markovian dynamics in a phenomenological way. Direct wavepacket dynamics [7] is possible, but may be limited in providing a general qualitative understanding. Cao, Messina, and Wilson [19] addressed the issue based on the system-bath Hamiltonian, by combining the weak field limit of the Wilson group formalism [20,21] with the Feynman-Vernon influence functional approach [22-24]. This approach provides a framework for investigating the role of the non-Markovian bath. However, practically, the use of the influence functional approach is limited to the harmonic bath, and the number of system degrees of freedom is small. A complementary approach is to apply the projection operator formalism, which can be applied to more general types of bath as long as certain criteria are met. This is the main subject of the present work. We first review the general formalism based on the Wilson group approach [20,21]. The resulting expression includes the gas phase system as a special case and also serves as a starting point for the explicit treatment of dissipative dynamics. Then, in the weak weak response limit, formally exact equations for the material response function and their second order perturbation approximations are derived. The article concludes with several comments on the implications of the present result.

Optimal Equation

Consider an ensemble of molecular systems with two electronic states within the optical range of interest. The effective system-bath Hamiltonian is written as

$$H_M = H_g |g\rangle\langle g| + (H_e + \epsilon_{eg})|e\rangle\langle e|$$

where H_g and H_e are nuclear Hamiltonians in the ground and excited states and ϵ_{eg} is the energy difference between the potential minima of the excited and ground states. Both nuclear Hamiltonians respectively consist of three terms, $H_g = H_g^0 + H_g^1 + H_b$ and $H_e = H_e^0 + H_e^1 + H_b$, where the terms with superscript 0 represent the system Hamiltonian and the terms with superscript 1 are those of the system-bath interaction. These expressions represent general situations where the system and the system-bath interaction Hamiltonians on the excited state surface are different from those on the ground state. In these definitions, the system Hamiltonians are renormalized in the sense that $Tr_b\{H_g^1 \rho_b\} = Tr_b\{H_e^1 \rho_b\} = 0$, where ρ_b is the canonical equilibrium density operator of the bath. In the presence of the optical field, the time dependent Hamiltonian is given by

$$H(t) = H_M - \mu(E^*(t)e^{i\epsilon_{eg}t}|g\rangle\langle e| + E(t)e^{-i\epsilon_{eg}t}|e\rangle\langle g|)$$

where the rotating wave approximation is invoked. We adopt the unit where $\hbar = 1$ through out this work.

Given a target state A and a time interval $\mathcal{T}$, the objective is to maximize the following average:

$$A(\mathcal{T}) = Tr\left\{AU(\mathcal{T},0)\rho(0)U^\dagger(\mathcal{T},0)\right\} \tag{1}$$

where $\rho(0)$ is the initial total density operator and $U(\mathcal{T},0)$ is the time evolution operator defined by $U(\mathcal{T},0) = \exp_{(+)}\left\{-i\int_0^{\mathcal{T}} dt\, H(t)\right\}$ with the subscript $(+)$ implying chronological time ordering. Various constraints might be necessary to conform to experimental reality, but here we assume the only constraint is the fixed total fluence. The optimal field according to the Wilson group formalism is then defined as a stationary solution of the functional $J[E(t)] = A(\mathcal{T}) - \lambda\int_0^{\mathcal{T}} dt|E(t)|^2$, where λ is the Lagrange multiplier to be determined. Taking functional derivative of $J[E(t)]$ with respect to $E^*(t)$ leads to the following optimal field equation:

$$\lambda E(t) = \frac{\delta A(\mathcal{T})}{\delta E^*(t)} \tag{2}$$

where the detailed form of the right hand side can be found by inserting the functional derivative of the time evolution operator, $\delta U(\mathcal{T},0)/\delta E^*(t) = ie^{i\epsilon_{eg}t}U(\mathcal{T},t)\mu|g\rangle\langle e|U(t,0)$, and that of the complex conjugate into the functional derivative of Eq. (1). The resulting expression is

$$\frac{\delta A(\mathcal{T})}{\delta E^*(t)} = iTr\left\{e^{i\epsilon_{eg}t}|g\rangle\langle e|U(t,0)\rho(0)U^\dagger(\mathcal{T},0)AU(\mathcal{T},t)\mu\right.$$
$$\left. -e^{i\epsilon_{eg}t}|g\rangle\langle e|U^\dagger(\mathcal{T},t)AU(\mathcal{T},0)\rho(0)U^\dagger(t,0)\mu\right\} \tag{3}$$

At this stage, we assume the initial density operator as the canonical equilibrium density operator

$$\rho(0) = |g\rangle\langle g|\rho_g^{eq} \tag{4}$$

where $\rho_g^{eq} = e^{-\beta H_g}/Z_g$ with $Z_g = Tr\left\{e^{-\beta H_g}\right\}$. Note that, in the ensemble represented by this equilibrium density operator, the system and bath degrees of freedom are entangled. As a result, the initial system-bath correlation cannot be disregarded. We assume that the target state on the electronic excited state is $A = \sigma_A|e\rangle\langle e|$, where σ_A is a reduced density operator defined in the Hilbert space of the system nuclear degrees of freedom. As a special case, σ_A can be an identity operator or the outer product of a pure state. With the specifications of Eq. (4) and the target state of $A = \sigma_A|e\rangle\langle e|$, Eq. (2) can be written as

$$\lambda E(t) = iTr\left\{e^{i\epsilon_{eg}t}\langle e|U(t,0)|g\rangle\rho_g^{eq}\langle g|U^\dagger(\mathcal{T},0)|e\rangle\sigma_A\langle e|U(\mathcal{T},t)|g\rangle\mu\right.$$
$$\left. -e^{i\epsilon_{eg}t}\langle e|U^\dagger(\mathcal{T},t)|e\rangle\sigma_A\langle e|U(\mathcal{T},0)|g\rangle\rho_g^{eq}\langle g|U^\dagger(t,0)|g\rangle\mu\right\} \tag{5}$$

which is the equation of the Wilson group formalism [20,21] valid for an arbitrary strength of the field. In general, solving Eq. (5) requires two difficult tasks: quantum dynamics of time dependent Hamiltonians and the nonlinear searching problem. The following section considers the weak response limit, where the problem reduces to quantum dynamics for time independent Hamiltonian and a time or frequency domain eigenvalue problem.

Weak Field Optimal Equation

To first order in the field strength, only the second term of Eq. (5) survives. Expanding $\langle e|U(\mathcal{T},0)|g\rangle$ with respect to $E(t)$ up to the first order, the weak field limit of Eq. (5) is written as

$$\lambda E(t) = \int_0^{\mathcal{T}} dt' \mathcal{M}(t,t') E(t') \tag{6}$$

Here, $\mathcal{M}(t,t')$, termed the material response function, is defined as

$$\mathcal{M}(t,t') = Tr\left\{ U_e^{\dagger}(\mathcal{T}-t)\sigma_A U_e(\mathcal{T}-t')\mu U_g(t')\rho_g^{eq}U_g^{\dagger}(t)\mu U_e^{\dagger}(\mathcal{T}-t) \right\} \tag{7}$$

where $U_{e(g)}$ is the time evolution operator for $H_{e(g)}$. Eqs. (6) and (7) correspond to the weak response limit of the Wilson group formalism. Examination of $\mathcal{M}(t,t')$ shows that the integral operator of Eq. (6) is Hermitian and that λ is always real. For an optimal field with unit fluence, λ is equal to the yield and corresponds to the second order approximation for $A(\mathcal{T})$ as follows:

$$\lambda = \int_0^{\mathcal{T}} dt \int_0^{\mathcal{T}} dt' E^*(t) \mathcal{M}(t,t') E(t') \approx A(\mathcal{T})$$

This relation implies that the desired optimal field is the eigenvector of Eq. (6) with the maximum eigenvalue λ. To simplify the formalism, we assume that the transition dipole operator is a constant, which is not affected by the trace operation. This simplification amounts to the Condon approximation. Then, the material response function of Eq. (7) can be written as

$$\mathcal{M}(t,t') = \mu^2 Tr_s\left\{ \sigma_A Tr_b\left\{ U_e(\mathcal{T}-t')\rho_g^{eq}U_g^{\dagger}(t-t')U_e^{\dagger}(\mathcal{T}-t) \right\} \right\} \tag{8}$$

where subscripts of s and b imply tracing over system and bath degrees of freedom respectively and preaveraging over the bath degrees of freedom has been taken. An equivalent expression can be found in Refs. 20 and 21. The averaging over the bath degrees in Eq. (8) suggests that approaches used in standard relaxation dynamics can be adopted here. For example, Cao *et al.* applied the Feynman-Vernon influence functional approach for the calculation of this material response function [19]. In the present paper, we apply the projection operator technique [25,26]. Before the derivation, we first define an analogue of the density operator, to which the projection operator is applied, and its basic relation to the material response function.

We define the operator within the trace over the bath in Eq. (7) as the two-time generalized density operator (TGDO):

$$G(t_2, t_1) \equiv U_e(t_2 + t_1)\rho_g^{eq} U_g^\dagger(t_1) U_e^\dagger(t_2) \quad , t_2, t_1 \geq 0$$

which is not a true density operator unless $t_1 = 0$. Taking the trace over the bath degrees of freedom leads to a reduced TGDO (rTGDO), $G_s(t_2, t_1) \equiv Tr_b\{G(t_2, t_1)\}$. Then, the material response function in Eq. (7) is expressed as

$$\mathcal{M}(t, t') = \left\{ \begin{array}{ll} Tr_s\{\sigma_A G_s(\mathcal{T} - t, t - t')\} & , t \geq t' \\ Tr_s\{\sigma_A G_s^\dagger(\mathcal{T} - t', t' - t)\} & , t < t' \end{array} \right. \tag{9}$$

where $G_s^\dagger$ is the Hermitian conjugate of G_s. The use of two different expressions in Eq. (9), depending on the time ordering of t and t', is necessary because the trace over the bath and the ensuing approximations break the time reversal symmetry of the full Hamiltonian dynamics.

The eigenvalue equation and the corresponding expression for the material response function can be transformed into the frequency domain. Taking the Fourier expansion of the electric field,

$$E(t) = \sum_n c_n e^{-i\Omega_n t} \quad , \Omega_n = \frac{2\pi n}{\mathcal{T}}$$

Eq. (6) can be transformed into

$$\sum_m \mathcal{M}_{nm} c_m = \lambda c_n$$

where

$$\begin{aligned} \mathcal{M}_{nm} &\equiv \int_0^{\mathcal{T}} dt \int_0^{\mathcal{T}} dt' e^{i(\Omega_n t - \Omega_m t')} \mathcal{M}(t, t') \\ &= \int_0^{\mathcal{T}} dt e^{i(\Omega_m - \Omega_n)t} \\ &\quad \times Tr_s\left\{\sigma_A \int_0^{\mathcal{T}-t} dt' \left(e^{i\Omega_m t'} G_s(t, t') + e^{-i\Omega_n t'} G_s^\dagger(t, t')\right)\right\} \end{aligned}$$

Master Equation for Reduced Two-time Generalized Density Operator (rTGDO)

We now use the projection operator technique [25,26] to derive a formally exact equation for $G_s(t_2, t_1)$, which is then reduced to a set of approximate equations valid to the second order of the system-bath interaction. First, we define the following interaction picture TGDO:

$$\tilde{G}(t_2, t_1) \equiv e^{i(H_e^0 + H_b)t_2} G(t_2, t_1) e^{-i(H_e^0 + H_b)t_2} = e^{i(\mathcal{L}_e^0 + \mathcal{L}_b)t_2} G(t_2, t_1)$$

which satisfies the quantum Liouville equation:

$$\frac{\partial}{\partial t_2} \tilde{G}(t_2, t_1) = -i[\tilde{H}_e^1(t_2), \tilde{G}(t_2, t_1)] \equiv -i\tilde{\mathcal{L}}_e^1(t_2)\tilde{G}(t_2, t_1)$$

where $\tilde{H}_e^1(t_2) = e^{i(\mathcal{L}_e^0 + \mathcal{L}_b)t_2} H_e^1$. The projection operator is defined as $\mathcal{P}(\cdot) \equiv \rho_b Tr_b(\cdot)$ and its complement is denoted as $\mathcal{Q} = 1 - \mathcal{P}$. Under the conditions of $Tr_b\{H_{g(e)}^1 \rho_b\} = 0$, one can show that

$$\frac{\partial}{\partial t_2} \mathcal{P}\tilde{G}(t_2, t_1) = -i\mathcal{P}\tilde{\mathcal{L}}_e^1(t_2)\mathcal{Q}\tilde{G}(t_2, t_1) \tag{10}$$

$$\frac{\partial}{\partial t_2} \mathcal{Q}\tilde{G}(t_2, t_1) = -i\mathcal{Q}\tilde{\mathcal{L}}_e^1(t_2)\mathcal{Q}\tilde{G}(t_2, t_1) - i\mathcal{Q}\tilde{\mathcal{L}}_e^1(t_2)\mathcal{P}\tilde{G}(t_2, t_1) \tag{11}$$

The formal solution of Eq. (11) is

$$\mathcal{Q}\tilde{G}(t_2, t_1) = \exp_{(+)} \left\{ -i \int_0^{t_2} dt_2' \mathcal{Q}\tilde{\mathcal{L}}_e^1(t_2') \right\} \mathcal{Q}\tilde{G}(0, t_1)$$
$$- \int_0^{t_2} dt_2' \exp_{(+)} \left\{ -i \int_{t_2'}^{t_2} dt_2'' \mathcal{Q}\tilde{\mathcal{L}}_e^1(t_2'') \right\} i\mathcal{Q}\tilde{\mathcal{L}}_e^1(t_2')\mathcal{P}\tilde{G}(t_2', t_1)$$

Inserting this equation into Eq. (10), we have

$$\frac{\partial}{\partial t_2} \mathcal{P}\tilde{G}(t_2, t_1) = -i\mathcal{P}\tilde{\mathcal{L}}_e^1(t_2) \exp_{(+)} \left\{ -i \int_0^{t_2} dt_2' \mathcal{Q}\tilde{\mathcal{L}}_e^1(t_2') \right\} \mathcal{Q}\tilde{G}(0, t_1)$$
$$- \int_0^{t_2} dt_2' \mathcal{P}\tilde{\mathcal{L}}_e^1(t_2) \exp_{(+)} \left\{ -i \int_{t_2'}^{t_2} dt_2'' \mathcal{Q}\tilde{\mathcal{L}}_e^1(t_2'') \right\}$$
$$\times \mathcal{Q}\tilde{\mathcal{L}}_e^1(t_2')\mathcal{P}\tilde{G}(t_2', t_1) \ . \tag{12}$$

The explicit expression for rTGDO is $G_s(t_2, t_1) = e^{-i\mathcal{L}_e^0 t_2} Tr_b\left\{ \tilde{G}(t_2, t_1) \right\}$. The evolution equation of this operator with respect to t_2, obtained from

Eq. (12), reads

$$\frac{\partial}{\partial t_2} G_s(t_2, t_1) = -i\mathcal{L}_e^0 G_s(t_2, t_1)$$

$$- Tr_b \left\{ i\mathcal{L}_e^1 e^{-i(\mathcal{L}_e^0 + \mathcal{L}_b)t_2} \exp_{(+)} \left\{ -i \int_0^{t_2} dt_2' \mathcal{Q}\tilde{\mathcal{L}}_e^1(t_2') \right\} \mathcal{Q}G(0, t_1) \right\}$$

$$- \int_0^{t_2} dt_2' Tr_b \left\{ \mathcal{L}_e^1 e^{-i(\mathcal{L}_e^0 + \mathcal{L}_b)t_2} \right.$$

$$\left. \times \exp_{(+)} \left\{ -i \int_{t_2'}^{t_2} dt_2'' \mathcal{Q}\tilde{\mathcal{L}}_e^1(t_2'') \right\} \mathcal{Q}\tilde{\mathcal{L}}_e^1(t_2')\rho_b \right\} e^{i\mathcal{L}_e^0 t_2'} G_s(t_2', t_1) \quad (13)$$

where t_1 is a fixed parameter. The solution of Eq. (13) requires both $\mathcal{Q}G(0, t_1)$ and $\mathcal{P}G(0, t_1) = \rho_b G_s(0, t_1)$. Employing a similar projection operator formalism, the former can be expressed in terms of the latter and the time evolution operator for the latter can also be obtained. Unlike the propagation with respect to t_2, the propagation with respect to t_1 couples the excited and the ground state Hamiltonian. Except for this difference, a similar procedure can be taken. For this purpose, we define the mixed interaction picture operator:

$$\bar{G}(0, t_1) \equiv e^{i(H_e^0 + H_b)t_1} G(0, t_1) e^{-i(H_g^0 + H_b)t_1} \equiv e^{i(\mathcal{L}_{eg}^0 + \mathcal{L}_b)t_1} G(0, t_1)$$

The time derivative of this operator is given by

$$\frac{\partial}{\partial t_1} \bar{G}(0, t_1) = -i\tilde{H}_e^1(t_1)\bar{G}(0, t_1) + i\bar{G}(0, t_1)\tilde{H}_g^1(t_1)$$

$$\equiv -i\bar{\mathcal{L}}_{eg}^1(t_1)\bar{G}(0, t_1) \quad (14)$$

where $\tilde{H}_e^1(t_1) = e^{i(\mathcal{L}_e^0 + \mathcal{L}_b)t_1} H_e^1$ and $\tilde{H}_g^1(t_1) = e^{i(\mathcal{L}_g^0 + \mathcal{L}_b)t_1} H_g^1$. Applying the projection operator and its complement to Eq. (14), one can show that

$$\frac{\partial}{\partial t_1} \mathcal{P}\bar{G}(0, t_1) = -i\mathcal{P}\bar{\mathcal{L}}_{eg}^1(t_1)\mathcal{Q}\bar{G}(0, t_1) \quad (15)$$

where

$$\mathcal{Q}\bar{G}(0, t_1) = \exp_{(+)} \left\{ -i \int_0^{t_1} dt_1' \mathcal{Q}\tilde{\mathcal{L}}_{eg}^1(t_1') \right\} \mathcal{Q}G(0, 0)$$

$$- \int_0^{t_1} dt_1' \exp_{(+)} \left\{ -i \int_{t_1'}^{t_1} dt_1'' \mathcal{Q}\bar{\mathcal{L}}_{eg}^1(t_1'') \right\} i\mathcal{Q}\bar{\mathcal{L}}_{eg}^1(t_1')\mathcal{P}\bar{G}(0, t_1') \quad (16)$$

When transformed back into the Schrödinger picture, Eq. (15) become

$$\frac{\partial}{\partial t_1} \mathcal{P}G(0, t_1) = -i\mathcal{L}_{eg}^0 \mathcal{P}G(0, t_1) - i\mathcal{P}\mathcal{L}_{eg}^1 \mathcal{Q}G(0, t_1) \quad (17)$$

where

$$\mathcal{Q}G(0,t_1) = e^{-i(\mathcal{L}^0_{eg}+\mathcal{L}_b)t_1} \exp_{(+)}\left\{-i\int_0^{t_1} dt'_1\,\mathcal{Q}\bar{\mathcal{L}}^1_{eg}(t'_1)\right\}\mathcal{Q}G(0,0)$$

$$-\int_0^{t_1} dt'_1\; e^{-i(\mathcal{L}^0_{eg}+\mathcal{L}_b)t_1}\exp_{(+)}\left\{-i\int_{t'_1}^{t_1} dt''_1\,\mathcal{Q}\bar{\mathcal{L}}^1_{eg}(t''_1)\right\}$$

$$\times\, i\mathcal{Q}\bar{\mathcal{L}}^1_{eg}(t'_1)\mathcal{P}\bar{G}(0,t'_1) \tag{18}$$

Eqs. (13), (17), and (18) along with the initial conditions, $\mathcal{P}G(0,0)$ and $\mathcal{Q}G(0,0)$, form *a closed set of formally exact equations necessary for calculating the material response function.* These equations need to be solved sequentially, first with respect to t_1 and then with respect to t_2.

For practical purposes, it is necessary to make approximations in order to derive solvable equations. Due to the initial condition $\rho(0) = \rho^g_{eq}$ and the fact that bath average of the system-bath interaction Hamiltonians are zero, the leading term of $\mathcal{Q}G(0,0)$ is first order in system-bath interaction. Thus a consistent second-order approximation for Eqs. (13), (17), and (18) can be found by replacing all the time ordered exponential operators involving $\mathcal{Q}\tilde{\mathcal{L}}^1_e$ and $\mathcal{Q}\bar{\mathcal{L}}^1_{eg}$ with unity, and approximating $\mathcal{P}G(0,0)$ and $\mathcal{Q}(0,0)$ by their leading terms. Making the former approximations first, the time evolution equation of rTGDO with respect to t_2 becomes

$$\frac{\partial}{\partial t_2}G_s(t_2,t_1) \approx -i\mathcal{L}^0_e G_s(t_2,t_1) - S(t_2,t_1)$$

$$-\int_0^{t_2} dt'_2\,Tr_b\left\{\mathcal{L}^1_e e^{-i(\mathcal{L}^0_e+\mathcal{L}_b)(t_2-t'_2)}\mathcal{Q}\mathcal{L}^1_e\rho_b\right\}G_s(t'_2,t_1) \tag{19}$$

where

$$S(t_2,t_1) = Tr_b\left\{i\mathcal{L}^1_e e^{-i(\mathcal{L}^0_e+\mathcal{L}_b)t_2}e^{-i(\mathcal{L}^0_{eg}+\mathcal{L}_b)t_1}\mathcal{Q}G(0,0)\right\}$$

$$-\int_0^{t_1} dt'_1\,Tr_b\left\{\mathcal{L}^1_e e^{-i(\mathcal{L}^0_e+\mathcal{L}_b)t_2}e^{-i(\mathcal{L}^0_{eg}+\mathcal{L}_b)(t_1-t'_1)}\mathcal{Q}\mathcal{L}^1_{eg}\rho_b\right\}G_s(0,t'_1)$$

The time evolution equation of rTGDO with respect to t_1 becomes

$$\frac{\partial}{\partial t_1}G_s(0,t_1) \approx -i\mathcal{L}^0_{eg}G_s(0,t_1) - Tr_b\left\{i\mathcal{L}^1_{eg}e^{-i\mathcal{L}^0_{eg}t_1}\mathcal{Q}G(0,0)\right\}$$

$$-\int_0^{t_1} dt'_1\,Tr_b\left\{\mathcal{L}^1_{eg}e^{-i\mathcal{L}^0_{eg}(t_1-t'_1)}\mathcal{Q}\mathcal{L}^1_{eg}\rho_b\right\}G_s(0,t'_1) \tag{20}$$

Equations (19) and (20) are obtained by inserting Eq. (18) into Eqs. (13) and (17) and making second order approximations as prescribed above. In both Eqs. (19) and (20), the inhomogeneous terms involve $QG(0,0)$. Also, the solution of Eq. (20) requires the initial condition of $G_s(0,0) = Tr_b\{\rho_g^{eq}\}$. In order to find the leading terms of these two initial conditions, we use the following first order approximation for the canonical density operator:

$$e^{-\beta(H_g^0+H_g^1+H_b)} \approx e^{-\beta(H_g^0+H_b)} - \int_0^\beta d\tau \, e^{-(\beta-\tau)(H_g^0+H_b)} H_g^1 e^{-\tau(H_g^0+H_b)}$$

Then, one can show

$$G_s(0,0) \approx \frac{1}{Z_g^0} e^{-\beta H_g^0} \quad , \quad Z_g^0 = Tr_b\{e^{-\beta H_g^0}\} \tag{21}$$

$$QG(0,0) \approx -\frac{1}{Z_g^0 Z_b} \int_0^\beta d\tau \, e^{-(\beta-\tau)(H_g^0+H_b)} H_g^1 e^{-\tau(H_g^0+H_b)} \tag{22}$$

Equations (19)-(20) along with the initial conditions of Eqs. (21) and (22) form a closed set of equations which is valid up to the second order of the system bath interaction. Unlike the usual Bloch-Redfield equation, these are equations in two dimensional time space of t_1 and t_2 and the integration over t_1 needs to be carried out first. For the special case where σ_A is an identity operator, no specification is made for the excited nuclear state, and the information of $G_s(0, t_1)$ is sufficient to specify the material response function. However, for general cases where a target excited state is specified in a nontrivial way, additional time evolution with respect to t_2 becomes necessary.

Concluding Remarks

Starting from the optimal quantum control formalism developed in the Wilson group, we have derived a set of formally exact equations valid in the weak response limit, with the use of the projection operator technique. In the weak system-bath interaction limit, we have obtained second order perturbative approximations. These equations need to be solved sequentially, as detailed in the last section. The solution of these equations provides the reduced two-time generalized density operator, as a function of two time arguments, which produces the material response function when overlapped with a target excited state nuclear density operator. The optimal field and the optimal yield are obtained by solving an eigenvalue problem for this material response function.

Our formalism does not depend on the detailed nature of the system-bath interaction, the measurement time interval $\mathcal{T}$, and the nature of the target state σ_A. As a result, the derived equations of motion are general enough to serve as a systematic framework for studying how such factors affect the qualitative nature of quantum control. In traditional quantum control studies, the issue has been finding the optimal field for predetermined Hamiltonian and the target state. However, in condensed phase, different issues can arise. For example, in the subject of quantum computation, an important issue is simply to find the best combinations that can nullify the effect of decoherence and dissipation as much as possible. In this case, the theoretical objective becomes finding the nature of system-bath interaction and the class of σ_A as well as the optimal field which can cooperatively maximize the eigenvalue of the optimal equation. Our formalism seems suitable for addressing these new issues as well as more traditional issues of quantum control.

Acknowledgments

This work is dedicated to the late Kent Wilson for his inspiring leadership in the field of quantum coherence control. One of us (J. C.) gratefully acknowledges Professor Wilson for his generosity, encouragement, and friendship. The research is supported by the Petroleum Research Fund administrated by the American Chemical Society.

References

1. Rice, S. A. ; Zhao, M. *Optical Control of Molecular Dynamics* ; Wiley : NewYork, 2000.
2. Gordon, R. J. ; Zhu, L. ; Seideman, T. *Acc. Chem. Res.* **1999**, *32*, 1007.
3. Shapiro, M. ; Brumer, P. *Adv. At. Mol. Opt. Phys.* **2000**, *42*, 287.
4. Rabitz, H. ; Zhu, W. *Acc. Chem. Res.* **2000**, *33*, 572.
5. C. J. Bardeen, V. V. Yakovlev, K. R. Wilson, S. D. Carpenter, P. M. Weber, and W. S. Waren *Chem. Phys. Lett.* **1997**, *280*, 151.
6. C. J. Bardeen, V. V. Yakovlev, J. A. Squier , and K. R. Wilson *J. Am. Chem. Soc.* **1998**, *120*, 13023.
7. C. J. Bardeen, K. R. Wilson, V. V. Yakovlev, V. V. Apkarian, C. C. Martens, R. Zadoyan, B. Kohler, and M. Messina *J. Chem. Phys.* **1997**, *106*, 8486.
8. Levis, R. J. ; Menkir, G. M. ; Rabitz, H. *Science* **2001**, *292*, 709.

9. Kennedy, S. P. ; Garro, N. ; Phillips, R. T. *Phys. Rev. Lett.* **2001**, *86*, 4148.

10. Cao, J. *J. Lumin.* **2000** *87-89*, 30.

11. M. Grifoni and P. Hänggi *Phys. Rep.* **1998** *304*, 219.

12. Agarwal, G. S. *Phys. Rev. A* **1999** *61*, 013809.

13. M. Thorwart, L. Hartmann, I. Goychuck, and P. Hänggi *J. Mod. Opt.* **2000**, *47*, 2905.

14. Luis, A. *Phys. Rev. A* **2001** *63*, 052112.

15. *Decoherence: Theoretical, Experimental, and Conceptual Problems*; Ph. Blanchard *et al.*, Eds.; Proceedings, Bielefeld, Germany, 1998; Springer; Berline, 1998.

16. Meier, C. ; Tannor, D. J. *J. Chem. Phys.* **1999** *111*, 3365.

17. Tang, H. ; Kosloff, R. ; Rice, S. A. *J. Chem. Phys.* **1996** *104*, 5457.

18. Kosloff, R. ; Ratner, M. A. ; Davis, W. B. *J. Chem. Phys.* **1997** *106*, 7046.

19. Cao, J. ; Messina, M. ; Wilson, K. R. *J. Chem. Phys.* **1997** *106*, 5239.

20. Y. Yan, R. E. Gillilan, R. M. Whitnell, and K. W. Wilson *J. Phys. Chem.* **1992** *97*, 2320.

21. J. L. Krause, R. M. Whitnell, K. R. Wilson, and Y. J. Yan, in *Femtosecond Chemistry*; J. Manz and L. Woste, Eds.; VCH Publishers; Weihneim; 1995.

22. Feynman, R. P. ; Hibbs, A. R. *Quantum Mechanics and Path Integrals*; McGraw-Hill Book Company; New York, 1965.

23. Feynman, R. P. *Statistical Mechanics*; Addison-Wesley Publishing Company; New York, 1972.

24. Weiss, U. *Series in Modern Condensed Matter Physics Vol. 2 : Quantum Dissipative Systems*; World Scientific; Singapore, 1993.

25. R. Zwanzig, in *Lectures in Theoretical Physics*, Boulder 1960, Vol. 3; W. E. Brittin, Eds.; Interscience; New York, 1961.

26. Seke, J. *Phys. Rev. A* **1987** *36*, 5841.

Chapter 10

Manifestation of Coherence on Laser Pulse Propagation: Propagation Dynamics of Phase-Controlled Lasers

Takashi Nakajima

Institute of Advanced Energy, Kyoto University, Gokasho, Uji, Kyoto 611–0011, Japan

We numerically study the propagation effects of phase-controlled two-color lasers in a two-level medium. By solving the coupled differential equations for atoms and fields simultaneously, it is found that the effects of propagation are striking in terms of the pulse shape and the yield.

INTRODUCTION

Since the first experimental observation [1] of the modulation of photoionization yield by controlling the relative phase of the fundamental and its third harmonic fields, people are technically convinced that the phase of laser field can indeed be manipulated to control photoabsorption products. The underlying idea is to induce the quantum mechanical interference between the three- and single-photon transitions, and the proto-type of the scheme was proposed by Brumer and Shapiro some years ago [2, 3]. Many aspects of phase-sensitive effects have been theoretically [4-10] as well as experimentally [11-20] investigated by various groups.

We should note that most of the related works have been limited to the response of single-atom or -molecule. If the purpose of manipulating the relative phase of lasers is to obtain more product yields in the desired state, one must definitely go beyond the single-atom or -molecule response. Namely the propagation effects must be taken into account. We are aware of only a few works which are dealing with the propagation effects of phase-controlled lasers. Experimental study has been carried out by Chen and Elliott [21] along such a context. More recently Petrosyan and Lambropoulos [22] have theoretically examined the propagation effects of phase-controlled lasers in an optically dense medium of Xe.

In this paper we investigate the time- and space-evolution of phase-controlled two-color lasers in a two-level medium by numerically solving the coupled differential equations for atoms and fields. The purpose is to understand the propagation effects which are often overlooked in the field of coherent control. Given the situation above, we have had the following questions in mind. First, concerning the properties of the fields, how are the temporal profiles of the phase-controlled pulse pair affected by the medium during the propagation? Can the phase difference be well-maintained throughout the medium? Needless to say, these are of essential importance for coherent control. Our second question is related to the atomic response itself. Since the fields drive atoms and vice versa, the atomic response must be necessarily altered during the propagation, if the properties of the fields are altered at all in the medium. Of course the dynamics of the fields and atoms are coupled to each other. Therefore, there should be a consistent picture for the understanding of the whole process, and that is what we would like to clarify in this work.

MODEL

The system we consider in this paper is shown in Figure 1, which is nothing but a two-level system, a proto-type of the phase control scheme. When this system is subject to two radiation fields, a fundamental field with frequency ω_1 and its third harmonic with frequency $\omega_3 (= 3\omega_1)$ in this case, the photoabsorption process can be controlled by externally changing the phase difference between two fields.

Needless to say, this process is based on a quantum mechanical interference caused by the three- and single-photon transitions of the fundamental and its third harmonic fields, respectively. The quantities we will look at here are the temporal as well as spatial variation of the pulse shape, the phase difference between two fields, and the photoabsorption yield. For a complete description of the system dynamics, we now construct the coupled differential equations for atoms and fields. Due to the numerical

146

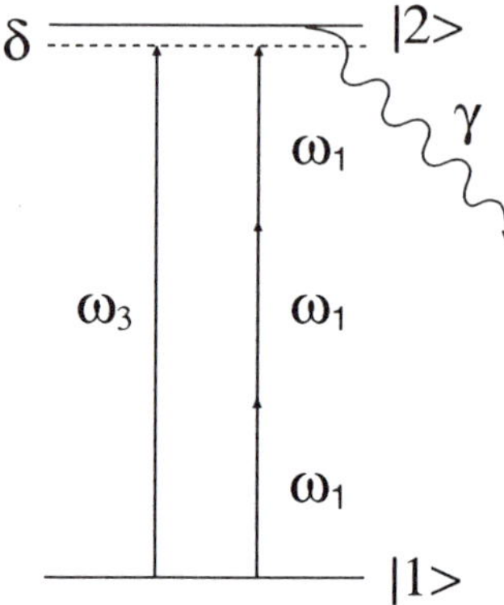

Figure 1: *Two-level system considered in this paper.*

convenience, we employ a local frame (ζ, τ) rather than a lab frame (z, t) throughout this paper. Obviously these two frames are connected through the relations $\tau = t - z/c$ and $\zeta = z$ where c is the speed of light in the vacuum. Using amplitude equations the response of the atom can be described as,

$$\frac{\partial}{\partial \tau} u_1 = i \left[\Omega^{(3)*}(f(\zeta, \tau)^*)^3 + \Omega^* h^*(\zeta, \tau) \right] u_2 \tag{1}$$

$$\frac{\partial}{\partial \tau} u_2 = i \left[-\left(\Delta - i\frac{\gamma}{2} \right) + (S_2 - S_1)|f(\zeta, \tau)|^2 \right] u_2$$
$$+ \left[\Omega^{(3)}(f(\zeta, \tau))^3 + \Omega h(\zeta, \tau) \right] u_1 \tag{2}$$

with u_1 and u_2 being the probability amplitudes of states $|1\rangle$ and $|2\rangle$, respectively, and Δ and γ are the detuning and the radiative decay rate out of the system. $\Omega^{(3)}$ and Ω stand for the three- and single-photon Rabi frequencies at the peak intensities by the lasers with frequencies ω_1 and ω_3, respectively. The *complex* field amplitudes with frequencies ω_1 and ω_3 are represented by $f(\zeta, \tau)$ and $h(\zeta, \tau)$, respectively, which have been normalized at the peak field amplitudes at the entrance to the medium. Note that they are functions of both time and space. For simplicity we restrict ourselves to the one-dimensional propagation along the ζ-axis in this work. S_1 and S_2 represent ac Stark shifts of states $|1\rangle$ and $|2\rangle$, at the peak intensities of the fundamental field at the entrance to the medium, respectively.

The polarizations of the medium with frequencies ω_1 and ω_3 can be obtained, after some algebra, as,

$$P_1 = 2N \left[2\hbar(s_1|u_1|^2 + s_2|u_2|^2)\varepsilon_{10} + 3\mu^{(3)}(\varepsilon_{10}^*)^2 u_2 u_1^* \right] \tag{3}$$

$$P_3 = 2N\mu u_2 u_1^* \tag{4}$$

where we have introduced the reduced ac Stark shifts $s_1 = S_1/|\varepsilon_{10}|^2$ and $s_2 = S_2/|\varepsilon_{10}|^2$, peak field amplitudes ε_{10} and ε_{30} for the fundamental and its third harmonic, and atom density N.

In order to study the propagation effects, we also need equations for the fields. The wave equations for the fundamental and its third harmonic fields are found to be written as,

$$\frac{\partial}{\partial\zeta}f(\zeta,\tau) = \; i\left(\frac{\omega_1 N}{c\epsilon_0 \varepsilon_{10}}\right)\left[2\hbar(s_1\,|u_1|^2 + s_2\,|u_2|^2)\varepsilon_{10}f(\zeta,\tau)\right.$$
$$\left. +3\mu^{(3)}(\varepsilon_{10}^*)^2(f(\zeta,\tau)^*)^2 u_2 u_1^* \right] \tag{5}$$

$$\frac{\partial}{\partial\zeta}h(\zeta,\tau) = i\left(\frac{\omega_3 N}{c\epsilon_0 \varepsilon_{30}}\right)\mu u_2 u_1^* \tag{6}$$

where ϵ_0 is the permittivity of free space. From these equations it is clear now that the two fields can communicate through the medium. Perhaps it is more precise to say that the harmonic field can communicate with, and is influenced by the fundamental field, since the right hand side of Equation (6) obviously depends on both fundamental and harmonic fields, as can be seen from Equations (1) and (2). On the other hand, the fundamental field is hardly affected by the harmonic field, since, although there appears a term containing $u_2 u_1$ on the right hand side of Equation (5), it is a higher order term. The dominant contribution on the right hand side of Equation (5) comes from the first term which is associated with the induced polarizability.

NUMERICAL RESULTS

We now present a few representative results obtained by numerically solving the coupled differential equations given by Equations (1), (2), (5), and (6). Before presenting results we describe the various assumptions made for the calculations. The temporal profiles of the two fields at the entrance to the medium are assumed to be in the Gaussian form, i.e.,

$$f(\zeta = 0, \tau) \equiv f_0(\tau) = exp\left[-4ln2\left(\frac{\tau}{\tau_1}\right)^2\right] \tag{7}$$

$$h(\zeta = 0, \tau) \equiv h_0(\tau) = exp(i\phi_0)exp\left[-4ln2\left(\frac{\tau}{\tau_3}\right)^2\right] \tag{8}$$

where ϕ_0 is an initial phase difference between two fields at the entrance to the medium, and τ_1 and τ_3 are the pulse durations (FWHM) of the fundamental and harmonic fields, respectively, with the relation $\tau_1 = \sqrt{3}\tau_3$. Note that this relation is necessary for the complete temporal overlap of the three- and single-photon excitations by the fundamental and harmonic at the entrance to the medium, respectively. It is also assumed that the three- and single-photon Rabi frequencies have initially the same amplitudes, i.e., $\Omega^{(3)} = \Omega$. Obviously these conditions for the pulse profile and the Rabi frequencies guarantee that the interference is maximum at the entrance to the medium.

For simplicity it is further assumed that the detuning has been taken to be zero, and the ac Stark shifts S_1 and S_2 have been neglected throughout this work. It is now convenient to introduce the absorption coefficients α for the single-photon excitation, which can be expressed as,

$$\alpha = 4\Omega\mu\gamma^{-1}\frac{\omega_3 N}{c\epsilon_0\varepsilon_{30}} \tag{9}$$

As usual, the optical depth for the single-photon excitation can be defined as $\alpha\zeta$. If we are to describe the system in terms of the parameters τ_1, τ_3, $\Omega^{(3)}$, Ω, γ, δ, and α, the ratio of $(\omega_1 N/c\epsilon_0\varepsilon_{10})3\mu^{(3)}(\varepsilon^*)^2$ and $(\omega_3 N/c\epsilon_0\varepsilon_{30})\mu$ must be set to a certain value. The reason for this is obvious by comparing Equations (5) and (6). Using the assumption $\mu^{(3)}\varepsilon_{10}^3 = \mu\varepsilon_{30}$ which is equivalent to $\Omega^{(3)} = \Omega$, this ratio can be simplified as

$$\frac{3(\omega_1 N/c\epsilon_0\varepsilon_{10})\mu^{(3)}(\varepsilon_{10}^*)^2}{(\omega_3 N/c\epsilon_0\varepsilon_{30})\mu} = \left(\frac{\varepsilon_{30}}{\varepsilon_{10}}\right)^2 \tag{10}$$

A typical ratio of the intensities of both lasers is $10^{-7} \sim 10^{-8}$ for a dipole transition of a neutral atom, and therefore we have specifically chosen the ratio given by Equation (10) to be 10^{-8}. We should mention that whether the ratio given in Equation (10) is 10^{-7} or 10^{-8} does not make any difference in our numerical results. In all the numerical results presented in this paper, we refer to the optical depth with respect to the single-photon process, since, within the optical depth considered here the depletion of the fundamental field does not occur. In other words, whether the ratio given in Equation (10) is 10^{-7} or 10^{-9} does not make any difference in our numerical results.

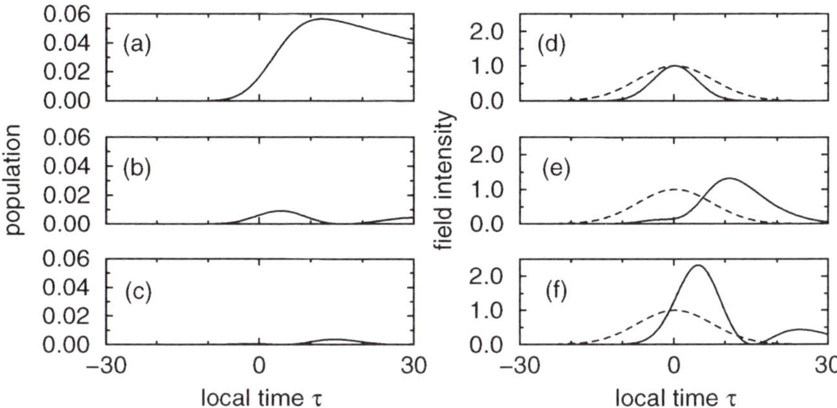

Figure 2: *Temporal evolution of the upper state population (left column) and the fundamental (dashed line in right column) and harmonic (solid line in right column) field amplitude as a function of local time τ at different optical depths $\alpha\zeta = 0$ ((a) and (d)), 20 ((b) and (e)), and 60 ((c) and (f)) under the presence of both fundamental and harmonic fields. The initial phase difference is taken to be $\pi/4$. All the parameters are given in the text.*

In summary, the parameters employed for the calculations are $\tau_3 = 10$, $\tau_1 = 10\sqrt{3}$, $\Omega^{(3)} = \Omega = 0.01$, $\gamma = 0.02$, $\Delta = 0$, $\alpha = 0.1$.

In Figure 2 we show the temporal variation, at different optical depths $\alpha\zeta = 0, 20$, and 60, of the population in the upper state $|2\rangle$ (left column) and the pulse profiles of the fundamental and its third harmonic (right column) for the initial phase difference $\phi_0 = \pi/4$. Since the pulse area is rather small, i.e., $\Omega\tau_3 \sim 0.1$, no Rabi oscillation is seen. As the pulse pair propagates in the medium, the population in the upper state $|2\rangle$ gradually decreases up to $\alpha\zeta \sim 10$ (not shown here) because of the depletion of the harmonic field. At $\alpha\zeta > 10$, however, the harmonic field regains its intensity. Obviously the energy of the harmonic field is provided by the much more intense fundamental field. Needless to say the fundamental field is practically intact in terms of the pulse shape and intensity, since the three-photon absorption cross section is very small and the fundamental field is very intense. Similar results are presented in Figure 3 for the initial phase difference $\phi_0 = 3\pi/4$. The variation of the both population and harmonic field profile is not as prominent as that in Figure 2, since there is already a more destructive interference taking place for $\phi_0 = 3\pi/4$ than $\phi_0 = \pi/4$. Recall that the complete destructive interference takes place for $\phi_0 = \pi$, where the upper state is never populated at any time at any depth in the medium. It should be noted that, in both Figures 2 and 3, the upper state is barely populated even after the harmonic field has regained its

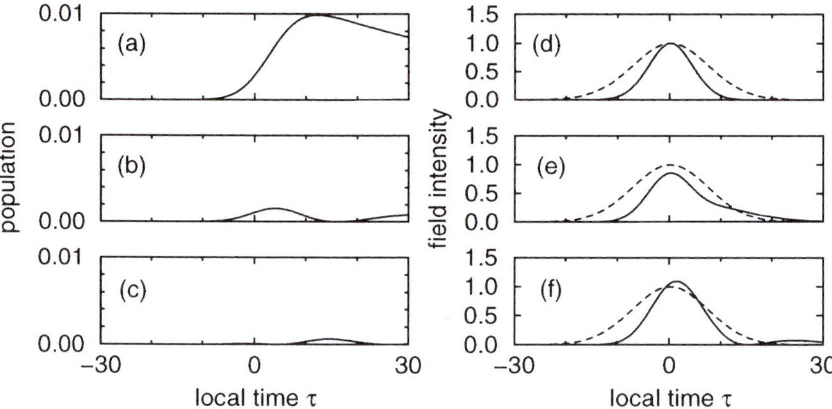

Figure 3: *Temporal evolution of the upper state population (left column) and the fundamental (dashed line in right column) and harmonic (solid line in right column) field amplitude as a function of local time τ at different optical depths $\alpha\zeta = 0$ ((a) and (d)), 20 ((b) and (e)), and 60 ((c) and (f)) under the presence of both fundamental and harmonic fields. The initial phase difference is taken to be $3\pi/4$. All the parameters are given in the text.*

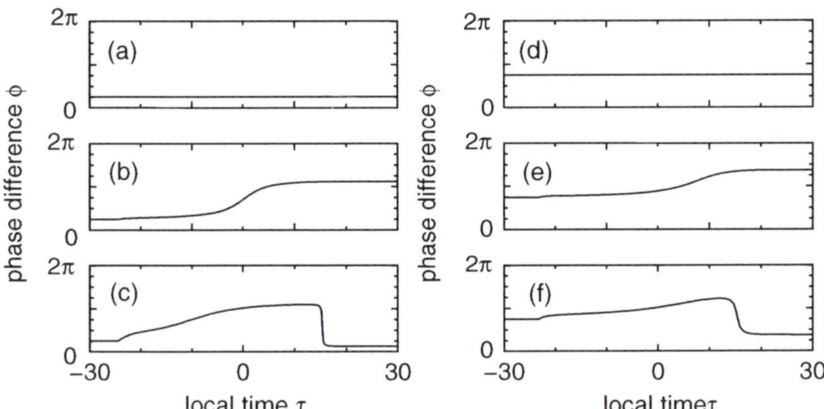

Figure 4: *Temporal evolution of the phase difference for the initial phase difference $\phi = \pi/4$ ((a)-(c)) and $3\pi/4$ ((d)-(f)) at different optical depths $\alpha\zeta = 0$ ((a) and (d)), 20 ((b) and (e)), and 60 ((c) and (f)). Except for the initial phase difference, all the parameters are the same with those employed for Figures 2 and 3.*

intensity. In order to understand this, we plot in Figure 4 the variation of the phase difference at the corresponding optical depths $\alpha\zeta = 0, 20$, and 60. It can be seen that the phase difference tends to change toward π around the trailing edge of the pulse. This means that, as the leading edge of the pulse pair interacts with atoms, the dipole is induced in them, leading to the feedback from atoms to the fields. As a result, the modification of the phase difference starts at the leading edge of the pulse. It should also be noted that the phase difference approaches toward π as the pulse pair propagates deeper into the medium. Therefore the phase difference in graph (c) of Figures 2 and 3 are not $\pi/4$ or $3\pi/4$ anymore, but close to the value of π, which means that the destructive interference becomes automatically maximum during the propagation. This explains why the upper state is barely populated even after the harmonic field has regained its intensity.

In Figure 5 we plot the time- and space-integrated signal, defined as $\int d\tau d\zeta (1 - |u_1(\zeta, \tau)|^2)$, as a function of optical depth $\alpha\zeta$ for the initial phase differences $\phi_0 = 0, \pi/4, \pi/2, 3\pi/4$, and π. Saturation is clearly observed. Needless to say, saturation occurs because the phase difference tends to approach π as the pulse pair propagates further into the medium, leading

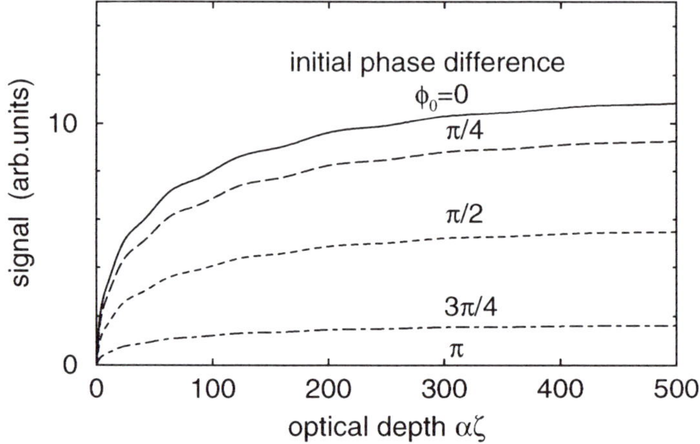

Figure 5: *Variation of the time- and space-integrated signal as a function of optical depth for different initial phase differences* $\phi_0 = 0(solid), \pi/4(dashed), \pi/2(dotted), 3\pi/4$ *(dot-dashed), and* π *(thin solid).*

to the complete destructive interference at sufficiently large $\alpha\zeta$. Our result suggests that, for the purpose of obtaining more signals, the increment of the interaction length along the propagation direction would be of some help up to some length. But a further increase of the interaction length is of no use, since the phase difference and the intensity of the pulse pair are automatically adjusted during the propagation in such a way that the complete destructive interference takes place, contributing to no increase of the signal.

SUMMARY

In summary we have numerically studied the propagation effects of phase-controlled lasers in a two-level medium by solving the set of coupled differential equations. Due to the coupling between atoms and fields, the properties of the third harmonic field are significantly modified during propagation in terms of the temporal pulse shape, amplitude, and phase, while those of the fundamental field are practically unchanged. This is because the fundamental field is much more intense, typically by 7-9 orders of magnitude for the dipole transition of a neutral atom, compared with the harmonic, resulting in no depletion during propagation. Regardless of the initial phase difference between two fields at the entrance to the medium, the pulse pair tends to be completely out of phase after sufficiently long propagation. The alterations of the harmonic field, in terms of the phase and the amplitude, lead to the complete destructive interference starting from *any* initial phase difference. Our results suggest that the inclusion of propagation effects is not always favorable for coherent control, since the saturation of the product yield takes place at a certain optical depth.

This work was supported by the Grant-in-Aid for scientific research from the Ministry of Education, Culture, Sports, Science and Technology of Japan.

References

[1] Chen, Ce; Yin, Y.-Y.; Elliott, D.S. *Phys. Rev. Lett.* **1990**, 64, 507.

[2] Shapiro, M.; Hepburn, J.W.; Brumer, B. *Chem. Phys. Lett.* **1988**, 149, 451.

[3] Shapiro, M.; Brumer, B. *J. Chem. Soc. Faraday Trans.* **1987**, 93, 1263.

[4] Nakajima, T.; Lambropoulos, P. *Phys. Rev. Lett.* **1993**, 70, 1081.

[5] Nakajima, T.; Lambropoulos, P. *Phys. Rev. A* **1994**, 50, 595.

[6] Nakajima, T; Zhang, J.; Lambropoulos, P. *J. Phys. B* **1997**, 30, 1077.

[7] Lambropoulos, P;Nakajima, T. *Phys. Rev. Lett.* **1999**, 82, 2266.

[8] Nakajima, T. *Phys. Rev. A* **2000**, 61, 041403(R).

[9] Schafer, K.J.; Kulander, K.C. *Phys. Rev. A* **1992**, 45, 8026.

[10] Charron, E.; Giusti-Suzor, A.; Mies, F.H. *Phys. Rev. Lett.* **1993**, 71, 692.

[11] Chen, Ce; Elliott, D.S. *Phys. Rev. Lett.* **1990**, 65, 1737.

[12] Yin, Y.-Y. ;Chen, Ce; Elliott, D.S. *Phys. Rev. Lett.* **1992**, 69, 2353.

[13] Park, S.-M.; Lu, S.-P.; Gordon, R.J. *J. Chem. Phys.* **1991**, 94, 8622.

[14] Zhu, L.;Kleiman, V. Li, X.; Lu, S.;Trentelman, K.; Gordon, R.J. *Science* **1995**, 270, 77.

[15] Wang, X.; Bersohn, R.;Takahashi, K.;Kawasaki, M.; Kim, H.-L. *J. Chem. Phys.* **1996**, 105, 2992.

[16] Dupont, E; Corkum, P.B.; Liu, H.C.; Buchanan, M; Wasilewski, Z.R. *Phys. Rev. Lett.* **1995**, 74, 3596.

[17] Kim, H.-L.; Bersohn, R; *J. Chem. Phys.* **1997**, 107, 4546.

[18] Xenakis, D; Karapanagioti, N.E.; Charalambidis, D. *Phys. Rev. A* **1999**, 59, 4840.

[19] Papastathopoulos, E; Xenakis, D; Faucher, O;, Hertz, E.; Charalambidis, D. *Phys. Rev. A* **1999**, 59, 4840.

[20] Schumacher, D.W.; Weihe, F.; Muller, H.G.; Bucksbaum, P.H. *J. Phys. B* **1999**, 32, 341.

[21] Chen, Ce; Elliott, D.S. *Phys. Rev. A* **1996**, 53, 272.

[22] Petrosyan, D.; Lambropoulos, P. *Phys. Rev. Lett.* **2000**, 85, 1843.

Chapter 11

Double-Pulse Manipulation of Vibrational Wavepackets in HgAr van der Waals Complex

Y. Sato[1,2], H. Chiba[1,2], M.Nakamura[1], and K. Ohmori[1,2]

[1]Institute of Multidisciplinary Research for Advanced Materials,
Tohoku University, Katahira 2–1–1, Aobaku, Sendai 980–8577, Japan
[2]CREST, Japan Science and Technology Corporation

Interferometric measurements and model calculations have been performed on the interaction of two nuclear wavepackets formed by two temporally separated 300 femtosecond (fs) laser pulses on the $A\,^3 0^+$ state of Hg-Ar van der Waals (vdW) dimer. The center wavelength of the fs laser is in the UV region around 254nm, which gives a field oscillation cycle at 848 attoseconds. The 300 fs laser pulse generates a wavepcket of the periodic nuclear motion of the Hg-Ar stretching vibration, with a period about $T_{vib} \approx 1$ ps, by a coherent superposition of the low-lying vibrational (v' = 3, 4 and 5) eigenstates. The model calculation shows how different spatial wave modes are formed by the coherent interaction of the two wavepackets. The wave mode changes drastically depending on the change by 424 attoseconds in the inter-pulse-delay τ (π -difference in the laser field phase) and also on the coarse change of τ by an integer or a half integer of T_{vib} (π -difference in the vibrational phase). The wavepacket interferometry has been measured with a sub-fs accuracy for $\tau \approx 2 T_{vib}$.

Coherent control of a molecular wavepacket (WP) is largely classified into two schemes (*1, 2*): a static or Brumer-Shapiro (*3*) scheme; and a dynamic or Tanner-Rice-Kosloff-Rabitz (*5, 6, 7*) scheme. Both schemes have, however, a common situation analogous to the Young's two-slit experiment. The static scheme has been experimentally implemented by using two or more lasers with narrow frequency bands, different colors, and temporally overlapped. (*1, 2*) The laser beams are passed through a gas to vary the phase difference between laser fields, which was introduced to control atomic processes by Elliot and coworkers (*7, 8*) and then extended to molecules, first by Gordon and coworkers. (*9-11*)

The dynamic scheme, on the other hand, is often called "pump/dump scheme" or "temporal coherent control". This is related to the bound state wavepacket interferometry (*12*) or the Ramsey fringes. (*13,14*) The dynamic scheme control employs two ultra short laser pulses, which are separated in time, and the time delay between the pulses is varied with interferometric precision. Each of the pulses creates a wavepacket (WP) and the two WPs interfere constructively or destructively depending on their relative phase determined by the delay time between the two laser fields. Pulse shaping is sometimes combined to the latter scheme in the case of strong field excitation. (*15*)

We describe here an interferometric study on the dynamic coherent control of the vibrational WP propagating on the $A\,^3 0^+$ state of the Hg-Ar van der Waals (vdW) dimer. The vibrational WP interferometry, by means of the inter-pulse-delay method, has been studied earlier by the phase-locked detection technique (*16*), and more recently by the method of three (pump, control, and probe) ultra short pulses. (*17*) A preliminary experimental result of our inter-pulse-delay method has been reported previously. (*18*) Main purpose of the present work is to extend the wavepacket interferometry (WPI) to UV or VUV wavelength region. This would extend application range of the coherent control over much wider classes of target materials. Furthermore, since the coherent control occurs in harmony with the optical cycle of the laser field, these short wavelengths would result attosecond (as) clocking time, indicating an ultra fast engine to drive a coherent control. However, WPI remains still to be a technical challenge in these short wavelengths. The optical interference technique itself appears to be more and more difficult in going to shorter wavelengths and we need many careful considerations to the optical settings in order to attain stable performance.

For obtaining an inter-pulse delay with attoseconds precision, we use a gas cell, instead of the angle-variable plates (*14, 17, 19*), placed in one of the arms of a Michelson-type delay stage. Our method will be described in some detail elsewhere. (*20*) This is a combination of the methods that have been used in the static phase control and the dynamic time-domain control schemes. The phase difference and the time delay between two identical laser fields have essentially the same meaning for the interferometric measurements. In what follows, we discuss first characteristic features of the vibrational WP and then discuss four different types of coherent interaction of two WPs. The discussion is based on model calculations on the HgAr system. We then present experimental results of the WPI observed by means of the double-pulse-delay method with an accuracy better than 100 attoseconds.

Vibrational wavepacket simulations

We consider two electronic states of the HgAr vdW complex, the ground state ($X^1 0^+$) and the excited state ($A^3 0^+$) interacting each other through the electric field of a laser pulse. The single pulse field is assumed to have a Gaussian envelope with a half-width δ and a center frequency ω_L:

$$E_{SP}(t) = E_0 \exp\left[-2\ln 2\left(\frac{t}{\delta}\right)^2\right]\cos(\omega_L t). \qquad (1)$$

The double pulse field is then given as (9),

$$E_{DP}(t) = E_{SP}(t) + E_{SP}(t-\tau), \qquad (2)$$

where τ is the delay time between the two pulses.

The HgAr complex is assumed to be a linear oscillator. Following Tanner, Kosloff, and Rice (21), the time-dependent Schrödinger equation for the vibrational motion is written

$$i\frac{\partial}{\partial t}\begin{bmatrix}\psi_X(t,R)\\\psi_A(t,R)\end{bmatrix}=\begin{bmatrix}T(R)+V_X(R) & -\mu_{XA}E(t)\\-\mu_{AX}E(t) & T(R)+V_A(R)\end{bmatrix}\begin{bmatrix}\psi_X(t,R)\\\psi_A(t,R)\end{bmatrix} \qquad (3)$$

with
$$T(R) = -\frac{\hbar^2 d^2}{2m\,dR^2} \quad,$$

where R is the Hg–Ar internclear distance and V_i is the potential energy curve for electronic state i. μ_{XA} is the X–A transition dipole moment, which we assume to be independent of R. Following our experiments, the laser frequency ω_L is 7.407 fs^{-1} corresponding to the laser center wavelength $\lambda = 254.3$ nm, and the laser pulse width δ is 300 fs. We have carried out numerical integration of the coupled equations, eq 3, by a grid method, which is a modification of the method by Tanner, Kosloff, and Rice.(*17*) The computational detail will be described elsewhere. The potential curves of the HgAr X and A states determined by Yamanouchi *et al.* (*18*) are used in the calculation. The HgAr is initially at the vibrational level $v'' = 0$ of the X state.

Wavepacket characteristics

Figure 1(a) is a 3D picture of the vibrational WP formed on the A state by the single pulse field given in eq 1. The calculated probability density $|\psi_A(t,R)|^2$ is displayed against t and R. Figure 1(b) is a 2D map representation of the same $|\psi_A(t,R)|^2$ expanded for smaller time range 0–10 ps, which shows characteristic features of the spatial mode (nodal structure) of the WP. The WP is first generated near the outer (attractive) wall of the A state potential, starts to move inwards, and oscillates between the inner and outer turning points with a period about $T_{vib} \approx 1$ ps. The WP is formed by a coherent superposition of mainly three vibrational eigenstates $v' = 3$, 4, and 5 of the A state. The oscillation period $T_{vib} \approx 1$ ps corresponds to the vibrational level spacing (31.6 cm^{-1} for $v' = 3$–4 spacing, and 29.2 cm^{-1} for $v' = 4$–5 spacing). Since the laser pulse width (300 fs) is comparable to T_{vib}, the WP has a broad feature with a mixed character of localized (non-stationary) and delocalized (stationary) distributions. At the earliest times of generation, the WP mimics the spatial distribution of the $v'' = 0$ vibrational wavefunction of the X state and then starts to spread to form nodal waves due to the reflection effect by the potential walls. Prominent peaks are formed at the turning point regions. It will be shown in the following sessions that the broad character of the present WP is rather fortunate to study the interference effects.

Anharmonicity of the vibrational levels gives the collapse and the revival of the wavepacket (*13, 14, 23, 24*), which appear clearly as a periodic change of the amplitude near the inner and the outer turning point regions. The collapse is partial because of the small number of the levels involved. The expanded picture in Figure 1(b) shows that, in the midst of the collapse, the peaks split into two and the spatial mode changes so that a non-classical nature appears in the vibrational movement. This is owing to that the WP involves two different vibrational periods corresponding to the 3–4 and 4–5 spacing. It is obvious that the collapse is by no means de-coherence and it is not proper to say that the WP spreads in time during the collapse.

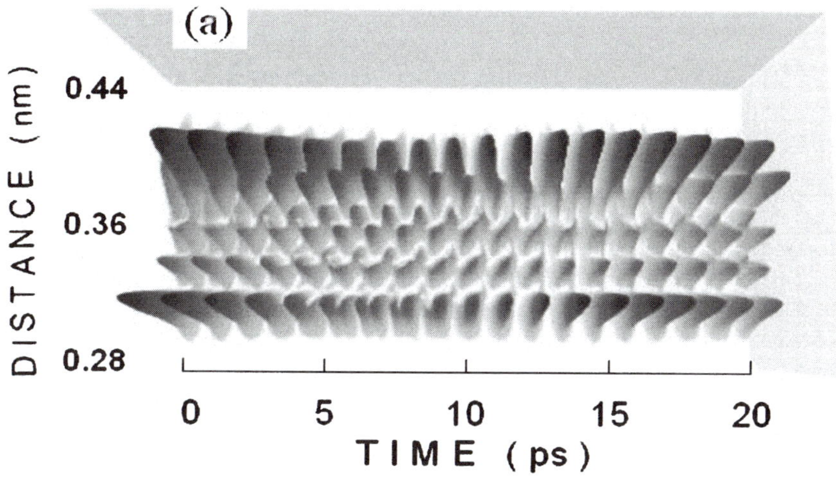

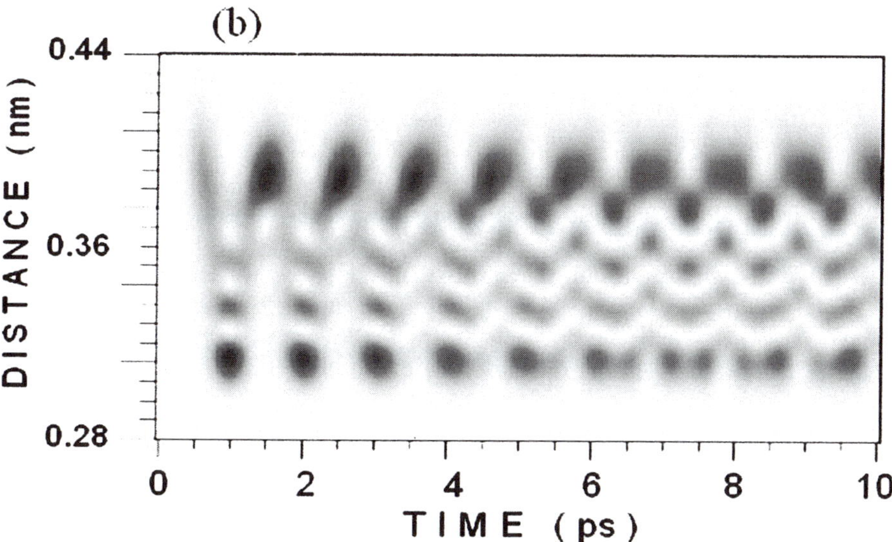

Figure 1. Calculation of the wavepacket generated by a 300 fs pulse on the A state of HgAr. The probability density density $|\psi_A(t,R)|^2$ is shown by a 3D picture in (a) and by a 2D image map in (b). Note that time range is 0 – 20 ps in (a) and 0 – 10 ps in (b). The center time of the laser pulse is 0.5 ps.

Control at the time of recurrence to the birthplace.

Calculations to simulate the present experiments are performed for the double-pulse field given in eq 2. The inter-pulse-delay τ is taken around 2 ps, which is about $2T_{vib}$, so that the second pulse generates a new WP at about the time when the old WP formed by the first pulse comes back to the outer wall of the potential well, that is, the initial $(X, v=0) - (A, v'=3,4,5)$ Franck-Condon region. The interference between the two WPs occurs constructively or destructively depending on the relative phase difference. Calculations are shown in Figure 2 for two delay times differ by only 0.424 fs. One of the mostly constructive interferences occurs at τ = 2010.115 fs, as shown in Figure 2 (a), while almost completely destructive one occurs at τ = 2010.539 fs as shown in Figure 2 (b). The WP density is three-times more intensified in (b) than in (a) for a better contrast in (b). The difference in the delay time, 0.424 fs, is just a half of the optical cycle ($T_{Laser}/2 = \pi/\omega_L$) of the laser field, *i.e.*, the phase difference π. The constructive and destructive interferences appear alternately at the delay time interval of 0.424 fs over a long span of τ (about $\pm$ 150 fs) around 2 ps. This is because the laser pulse width is broad in the present case. The constructively superposed WP has quite the same characteristics as the single-pulse-WP shown in Figure 1 and reproduces the collapse and revival quite in the same manner.

Control at a half way of the vibrational recurrence

When the second WP is generated without space overlapping with the first one, interaction of two WPs is quite different from those shown in Figure 2. Figure 3 shows how the two WPs interact when the delay time is set at about $1.5\,T_{vib}$. The second WP is, therefore, generated while the old one travels around the other side (the inner wall region) of the potential well. This setting of τ at a half integer of the vibrational recurrence time does not result in such enhancement and vanish (dump) of WP as seen in Figure 2, but produces quite different types of constructive and destructive interferences to result in new spatial modes of WP. Figure 3 shows two different spatial modes formed depending on the delay time difference of 0.424 fs (π -difference in the laser field phase). Figure 3(a) is a constructive interference at τ = 1500.803 fs and (b) is a destructive one at τ = 1501.227 fs. The WP density is three-times more intensified in (b) than in (a) to give a better contrast image in (b). Since the WP is broad in both space and time, the two WPs start to interact at an early time of the second WP generation. Even when the spatial overlap of the two WPs is small,

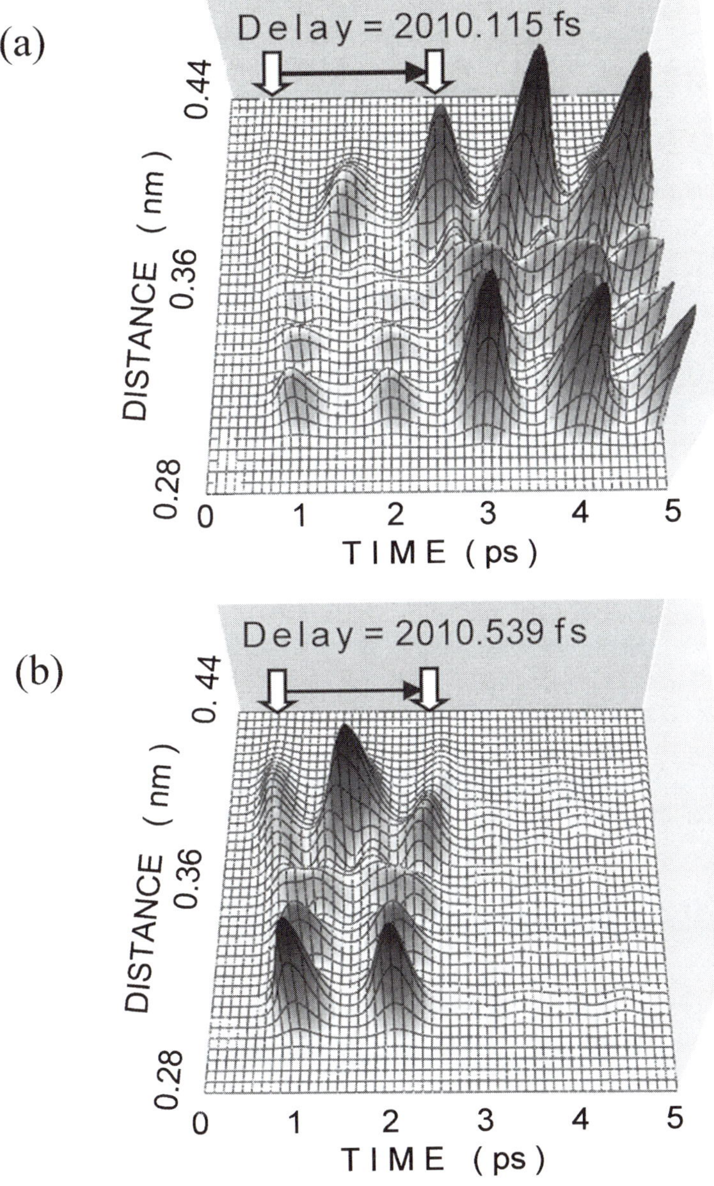

Figure 2. Calculation of the double-pulse control of the vibrational wavepacket on the A state of HgAr. The probability density is shown for the inter-pulse-delay 2010.115 fs in (a) and 2010.539 fs in (b). Constructive and destructive interferences appear in the delay-time difference of only 0.424 fs.

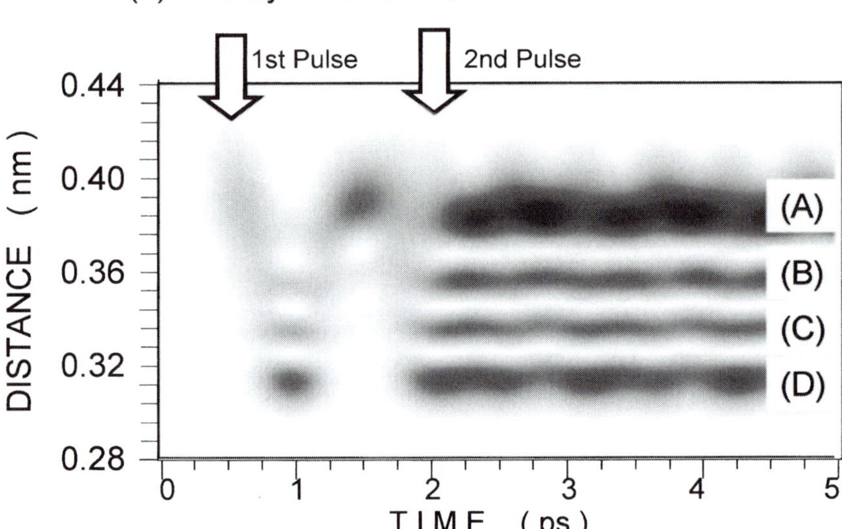

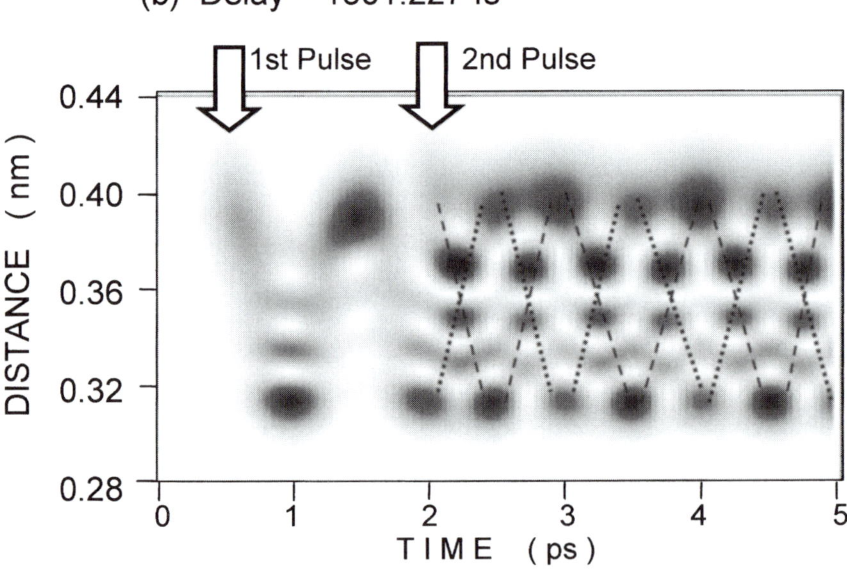

Figure 3. Interaction of the two vibrational wavepackets generated by two 300 fs pulses with the inter-pulse-delay $\tau \approx 1.5\, T_{vib}$. (a) Constructive interaction by inter-pulse-delay = 1500.803 f s. (b) Destructive interaction by inter-pulse-delay = 1501.227 fs. (A) and (D) denote the outer and inner turning point regions. (B) and (C) are humps in the constructive interaction and dips in the destructive interaction.

the effect of the interference prevails over the whole spatial range. This is illustrated in Figure 4(a), where the superposed wavepacket at the time of the birth of the second WP, *i.e.*, when the second laser field reaches its maximum amplitude, is shown for $\tau = 1500.803$ fs and $\tau = 1501.227$ fs. A difference between the constructive and destructive interaction is obvious near the outer wall. The destructive interaction makes a dip in this region. This is the region where the new WP overlaps with the small tail of the old one. However, the WP interaction extends over the region where the two WPs do not yet overlap.

A clear difference in the spatial mode appears at the crossing points of the two WPs, which are located at the middle between the outer and inner turning point regions as shown in Figure 3(b). Figure 4(b) shows a comparison of the cnstructive ($\tau = 1500.803$ fs) and destructive ($\tau = 1501.227$ fs) interaction at the time of the first crossing, the crossing located near 0.36 nm and at a time 2.25 ps in Figure 3(b). The destructive interaction results in null value of the WP density at the crossing, while the constructive one makes a sharp peak there, indicating that the spatial modes which are nearly in anti-phase relation each other are formed by the delay time difference of 0.424 fs.

Experiment – method and result –

Figure 5 shows the pump-control-probe scheme in the experiments and the relevant potential curves. Experimental setup is essentially the same as that used in the previous fs laser study on the Hg-CO vdW complex (*25*) except incorporation of a Michelson-type double-arm-delay stage to generate a pair of 300 fs laser pulses. Details of the optical delay stage will be described elsewhere. (*20*) Briefly, the output of a mode-locked Ti:Sapphire laser pumped by Ar[+] laser is amplified by a double-stage regenerative amplifier pumped by two 10Hz Nd:YAG lasers to give the final output around 760 nm. It is frequency tripled to generate a 300 fs pulse around 253 nm. The pulse is introduced into the Michelson-type delay generator, where the input pulse is divided by a half mirror and then recombined on another half mirror to generate coaxially a pair of 300 fs pulses. The pulse energy is about 10 μJ for each pulse. The inter-pulse-delay τ is coarsely tuned by sliding a mechanical delay stage placed in one of the arms. The minimum step of the coarse tuning is about 3 fs. An Ar gas cell is placed in the other arm of the Michelson stage, and its pressure variation gives a fine-tuning of τ with an accuracy better 50 as. The use of a gas is motivated from our long time experience on the optical dispersion measurements by the Mach-Zender interferometer operated with an incoherent light source (*26*), which has given high contrast optical dispersion spectrum around 253 nm. (*27*) The Ar pressure variation is calibrated against the delay time variation by measurements of the

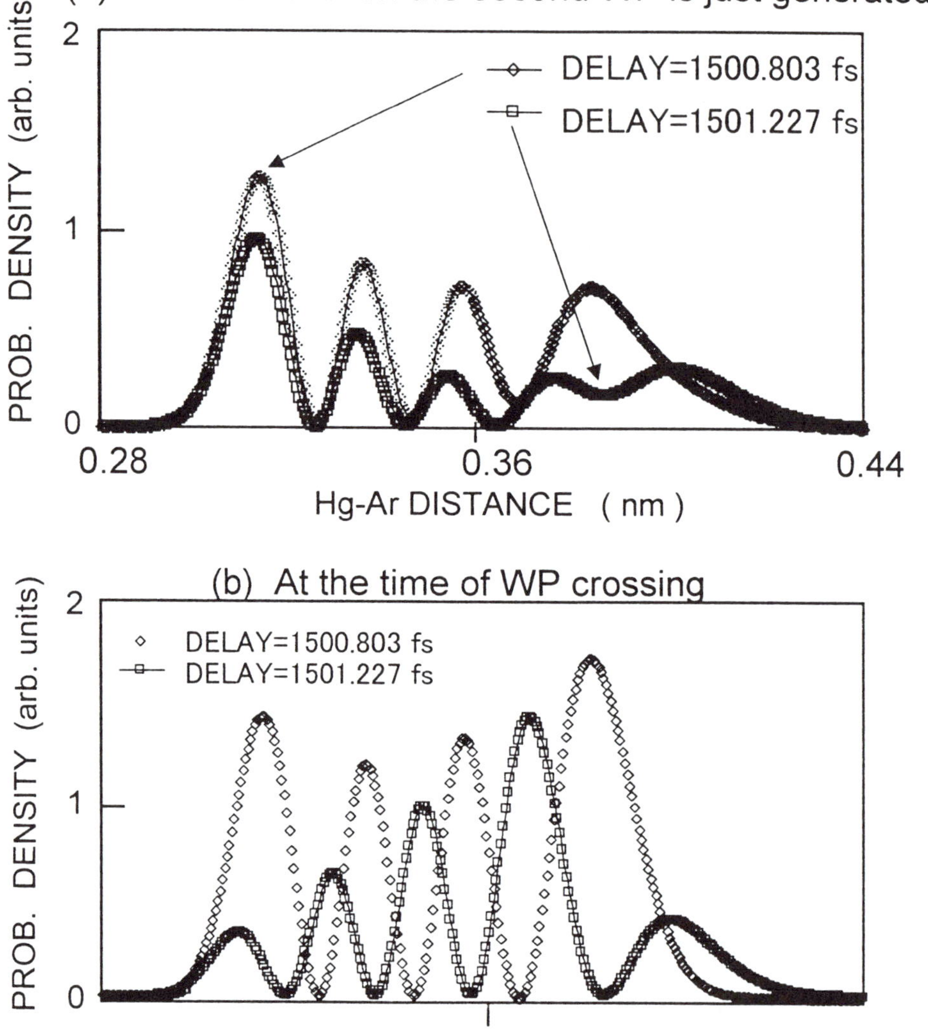

Figure 4. Wavepacket probability density profile (Cut of Fig. 3) at different times: (a) Time = 2.0 ps,The second wavepacket is generated at about this time. (b) Time = 2.25 ps. The first and the second wavepackets cross at this time. Solid curves: Inter-pulse-delay = 1500.803 fs, Dotted curves: Inter-pulse-delay = 1501.227 fs.

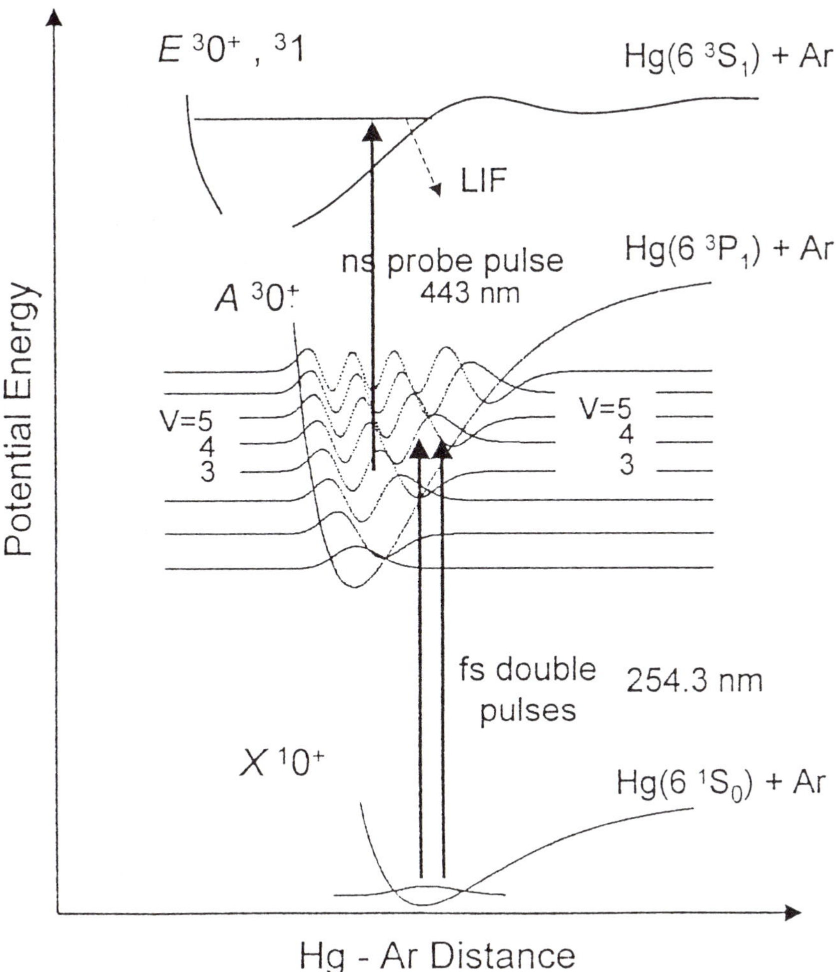

Figure 5. Pump-control-probe scheme and schematic sketch of the related potential curves for the present wavepacket interferometry. The vibrational wavepackets are generated on the A state of the HgAr van der Waals complex.

optical interference between the two 300 fs laser pulses with the coarse mechanical delay being set around 0.0 ps so that the two laser pulses are temporally overlapped. An example of the optical interferogram is shown in Figure 6(a) as a function of the Ar pressure. The period of the oscillation corresponds to the optical cycle 848 attoseconds of the laser field. We are modifying the optical setup and the optical interferogram is now being improved to attain a contrast more than 80%.

The cooled HgAr vdW dimers are produced by the supersonic jet expansion of heated Hg vapor (470 K) seeded in a high-pressure Ar gas into a vacuum chamber through a pulsed valve of 0.3 mm diameter. The HgAr complex is irradiated by the sequence of two identical 300 fs pulses. The wavelength of the pulse is tuned to 254.3 nm. The delay τ between the pulses is coarsely set around 2 ps so that the two fs pulses are well temporally separated. Since the 2 ps delay is nearly equal to $2T_{vib}$, the second WP is generated when the first one returns to the outer turning point region. Varying the Ar pressure enables fine-tuning of τ. A nanosecond laser pulse of the wavelength 443 nm is delayed by 30 ns from the fs pulses and is used for laser-induced-fluorescence (LIF) detection of the population of the vibrational level $v = 3$ of the $A\ ^3 0^+$ state. The LIF signal beats clearly against the Ar pressure variation, as shown in Figure 6(b). The period of the beat is the same as that observed on the optical interferogram in Figure 6(a), indicating that the constructive/destructive interactions of the two WPs occur at the deley interval of about 424 attoseconds in coincidence with the calculation.

Discussion

The two electronic states X and A of the HgAr vdW complex constitute a simple system to study the nuclear wave packet dynamics. The radiative $X \leftarrow A$ spontaneous decay is very slow at an order of 100 ns. (28,29) In contrast to the Hg-CO and Hg-N$_2$ vdW complexes (25, 29), the nonadiabatic decay of the A state to the nearby $a^3 0^-$ and $B\ ^3 1$ states is negligibly slow in the Hg-rare-gas complexes. (29) The vibrational WP generated on the $A\ ^3 0^+$ state has, therefore, a long Lifetime. This is a reason why the LIF detection of the A state is useful at a long time (30 ns) after the fs pulses for probing the results of the WP interaction. Since the HgAr complex is supplied by the jet-expansion method, it is produced at the lowest vibrational level of the X state with a very cold rotational temperature (ca. 6K). All these situations underlie the present simple model calculations of the WP dynamics.

We didn't pay attention to the phase relation between the two fs laser pulses in our experiments. However, a phase-locked relation between the two sequential laser pulses is not necessary a condition to obtain a high contrast interferometry.

Bucksbaum and coworkers (*12*) has demonstrated that even an incoherent light source gives a high contrast interferometry. We have recently modified our experimental setup to improve performance of the Michelson-type double-pulse generator and attained a high contrast WPI more than 90% peak-to-bottom ratio. Thus the data in Figure 6 are now much improved. Our new WPI data will be submitted elsewhere.

Considering the laser power used and the low oscillator strength for the *A-X* transition, we estimate that only 10% at most of the HgAr molecules are excited by the one shot of the laser pulse. It is important to note that, even though there could be much more chance for the photons in the second pulse to be absorbed by the molecules left at the ground state, most of the photons in the second pulse seem to be absorbed by the excited molecules to results in the present high-contrast WPI. This means that the particle picture cannot be used to understand the present WPI result, but only the quantum wave picture gives its proper explanation. The present high-contrast WPI in Figure 6(b) is a complete molecular version of the Young's two slits experiment. When an observation is made for an ensemble of non-interacting identical particles without distinguishing individual particles, the wave nature of one particle appears to represent the gross (macroscopic) wave nature of the ensemble. This is the reason why the present model calculations for one particle time-dependent Schrödinger equation explain properly the present WPI result. This situation bears resemblance to the Bose-Einstein condensation.

Use of a gas for the time delay control is found to be reliable to attain a time resolution in attoseconds. This means that the light velocity dispersion in a gas is not a problem for the present fs pulses. Even in our recent high-contrast WPI measurements, we have not checked what a phase relation exists actually between the two pulses separated far apart around 2 ps. The optical transition itself of the molecule seems to serve as a phase selector. (*12, 30*)

In conclusion, we have observed the nuclear WPI with about 70% peak/bottom contrast by controlling the WP phase difference in attoseconds region for the first time owing to the use of the fs laser pulses tuned to the UV wavelengths. The present high contrast WPI is a molecular version of the Young's two slits experiments, where the particle nature is depressed and the microscopic phase of a single molecule represents the macroscopic phase of the molecular ensemble.

The present calculations show that the double-pulse-control encodes four fundamental spatial modes on the vibrational wavepacket; depending on the combination of the laser-field-phase difference $P_L = 0$ or π and the vibrational phase difference $P_V = 0$ or π. The combination $(P_L, P_V) = (0, 0)$ enhances the normal mode WP, $(\pi, 0)$ gives the complete vanishment (dump) of WP, $(0, \pi)$ and (π, π) generates new WP modes which are in anti-phase relation each other.

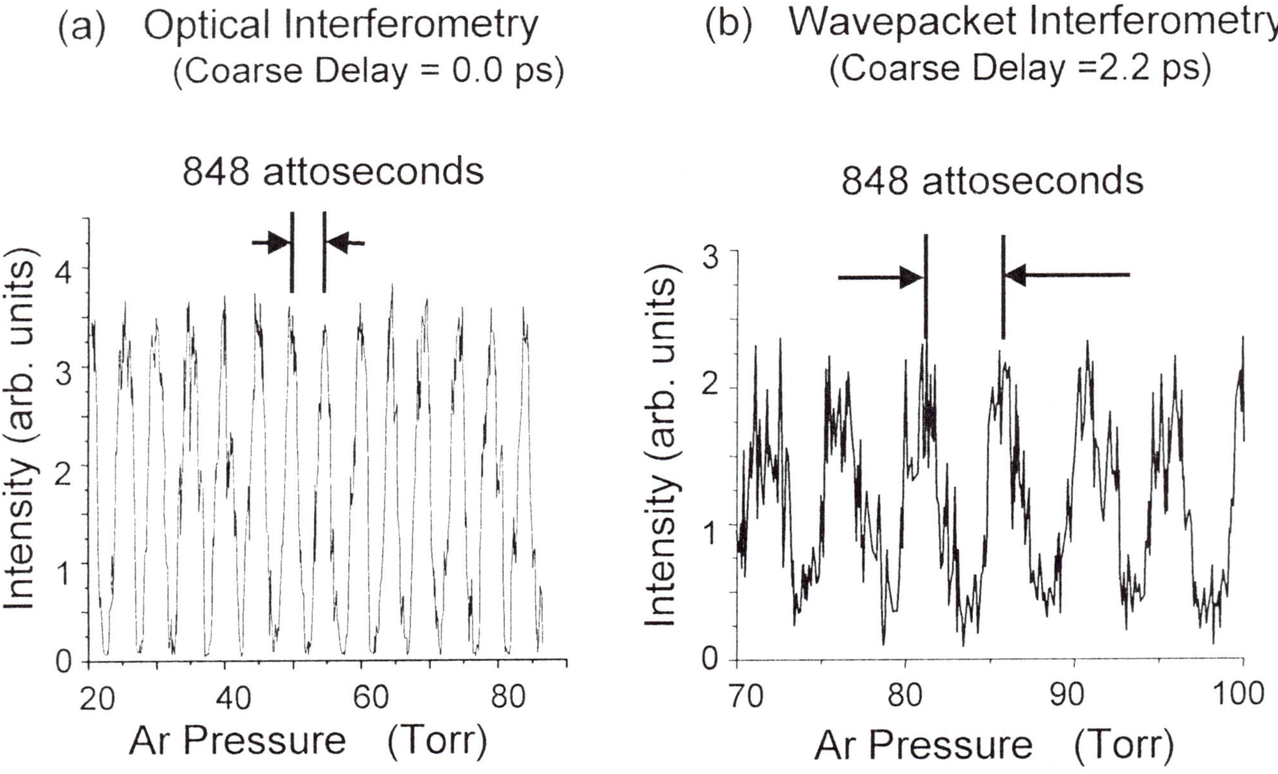

Figure 6. (a) Optical interferogram of two300 fs laser tuned to 253 nm.   (b) Experimental result of the wavepacket interferometry.

It would be an important next step to explore a method to decode these spatial modes written on the wavepacket.

Acknowledgments

The authors appreciate many contributions to the experiment and the calculation by K. Amano. The authors wish to thank Y. Fujimura, H. Kono, Y. Ohtsuki, R. J. Gordon, K. Misawa for helpful discussions. Y.S. is grateful to H. Lang for inviting our group to join his team in CREST. This work has been supported by the Grant-in Aid for Scientific Research on Priority Areas "Molecular Physical Chemistry", by the Grant-in Aid for Science (No. 13440171, No. 13440120) from the Ministry of Science and Technology. This work has been supported by CREST of JST.

References

1. Rice, S.A. *Adv. Chem. Phys.* **1997**, *101*, 213.
2. Gordon, R. J. ; Rice, S. A. *Ann. Rev. Phys. Chem.* **1997**, *48*, 601.
3. Brumer, P. ; Shapiro, M. *Chem. Phys. Lett.* **1986**, *126*, 541
4. Tannor, D. J. ; Rice, S. A. *J. Chem. Phys.* **1985**, *83*, 5013.
5. Kosloff, R.; Rice, S. A.; Gaspard, P.; Tersigni, S.; Tannor, D. J. *Chem. Phys.* **1989**, *139*, 201.
6. Shi, S.; Woody, A. ; Rabitz, H. *J. Chem. Phys.* **1988**, *88*, 6870.
7. Chen, C.; Yin, Y.-Y ; Elliot, D. S. *Phys. Rev. Lett.* **1990**, *64*, 507.
8. Chen, C. ; Elliot, D. S. *Phys. Rev. Lett.* **1990**, *65*, 1737.
9. Park, S. M.; Lu, S. P. ; Gordon, R. J. *J. Chem. Phys.* **1991**, *94*, 8622.
10. Lu, S. P.; Park, S. M.; Xie, Y. J.; Gordon, R. J. *J. Chem. Phys.* **1992**, *96*, 6613.
11. Zhu, L.; Kleiman, V.; Li, X.; Lu, S-O; Trentelman, K.; Gordon, R. J. *Science*, **1995**, *270*, 77.
12. Jones, R. R.; Schumacher, D. W.; Gallagher, T. F.; Bucksbaum, P. H., *J. Phys. B: At. Mol. Opt. Phys.* **1995**, *28*, L405.
13. Noordam, L. D.; Duncan, D. I.; Gallagher, T. F., *Phys. Rev. A*, **1992**, *45*, 4734.
14. Jones, R. R.; Raman, C. S.; Schumacher, D. W.; Bucksbaum, P. H. *Phys. Rev. Lett.* **1993**, *71*, 2575.
15. We would like to note Kent R. Wilson and coworkers' pioneering works, for examples, Kohler, B.; Yakovlev, V. V.; Che, J.; Krause, J. L.; Messina, M., Wilson, K. R.; Schwentner, N.; Whitnell, R. M.; Yan, Y. *Phys. Rev. Lett.* **1995**, *74*, 3360; Bardeen, C. J.; Che, J.; Wilson, K. R., Yakovlev, V. V.; Apkarian, V. A.; Martens, C. C.; Zadoyan, R.; Kohler, B.; Messina M. *J. Chem. Phys.* **1997**, *106*, 8486.

16. Scherer, N. F.; Carlson, R. J.; Matro, A.; Du, M.; Ruggiero, A. J.; Romerorochin, V.; China, J. A.; Fleming, G. R.; Rice, S. A. *J. Chem. Phys.* **1991**, *95*, 1487.

17. Blanchet, V.; Bouchene, M. A.; Giral B. *J. Chem. Phys.* **1998**, *108*, 4862.

18. Ohmori, K.; Amano, K.; Chiba, H.; Nakamura, M.; Okunishi, M.; Sato, Y. *Spectral Line Shapes*, ed. Seidel, J. AIP Press, **2001**, *Vol.11*,284.

19. Christian, J. F.; Broers, B.; Hoogenraad, J. H.; van der Zande, W. J.; Noordam, L. D. *Optics Commun.* **1993**, *103*, 79.

20. Ohmori, K.; Nakamura M.; Chiba, H.; Amano, K.; Okunishi, M.; Sato, Y. *J. Photochemistry Photobiology A: Chemistry*, **2001**, *accepted.*

21. Tannor, D. J.; Koslof, R.; Rice, S. A. *J. Chem. Phys.* **1986**, *85*, 5806.

22. Yamanouchi, K.; Isogai, S.; Okunishi, M.; Tsuchiya, S. *J. Chem. Phys.* **1988**, *88*, 205.

23. Yeazel, J. A.; Mallailieu, M.; Stroud Jr., C. R. *Phys. Rev. Lett.* **1990**, *64*, 2007.

24. Vrakking, M. J. J.; Fischer, I.; Villeneuve, D. M.; Stolow A. *J. Chem. Phys.* **1995**, *103*, 4538.

25. Ohmori, K.; Amano, K.; Chiba, H.; Okunishi, M.; Sato, Y. *J. Chem. Phys.* **2000**, *113*, 461.

26. Ueda, K.; Komatsu, T.; Sato, Y. *J. Chem. Phys.* **1989**, *91*, 4495.

27. Sato, Y.; Nakamura, T.; Okunishi, M.; Ohmori, K.; Chiba, H.; Ueda, K. *Phys. Rev. A* **1996**, *53*, 867.

28. The oscillator strength of the HgAr $X \leftarrow A$ transition can be approximated as 1/3 of the atomic Hg $6\,{}^3P_1 \rightarrow 6\,{}^1S_0$ oscillator strength f (= 0.0255, See Ref. 27) , which gives the fluorescence lifetime 114 ns of Hg $6\,{}^3P_1$. (See Ref.29).

29. Ohmori, K.; Kurosawa, T.; Chiba, H.; Okunishi, M.; Ueda, K.; Sato, Y. *J. Chem. Phys.* **1995**, *102*, 7341.

30. private communication with P. H. Bucksbaum and his group at the symposium.

Chapter 12

Real-Time Spectroscopy of Molecular Vibration Using Sub-5-fs Pulses

Takayoshi Kobayashi, Akira Shirakawa, and Takao Fuji

Department of Physics, Faculty of Science, University of Tokyo Hongo 7–3–1, Bunkyo-Ku, Tokyo 113–0033, Japan

Transform-limited (TL) visible pulses with as short a duration as 4.7 fs with a 5 μJ pulse energy have been generated for the first time from a novel noncollinear optical parametric amplifier (NOPA). Applications of the sub-5-fs pulse source to the real-time spectroscopy of conjugated polymers, dye molecules, and J-aggregates are described. Several new phenomena, namely mode coupling, dynamic Duschinsky rotation, and dynamic intensity borrowing are found to take place in a polydiacetylene cast film, cresyl violet doped in polymers, and J-aggregates of porphyrin toluene sulfonate in aqueous solution.

Introduction

Time-resolved spectroscopy has long been a powerful method of electronic structure in the exited molecules and photochemical intermediates. The time resolution has been improved from the microsecond regime in 50-60's to femtosecond regime in 90's. The pulsed light from mode-locked lasers has enabled to study ultrafast dynamics and determination of structures of molecules in their electronic excited state and intermediate states in photochemical reaction by the measurement of transient electronic and vibrational spectra. Zewail et al.(1, 2) have succeeded in the observation of the quantum mechanical tunneling of wavepacket in the photodissociation of ``quasi'' two-atom molecules like NaI. That study triggered the femtochemistry field and transition-state spectroscopy and rapid growth of the field is still in progress.

Recently, sub-5-fs visible pulse laser based on noncollinear optical prametric amplifier was constracted by our group.(3) The sub-5-fs pulses is very powerful for pump-probe real-time vibrational spectroscopy, since the oscillation period becomes as short as 11fs in the case of 3000 cm^{-1}. Information about the phase relation among vibrational modes and the initial phase of oscillation induced by photoexcitation through vibronic coupling can be obtained by sub-5-fs real-time spectroscopy described below. These informations cannot be obtained by any conventional stationary or time-resolved Raman spectroscopies. Therefore the applicability of such short pulses to the study of molecular vibrational systems is to be mentioned to show how useful the short pulse is demonstrated in several real systems.

A conjugated polymer: mode coupling

The ultrafast dynamics of a quasi-one-dimensional conjugated polymer polydiacetylene $(=CR-C \equiv C-CR'=)_n$ (R and R' are substituents) was studied by pump-probe spectroscopy using visible sub-5-fs pulses(4). The spectrally-resolved differential transmittance change in a thin film of a ladder polymer poly (5,7,17,19-tetracosatetraynylene bis(N-butoxycarbonylmethyl) carbamate) (PDA-4BCMU4A(8)) (5) at various wavelengths is shown in Fig. 1. The signal is dominated by a long-living multimode wave-packet motion including three intense stretching mode signals C–C ($\approx$ 1220 cm^{-1}, 27 fs), C=C ($\approx$ 1450 cm^{-1}, 23 fs), and C $\equiv$ C ($\approx$ 2080 cm^{-1}, 16 fs). Both features of the electronic and molecular dynamics indicate the relaxation from a 1^1B_u-free exciton to a geometrically-relaxed 2^1A_g state within 60-80 femtoseconds via vibronic coupling and internal conversion(6,7). This conclusion can be obtained from extremely weak fluorescence from the exciton state. The quantum yield is estimated to be lower than 10^{-5}.

The geometrical relaxation is clearly visualized by a real-time frequency analysis using a spectrogram(8). The spectrogram at 1.75 eV shown in Fig. 2

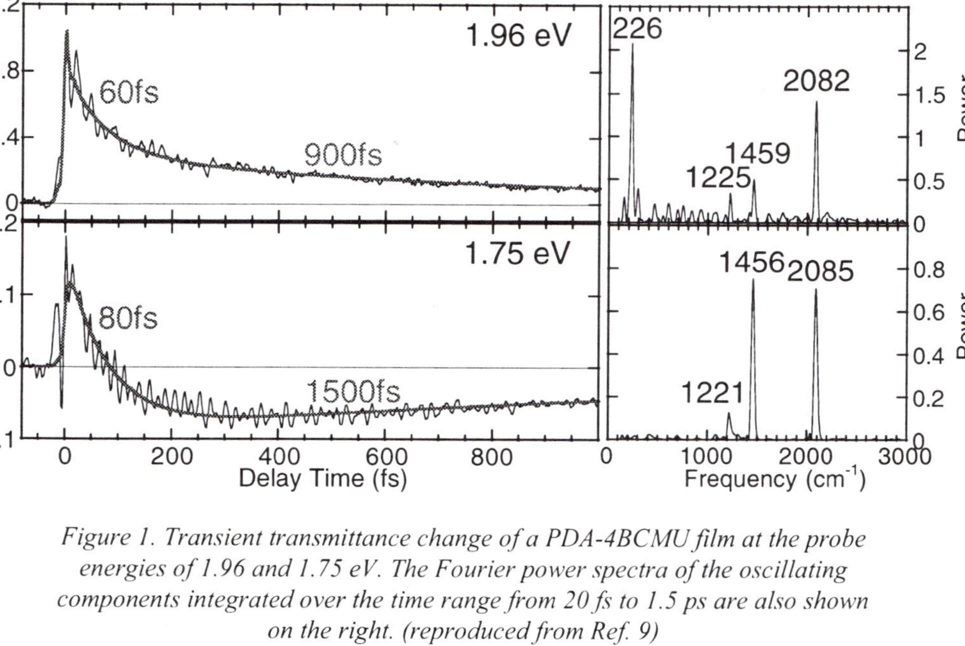

Figure 1. Transient transmittance change of a PDA-4BCMU film at the probe energies of 1.96 and 1.75 eV. The Fourier power spectra of the oscillating components integrated over the time range from 20 fs to 1.5 ps are also shown on the right. (reproduced from Ref. 9)

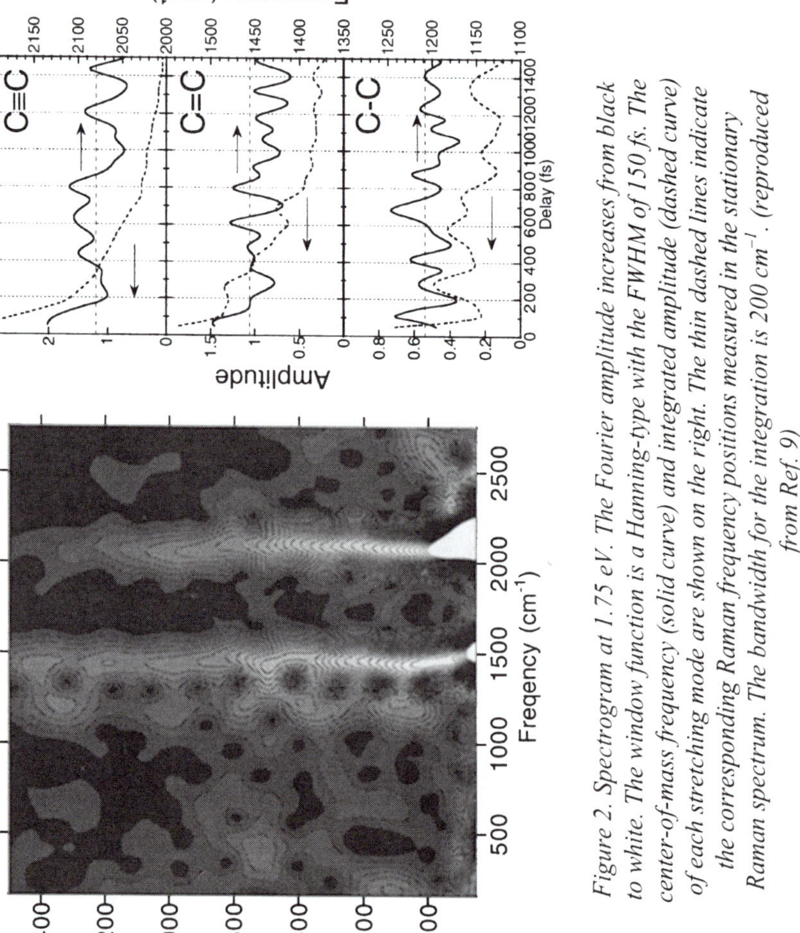

Figure 2. Spectrogram at 1.75 eV. The Fourier amplitude increases from black to white. The window function is a Hanning-type with the FWHM of 150 fs. The center-of-mass frequency (solid curve) and integrated amplitude (dashed curve) of each stretching mode are shown on the right. The thin dashed lines indicate the corresponding Raman frequency positions measured in the stationary Raman spectrum. The bandwidth for the integration is 200 cm⁻¹. (reproduced from Ref. 9)

evidently exhibits the highly vibronic non-equilibrium state characterized by the instantaneous mode frequency and amplitude modulations. The C–C and C=C stretching frequencies and amplitudes are π out-of-phase modulated with the period of 145 fs corresponding to the 230 cm^{-1} bending mode of the C–C=C bond. There is a possibility of a kind of artificial interference between two neighboring modes in the spectrogram analysis because of the limited spectral resolution determinced by the finite width of the window function. This effect will be discussed in detail elsewhere.(9) The feature of modulation is well explained by the coupling of the stretching modes via the bending motion in the geometrically-relaxed butatriene-like backbone. This is the first observation of such diabatic molecular motions which modify the vibrational frequencies of modes coupled to each other by the relevant mode.

A dye molecule: Duschinsky rotation

In order to clarify the mechanism of molecular vibrational mode coupling discovered in the conjugated polymers(4), we studied a simpler system. Cresyl violet (CV) was selected, since the Raman spectrum in literature(10), is shown to have only one intense Raman signal.

Figure 3(a) shows the normalized transient differential transmittance ($\Delta T(t)/T$) as a function of delay time t of the probe pulse after the pump pulse of CV doped in poly(vinyl alcohol) (PVA) against the pump-probe delay time at various wavelengths. Bleaching is observed throughout the probe wavelength region. The observed signal intensity does not decrease detectably within the probed delay time range of 1 ps because the lifetime of the excited singlet state, S_1, is reported to be as long as 3.2±0.1 ns(11). The oscillations observed at negative and near-zero delays are mainly due to pump-perturbed free-induction decay and coherent coupling(12,13) and also partially due to the interference between the probe and scattered pump. Special care was taken to reduce the artificial interference effect not to deteriorate the signal pattern resulting in the loss of information near zero delay time. It was performed by slightly increasing the angle between the pump and probe up to the level of a few degrees. The oscillation with a 57-fs period can be clearly seen for all the probe wavelengths between 600 and 640 nm. The oscillation period is corresponding to the ring-breathing mode and the period is corresponding to the frequency of 590 cm^{-1} observed in the Raman spectrum of CV(10). The Fourier transformation of the ΔT trace shown in Fig. 3(b) has a peak at this frequency. These vibrational

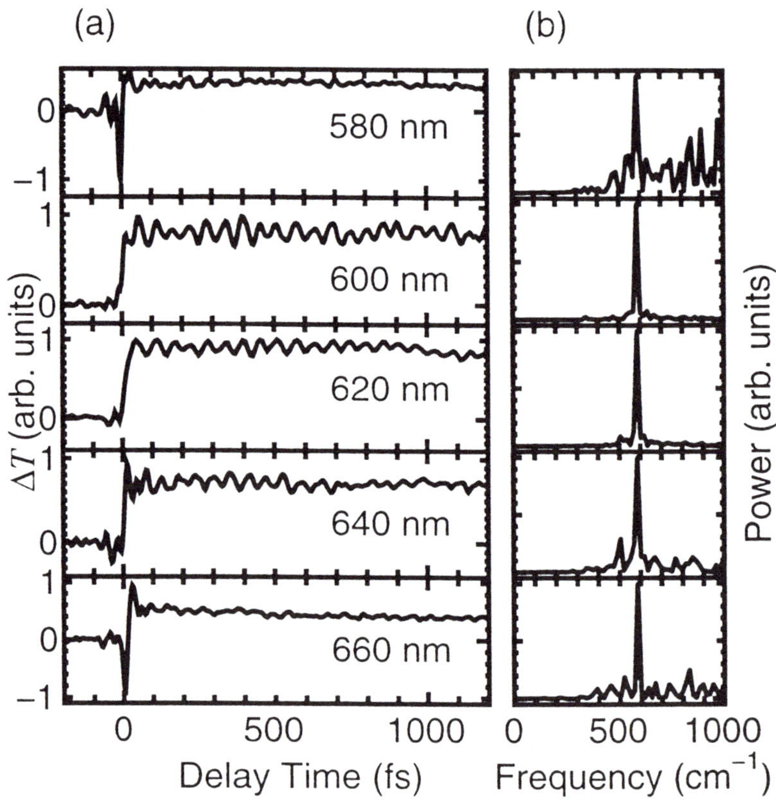

Figure 3. (a) Time dependence of the transmittance change of cresyl violet doped in poly(vinyl alcohol). The intensities are normalized at their positive or negative peaks. (b) Fourier power spectrum of time dependence of oscillating components of ΔT integrated over the delay time longer than 200 fs. The intensities are normalized at their peaks. (reproduced from Ref. 18)

signals are mostly due to the wave-packet motion of the excited state. In Ref. (14), the motion of the ground-state wave-packet is much more restricted than that of the excited state, because it is initially weakly displaced from the ground-state minimum and has a small initial momentum created by 6-fs-pulse excitation. In the present experiment the excitation pulse is shorter than 5 fs. Then the motion of the ground-state wave-packet is even more limited; and hence it can be concluded to be negligible and the vibrational structure can be assigned to the wave-packet motion on the excited-state potential surface.

In order to investigate the evolution of the instantaneous frequency of the ring-breathing mode, we analyze the transient differential transmittance using the spectrograms shown in Fig. 4. The amplitude and frequency of the peak of the spectrogram probed at 600 nm are shown in the upper part of Fig. 5. It clearly shows that the amplitude and frequency of the ring-breathing mode are modulated. The modulation period and corresponding frequency are estimated to be about 370 fs and 91 cm^{-1}, respectively, by Fourier transformation of the time-dependent amplitude and instantaneous frequency. The modulations of the amplitude and phase are out of phase by about π. The modulation depths of the amplitude and frequency are about 13 and 4%, respectively. The same features are also observed at a probe wavelength of 620 nm. Such time dependent modulations of the ring-breathing mode parameters and their phase relationship can never be obtained by frequency-domain spectroscopy, and it will be discussed in detail in this paper. These modulations indicate that the ring-breathing mode is coupled to another vibrational mode. At other probe wavelengths, the transmittance change due to the ring-breathing mode is so weak that these modulations of such a small transmittance is difficult to be detected by the spectrograms.

The power spectra of amplitude and phase modulations have two symmetric side bands on higher and lower sides of the center frequency. Therefore, it is impossible to discriminate the two cases only from Raman spectrum: (a) The two modes are coupled to each other through another mode with ω_m, and (b) the two modes are independent of each other. In the real-time experiment, each of the depth and phase of the modulations can be determined. From the experimental data shown in Fig. 5, the parameters of side bands can be obtained to be $m_{A0} = 0.18$ and $\varphi_m = 31\text{cm}^{-1}/91\text{cm}^{-1} = 0.34$. These are determined by standard deviations of the modulation data. Because the instantaneous amplitude and frequency are π out of phase from each other, the lower-frequency side band should be larger than that of higher frequency. Therefore, the time-dependent amplitude of the ring-breathing mode $A(t)$ can be given in the following equations.

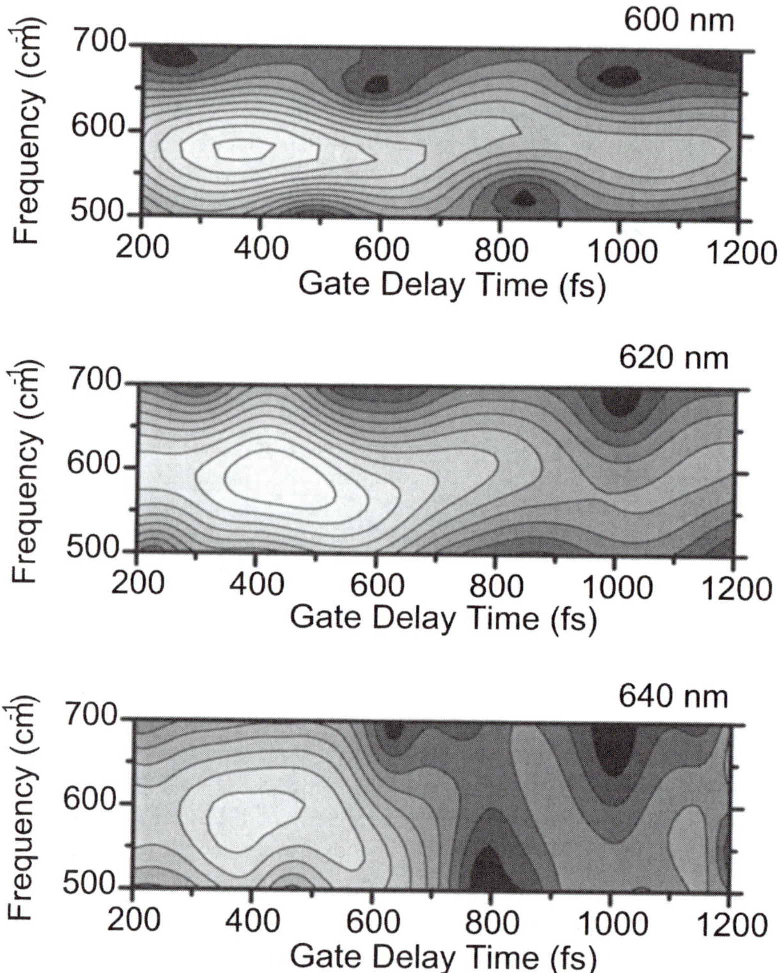

Figure 4. Spectrograms at three probe wavelengths calculated with a Hanning-type window function of 150 fs FWHM. The Fourier amplitude increases from darker to lighter with a step of 1/8 of the peak height. (reproduced from Ref. 18)

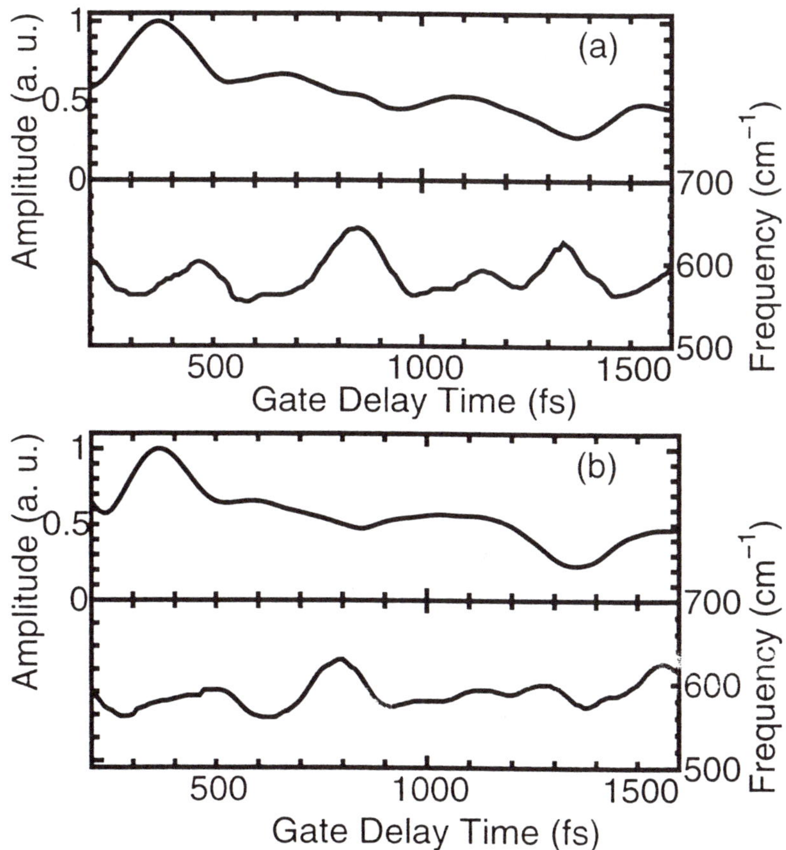

Figure 5. Instantaneous amplitudes and frequencies calculated for the ring-breathing mode in (a) CV/PVA and (b) CV/PMMA samples using data probed at 600nm. (reproduced from Ref. 18)

$$A(t) = A_0 \cos\omega_c t + \frac{1}{2}(A_0 m_{A0} - A_0\varphi_m)\cos(\omega_c + \omega_m)t$$

$$+ \frac{1}{2}(A_0 m_{A0} + A_0\varphi_m)\cos(\omega_c - \omega_m)t$$

$$= A_0\left(\cos\omega_c t - 0.08\cos(\omega_c + \omega_m)t + 0.26\cos(\omega_c - \omega_m)t\right). \tag{1}$$

These side bands are expected to appear in the Fourier transformation of the real-time pump-probe signal. Even though the intensity of lower-frequency mode the side band is 7% of that of the main peak, it is below the noise level in the spectrum shown in Fig. 3(b). Since the spectrogram analysis is very sensitive to the modulations of instantaneous frequency and amplitude in the real-time spectroscopy, these side bands can be detected much more easily than by Raman spectroscopy.

The modulation of the adiabatic harmonic potential of CV in the vibrational coordinates can be determined from the depths of the amplitude and frequency modulations. It is possible to explain the frequency modulation of the ring-breathing mode by assuming that the curvature of the harmonic potential of a breathing mode is modulated by the 91-cm^{-1} mode. The modulation of the curvature of the harmonic potential curve due to the effect is estimated to be 8% from the depth of frequency modulation of 4%. If the curvature of the potential is modulated, the amplitude of the $\Delta T/T$ signal due to the ring-breathing mode is also modulated even when the harmonicity of the potential curve is maintained because the Franck-Condon factor is modified by the time-dependent curvature. The Franck-Condon factor of the 0-0 transition, F_{00}, is estimated as 0.65 from the Stokes shift of the CV/PVA film and the Huang-Rhys factor of 0.43 estimated from the absorption spectrum. When the curvature is modulated by $\pm 8\%$, F_{00} corresponding the probe wavelength of 600 nm in the present experiment, is calculated to be modified by $\mp 2\%$. Therefore, the amplitude modulation of the $\Delta T/T$ signal due to the ring-breathing mode at 600 nm can be estimated to be 2%, which is smaller by 11% than the experimental results of 13%. Hence, it is necessary to invoke another mechanism of the potential curve modulation.

If the minimum point of the potential curve along the vibrational coordinate of breathing mode is assumed to displace periodically with the same modulation frequency as that of the curvature, then an additional modulation of the Franck-Condon factor is induced. For example, the modulation of F_{00} is calculated to be 11% when the potential minimum of the excited state is displaced by 12% of the amount of displacement of the potential minimum of the excited state from the ground-state minimum. Therefore, if the curvature of the potential is modulated by 4% of the central breathing frequency observed in Raman spectrum and 12% of the Franck-Condon displacement, then the displacement of the potential minimum modulated by the observed total amplitude modulation of 13% can be explained.

The above estimation of the modulation parameters can be used for the determination of two-dimensional (2-D) potential curve of the two relevant modes. The two axes spanning the 2-D potential surfaces are the coordinates of the ring-breathing mode, q_0, and another mode, q_1. At first we obtain the 2-D harmonic potential surface as

$$\frac{U_g(q_0, q_1)}{\hbar\omega} = \frac{1}{2}q_0^2 + \frac{1}{2}q_1^2 \tag{2}$$

$$\frac{U_g(q_0, q_1)}{\hbar\omega} = \frac{1}{2}(q_0 - a)^2 + \frac{1}{2}q_1^2 + \frac{\omega_{00}}{\omega}, \tag{3}$$

where U_g and U_e are the potential curves in the ground and excited states, respectively, ω is the frequency of ring-breathing mode. a is the parameter for the Stokes shift and ω_{00} is the 0-0 transition frequency. Associated with the curvature change and the relative displacement of potential surfaces can be represented by

$$\frac{U_g(q_0, q_1)}{\hbar\omega} = \frac{1}{2}\left(1 + \frac{q_1}{Q_1}\right)q_0^2 + \frac{1}{2}q_1^2 \tag{4}$$

$$\frac{U_e(q_0, q_1)}{\hbar\omega} = \frac{1}{2}\left(1 + \frac{q_1 + \Delta c Q_1}{Q_1}\right)\left((q_0 - a) - \frac{a\Delta a}{\Delta c Q_1}(q_1 + \Delta c Q_1)\right)^2$$
$$+ \frac{1}{2}(q_1 + \Delta c Q_1)^2 + \frac{\omega_{00}}{\omega} \tag{5}$$

where Δc is the fractional modulation of the curvature and Δa is the fractional displacement of potential minimum, and Q_1 is an arbitrary parameter. Here, a parameter set of $a = 0.9$, $\Delta c = 0.08$, $\Delta a = 0.12$, and $Q_1 = 1.0$ is used. The parameter a is determined by the energy difference between the peaks of absorption and fluorescence, 502 cm^{-1}.

Figure 6 shows the 2-D potential curve calculated by Eqs. (4) and (5). In Fig. 6(b) the 0-0 transition energy is subtracted. Such a rotation of the potential curve associated with the electronic transition from the ground to the excited states is

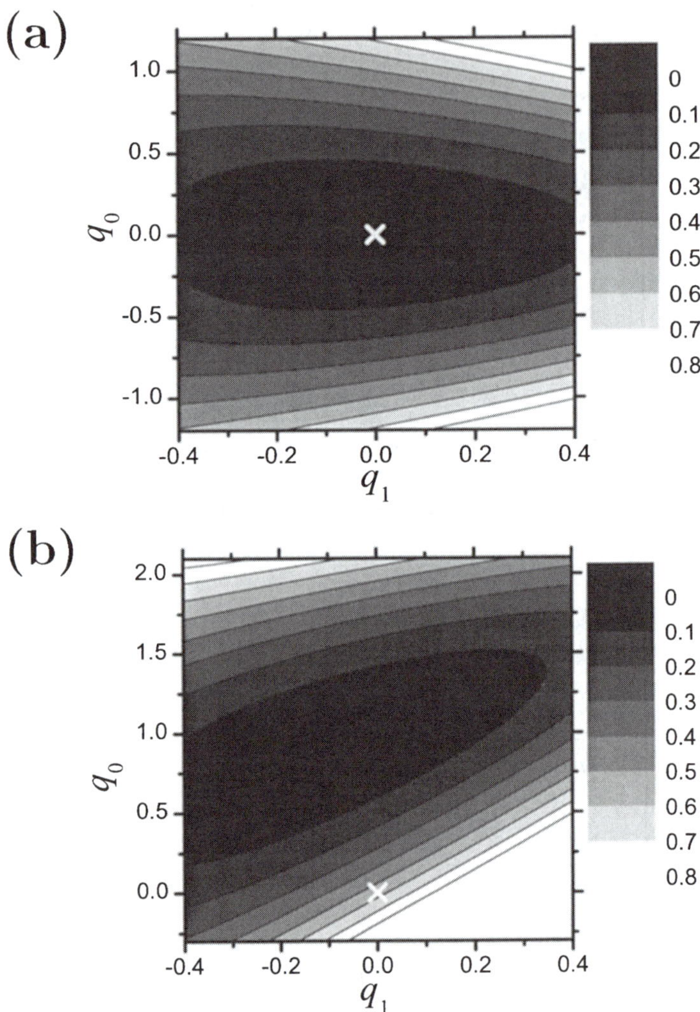

Figure 6. Two-dimensional potential curves of ground and excited states obtained from Eqs.(4) and (5), respectively. White crosses mark the points on the ground and excited states are the minimum and the Franck-Condon point, respectively, on their potential surfaces. (reproduced from Ref. 18)

known as the Duschinsky effect (15). This effect was observed by the difference between the frequency of the ground state and that of the excited state. For example, Book et al. (16) performed a pump-probe experiment using blue-copper proteins and observed modulation of the transmittance due to a molecular vibration which was not obtained by Raman spectroscopy. They assigned that the observed mode is to the Cu–S(cysteine) stretching mode in the excited state, of which frequency shifts are higher than that in the ground state, and ascribed this frequency shift to the Duschinsky rotation. In the present case, the Duschinsky rotation can be directly observed in real-time from the modulations of the vibrational mode. We would like to emphasize that this is the first dynamical observation of the Duschinsky effect by sub-5-fs real-time spectroscopy.

J-aggregates: intensity borrowing

Real-time spectroscopy of J-aggregates tetraphenyl porphyrin sulfonate was also studied using the sub-5-fs visible pulses. Time dependence of $\Delta T/T$ probed at several wavelengths are shown in Fig. 7(a). For the discussion of the molecular vibration, the oscillating components should be separated from the overall transient curves. The slowly varying decay component is subtracted from the transient signal. Fourier spectra shown in Fig. 7(b) clearly exhibits a strong peak at 244 cm^{-1}. The Fourier amplitude is drastically reduced at 1.77 eV at which the sign of the phase is changed. The peak at 241 cm^{-1} is also observed in the stationary Raman spectrum. The Raman signal of this mode is drastically enhanced by aggregation by a factor of more than 30(17). The phase of the oscillation is evaluated by the complex Fourier transformation. A π-phase jump is clearly observed around 1.77 eV, which is slightly higher than the peak photon energy of the Q-band (Fig. 8). At the probe photon energy of the phase jump, the signal changes from the bleaching (positive signal) to the photo-induced absorption (negative signal) and the oscillation amplitude is small. Although the oscillation in the bleaching and photo-induced absorption is π-phase shifted, relative values show in-phase oscillations. Associated with the ruffling mode of the aggregates transmittance change is observed at various probe photon energies covered by the shortest visible pulse as broad as 150 THz. The phase of the molecular vibration is found nearly constant, if we take into account the sign reversal of the signals between the bleaching region and photoinduced absorption region. Absorbance change is induced by either bleaching, induced

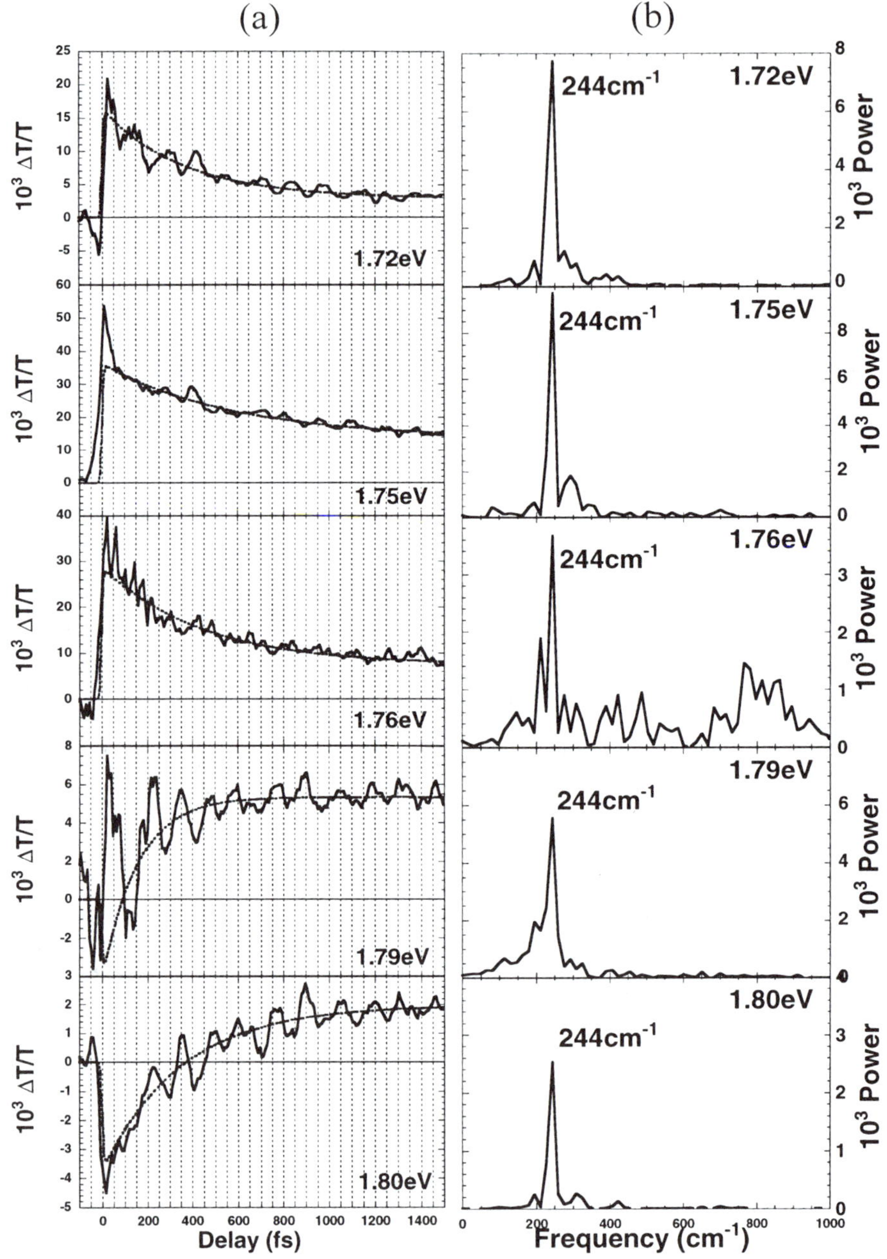

Figure 7. (a)Time-dependence of $\Delta T/T$ of J-aggregates of tetraphenylporphyrin sulfonate. (b) Fourier transform of oscilating components of $\Delta T/T$ integrated over the delay time from 100 fs to 1800 fs. (reproduced from Ref. 19)

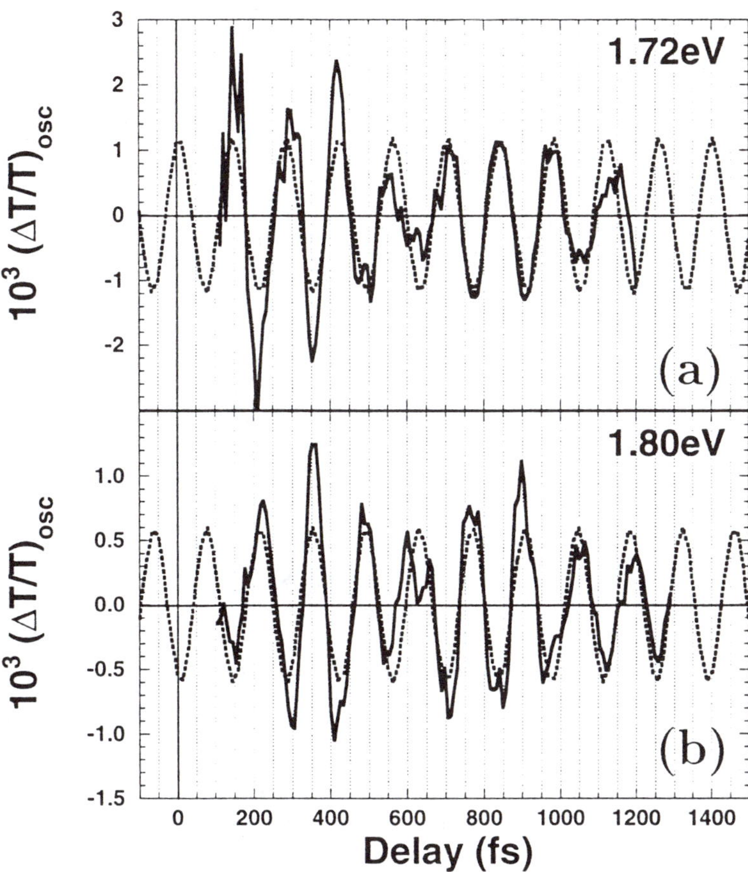

Figure 8. Oscilating components of $\Delta T/T$ at probe photon energies of (a)1.72 eV and (b)1.80 eV, respectively. It is clearly seen that the phase difference between them is out of phase by π . (reproduced from Ref. 19)

emission, and/or induced absorption. The former two are clearly related with the stationary ground-state absorption. The induced absorption is also considered to be related to the ground-state absorption. This is because the induced absorption is due to the transition from the $n = 1$ exciton state to $n = 2$ exciton state. It is expected to resemble that of $n = 0$ to $n = 1$ exciton state except for blue-shift due to the Pauli exclusion principle. The constant phase is different from expected continuously varying phase change along the probe photon energy. This unexpected result can be explained in terms of dynamical intensity borrowing. The intensity of absorption associated with the transition of the ground state to the Q-band is borrowed from that of Soret band by the configuration-interaction mechanism. Since the mode appears 30 times stronger than on a molecule dissolved in solution, it is concluded that the transition is strongly coupled to the Frenkel excitonic transition. The degree of the intensity borrowing is modified by vibrational motion of the ruffling mode by the exciton-phonon coupling. The coupling is not large enough to induce the potential minimum in the ground state from that of the ground state but large enough to modify the amount of intensity borrowing. It was found that even a very small change in the vibrational amplitude could introduce the change in the transition probability. From the experimental transmittance change, the intensity of the vibronic coupling is estimated.

Summary

Sub-5-fs real-time spectroscopy has been achieved using NOPA. Nonlinear mode coupling, was observed in a polydiacetylene cast film. Dushinsky rotation dynamics were observed in cresyl violet doped in polymers. Dynamic intensity borrowing was found in J-aggregate of porphyrin sulfonate. These phenomena cannot be detected by frequency-resolved spectroscopy but they can be detected by real-time spectroscopy. The result shown in this manuscript show that real-time spectroscopy with ultrashort pulses is very important for studying molecular vibrational dynamics.

Acknowledgements

The authors are grateful to the collaboration with T. Saito, H. Kano, M. Takasaka, and I. Sakane. This work was supported by a program of Research for the Future of Japan Society for the Promotion of Science (JSPS-RFTF-97P-00101).

References

1. Rose, T. S.; Rosker, M. J.; Zewail, A. H.; J. Chem. Phys. **1988**, 88, 6672.
2. Rose, T. S.; Rosker, M. J.; Zewail, A. H.; J. Chem. Phys. **1989**, 91, 7415.
3. Shirakawa, A.; Sakane, I.; Takasaka, M.; Kobayashi, T. *Appl. Phys. Lett.* **1999**, 74, 2268.
4. Kobayashi, T.; Shirakawa, A.; Matsuzawa, H.; Nakanishi, H. *Chem. Phys. lett.* **2000**, 321, 385.
5. Matsuzawa, H.; Okada, S.; Matsuda, H.; Nakanishi, H. *Chem. Lett.* **1997**, 11, 1105.
6. Kobayashi, T.; Yoshizawa, M.; Stamm, U.; Taiji, M.; Hasegawa, M. *J. Opt. Soc. Am. B* **1990**, 7, 1558.
7. Kobayashi, T.; Yasuda, M.; Okada, S.; Matsuda, H. Nakanishi, H. *Chem. Phys. Lett.* **1997**, 267, 472.
8. Vrakking, M. J. J.; Villeneuve, D. M.; Stolow, A.; *Phys. Rev. A* **1996**, 54, 37.
9. Kobayashi, T.; Shirakawa, A.; Saito, T.; Nakanishi, H. *J. Phys. Chem.* to be submitted.
10. Chang, T.-C.; Small, G. J. *Chem. Phys.* **1985**, 99, 479.
11. Bisht, P. B.; Fukuda, K.; Hirayama, S. *J. Chem. Phys.* **1996**, 105, 9349.
12. Brito Cruz, C. H.; Gordon, C. J. P.; Becker, P. C.; Fork, R. L.; Shank, C. V.; *IEEE J. Quantum Electronics*, **1988**, 24, 261.
13. Gardecki, J. A.; Constantine, S.; Zhou, Y.; Ziegler, L. D.; J. Opt. Soc. Am. B. **2000**, 17, 652.
14. Pollard, W. T.; Fragnito, H. L.; Bigot, J.-Y.; Shank, C. V.; Mathies, R. A. Chem. Phys. Lett. **1990**, 168, 239.
15. Duschinskii, F.; *Acta Physicochim. URSS* **1937**, 7, 551.
16. Book, L. D.; Arnett, D. C.; Hu, H.; Scherer, N. F. *J. Phys. Chem. A*, **1998**, 102, 4350.
17. Akins, D. L.; Özçelik, S.; Zhu, H.-R.; Guo, C. *J. Phys. Chem.* **1996**, 100, 14390.
18. Fuji, T.; Saito, T.; Kobayashi, T. *Chem. Phys. Lett.* **2000**, 332, 324.
19. Kano, H.; Saito, T.; Kobayashi, T. *J. Phys. Chem. B* **2001**, 105, 413.

Intense Fields

Chapter 13

Quantum Control by Adaptive Femtosecond Pulse Shaping

Application to CH$_2$BrCl in the Gas Phase and [Ru(dpb)$_3$]$^{2+}$ in the Solution Phase

Niels H. Damrauer and Gustav Gerber*

Physikalisches Institut, Universität Würzburg, Am Hubland, D–97074 Würzburg, Germany

Femtosecond adaptive pulse shaping is used to achieve bond-selective photochemistry for CH$_2$BrCl in the gas-phase as well as to control charge-transfer excitation of the [Ru(dpb)$_3$]$^{2+}$ complex in the solution phase (where dpb = 4,4'-diphenyl-2,2'-bipyridine). The technique exploits phase shaping of ultrafast broadband laser pulses within a 128-parameter search space. Because of the enormity of the variational space, a learning loop is implemented to find optimum pulse shapes. Successful applications of this technique to new systems in both the gas and solution phase emphasizes its generality and utility for controlling the outcome of light-matter interactions.

One of the central themes in modern chemistry is the struggle to control the outcome of reactions with the goal of creating useful and interesting products. As such, the modern synthetic chemist must have a broad knowledge of functionality and reaction conditions in order to create or break a variety of chemical bonds. Thermodynamic control is exercised with careful choice and design of starting materials followed by variation of external conditions such as temperature, pressure, and concentration. Kinetic control is implemented using catalysts designed to reduce barriers to desired product channels. With respect to these types of control, light can be thought of as having the potential to be very useful. In principle, excited-state potential energy surfaces could be exploited to change free-energy relationships between starting materials and products thus exercising thermodynamic control. Kinetic control, on the other hand, would be achievable if excess energy could be placed within molecular systems using tunable sources in order to overcome specific reaction barriers to product formation. Nonetheless, one has only to open a common journal or textbook to realize the basic shortcoming of photochemistry at achieving a prominent role in the methodology of synthetic control. While there are well understood and used photochemical routes to product formation, the numbers pale in comparison to the volumes of literature associated with more traditional synthetic methods. The responsibility for this shortcoming lies in the complexity of the excited-state electronic structure of many molecular reactants and the speed at which energy is statistically distributed within the molecule and to its surroundings following photo-excitation. Using typical laboratory photochemical variables such as intensity and wavelength, product yields are difficult to predict and even more so to control.

In the previous fifteen years, it has been shown that variations of spectral-temporal characteristics of applied coherent light (beyond simple variation of wavelength or intensity) can be used to exert control over the outcome of light-matter interactions by exploiting available interference phenomena.(*1-3*) Initial experiments subjected small molecules in the gas-phase to changes in a single control variable of the incident coherent light field. One example involved manipulation of the relative phase between two cw laser fields interacting with HCl(*4*) according to the Brumer-Shapiro control methodology.(*5*) A second example involved manipulation of the relative timing between two ultrashort laser pulses interacting with Na_2(*6*) according to the Tannor-Kosloff-Rice scheme.(*7*) For simple control problems, such single parameter changes can be sufficient. However, the approach becomes inadequate when the molecular and field-induced Hamiltonians are not known in sufficient detail to aid in the prediction of the control schemes (i.e. for complex molecular systems). In order to deal with this problem, theorists have applied optimal control theory to the problem of coherent control.(*8*) Allowing for variational freedom in the shape of the incident light field – for example, by permitting phase and/or amplitude modulation of the Fourier components of a broad band light pulse – it is generally determined that a large number of fields exist which can guide the temporal evolution of a system to target photoproduct distributions. Typically, however, such fields reflect the complexity of the dynamics available to the system subjected to control. For experimentalists interested in controlling polyatomic systems, the question that arises is how to generate such complex fields? This was answered by the technological development of femtosecond pulse shapers, which allowed rapid and sophisticated manipulation of broadband laser pulses within a large parameter space.(9,10) Nonetheless, combining optimal control theory with laser pulse shapers is not straightforward. First, theory cannot predict the optimal field when molecular excited-state potential energy surfaces are not known accurately enough. Second, it is not clear that fields predicted by theory can then be adequately reproduced under laboratory conditions. To address these problems, Judson and Rabitz suggested that the experimental outcome be used as feedback for actively optimizing the electric field.(8,11) Practically, this can be done using an iterative optimization (or learning) loop which implements (a) multi-parameter active control over E(t) using pulse shaping technology, (b) measurement of the experimental response to be used as feedback, and (c) a general search algorithm to evaluate and improve pulse shapes in the multi-parameter space. In this scheme, which is referred to as femtosecond adaptive pulse shaping, the molecules being subjected to control guide the search for an optimal E(t) within the constraints of the Hamiltonian and the experimental conditions. The methodology, thus, is ideally suited for complex control problems.

Application of adaptive femtosecond pulse shaping to molecular systems was first demonstrated by Wilson's group to control the excited-state population of a laser dye in solution(12) and by our group to control the product distribution of gas-phase photodissociation reactions.(13) Bucksbaum's group has used such a scheme to control the excitation of different vibrational modes in a molecular liquid,(14) and Motzkus' group has demonstrated control of vibrational dynamics in a four-wave mixing experiment.(15) This chapter discusses two of our most recent efforts. In the first case we have considered an old problem in the field of quantum control; namely, can light be used to undertake bond selective photochemistry. To address this, we have studied gas-phase CH_2BrCl molecules where the general goal was to cleave a specific carbon-halogen bond while leaving the other intact. In the second case, we have applied adaptive pulse shaping technology to control photophysical properties of a ruthenium(II) metal-to-ligand charge-transfer chromophore in solution. The results suggest that phase-shaped light fields can be used to preferentially excite a molecule according to its electronic structure but beyond simple wavelength consideration. This is a promising indication that these techniques can be used in the future to exert control over solution-phase photochemistry. These experiments, as well as previous demonstrations of adaptive femtosecond pulse shaping, show significant variety in the chemical systems considered as well as the control objectives. This emphasizes the generality and utility of the technique for controlling the outcome of light-matter interactions in complex molecular systems.

Photoreaction Control of Gas-Phase CH$_2$BrCl

For many years, chemists have been interested in using light for selectively cleaving particular bonds within polyatomic molecular systems. One approach has been to use narrow-band ultraviolet sources to generate localized excited-state populations in the hopes of promoting bond selective photodissociation reactions. In the context of this type of experiment, polyhaloalkanes (CH$_2$XY) have served as important model systems due to the presence of n(X)$\rightarrow\sigma^*$ (C-X) ultraviolet transitions for each available halogen atom. The idea was that wavelength tuning of the excitation source could access distinct repulsive surfaces of this type leading to different photoproduct distributions. Successful application of this type of control scheme has been shown by Butler et al. in studies of gas-phase CH$_2$BrI.(*16*) These workers demonstrated that excitation of the n(Br)$\rightarrow\sigma^*$ (C-Br) transition at 210nm resulted in exclusive fission of the stronger carbon-halogen (C-Br) bond while excitation to the red (248nm) resulted in photoproduct distributions dominated by cleavage of the weaker C-I bond. Interestingly, however, the bond selectivity observed for CH$_2$BrI is not a general condition for all polyhaloalkanes. Tzeng et al.(*17*) have shown this in studies of the closely related molecule CH$_2$BrCl excited at 248nm and 193nm. They found that at the lower energy excitation, the molecule dissociates exclusively (as expected) along the weaker carbon-bromine bond.(*18-20*) Higher energy excitation does activate the cleavage of the stronger carbon-chlorine bond, however, in contrast to CH$_2$BrI, the first photoreaction still dominates. These findings are consistent with studies of CF$_2$BrCl(*21*) as well as CBrCl$_3$.(*22,23*) In an effort to understand the conditions necessary for wavelength dependent control, Takayanagi and Yokoyama have used time-dependent quantum mechanical methods and demonstrate that increasing the non-adiabatic coupling between dissociative surfaces is sufficient to explain the experimental results.(*24*) These findings suggest that wavelength variation alone is not a general condition for successful control in these systems. We became interested in the possibility that femtosecond adaptive pulse shaping might be useful for demonstrating bond-specific photochemistry in the challenging CH$_2$BrCl system. These experiments are among the first to address bond-selective photochemistry using this method.

We implement an active control methodology within this system by combining adaptive femtosecond pulse shaping with mass spectrometry.(*13,25*) Photoproduct cations, to be used as feed-back in the adaptive learning algorithm, are generated within a reflectron time-of-flight mass spectrometer following interaction of a molecular beam of neutral species with the focused 800nm light. An example of a spectrum collected following excitation with a bandwidth-limited laser pulse (~80fs; ~12nm FWHM) is shown in Figure 1a. Two major fragment ions are observed corresponding to the loss of bromine in the dominant feature (CH_2Cl^+) and loss of chlorine in the next largest feature (CH_2Br^+). In both cases, the expected isotope pattern due to presence of the halogen is observed. Also seen are very minor contributions from CH_2^+, and the parent molecular ion CH_2BrCl^+. For implementation of the feedback optimization, attention was directed at the fragments CH_2Cl^+ and CH_2Br^+ resulting from carbon-halogen dissociation. Monitoring the peak areas arising from the single isotope fragments $CH_2{}^{35}Cl^+$ and $CH_2{}^{79}Br^+$, optimizations were run with the goal of maximizing and minimizing the product ratio CH_2Br^+/CH_2Cl^+. In this chapter, attention is focused on the maximization case, which is of particular interest because it attempts to find fields that preferentially cleave the carbon-chlorine bond, i.e., the stronger of the two carbon-halogen bonds. Further discussion of the minimization data is deferred because a strongly reduced total ion yield was observed following the optimization. We are concerned that systematic experimental errors associated with the smaller signal sizes may be responsible for driving the optimization to its endpoint. This will be discussed in detail elsewhere.(*26*) The fitness curve for maximization is shown in Figure 2a along with the value of the product ratio measured (throughout the optimization) using a short laser pulse that has not been modulated by the pulse shaper. In this particular optimization, such an unmodulated pulse serves as the initial guess.(*27,28*) Thus, we observe a significant improvement in the CH_2Br^+/CH_2Cl^+ product ratio from the starting place of a short laser pulse.

Using the spectral phase applied at the pulse-shaper and the measured laser spectrum, we can calculate time-dependent electric field shapes. Figure 3 shows such a field for the best individual of the last generation following maximization of the CH_2Br^+/CH_2Cl^+ product ratio. A complex structure is observed with decreased intensity with respect to a bandwidth-limited pulse. Because the optimization leads to a reduced intensity laser pulse, it is important to address whether the phase structure – and not simply the intensity change – is responsible for the selectivity obtained. One way that we have addressed this is by collecting a series of mass spectra as a function of pulse energy attenuation. This was done immediately before the optimization. The data (Figure 2b: filled circles) show that the product ratio decreases slightly as the energy of the short pulses is reduced. It is clearly seen that the maximized CH_2Br^+/CH_2Cl^+ product ratio achieved by the adaptive algorithm (dashed line) cannot be obtained with intensity variation of a short laser pulse by energy attenuation. We have also considered the effect of intensity variation of the laser pulse at constant energy (by pulse length variation) and find again that such changes cannot account for the ratio value obtained by the algorithm.(*26*)

The mass spectrum collected following maximization of the CH_2Br^+/CH_2Cl^+ product ratio shows substantial qualitative differences with respect to that

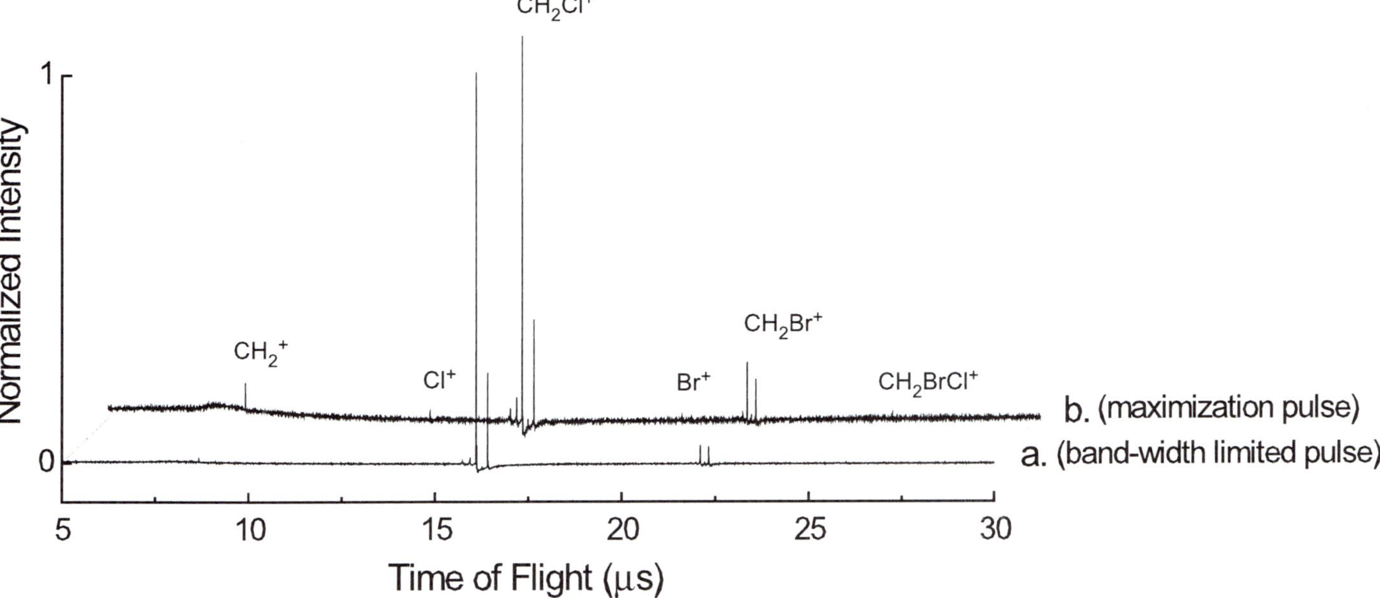

Figure 1. Time-of-flight mass spectrum collected for CH$_2$BrCl following multi-photon excitation with (a) a bandwidth-limited 800nm laser pulse (~80fs; ~230µJ/pulse), (b) a laser pulse of the same energy following adaptive maximization of the product ratio CH$_2$Br$^+$/CH$_2$Cl$^+$. The CH$_2$Cl$^+$ ion intensity in b. is 5.6 times smaller than in a.

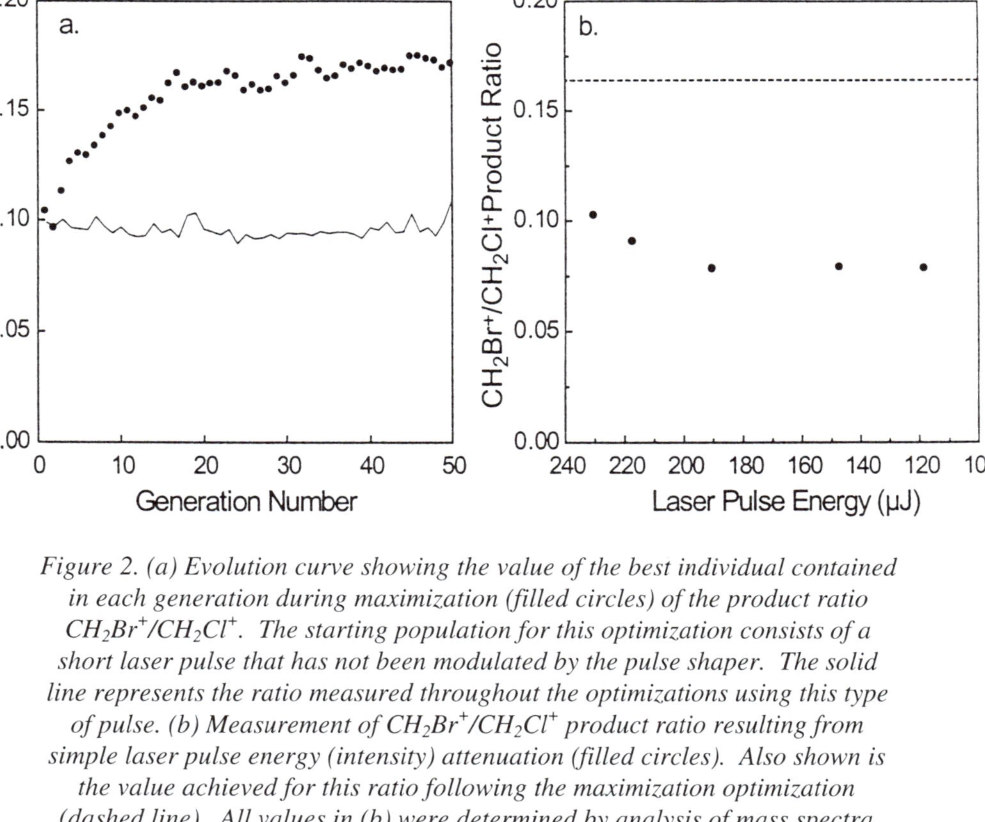

Figure 2. (a) Evolution curve showing the value of the best individual contained in each generation during maximization (filled circles) of the product ratio CH_2Br^+/CH_2Cl^+. The starting population for this optimization consists of a short laser pulse that has not been modulated by the pulse shaper. The solid line represents the ratio measured throughout the optimizations using this type of pulse. (b) Measurement of CH_2Br^+/CH_2Cl^+ product ratio resulting from simple laser pulse energy (intensity) attenuation (filled circles). Also shown is the value achieved for this ratio following the maximization optimization (dashed line). All values in (b) were determined by analysis of mass spectra collected during attenuator series and following the optimization.

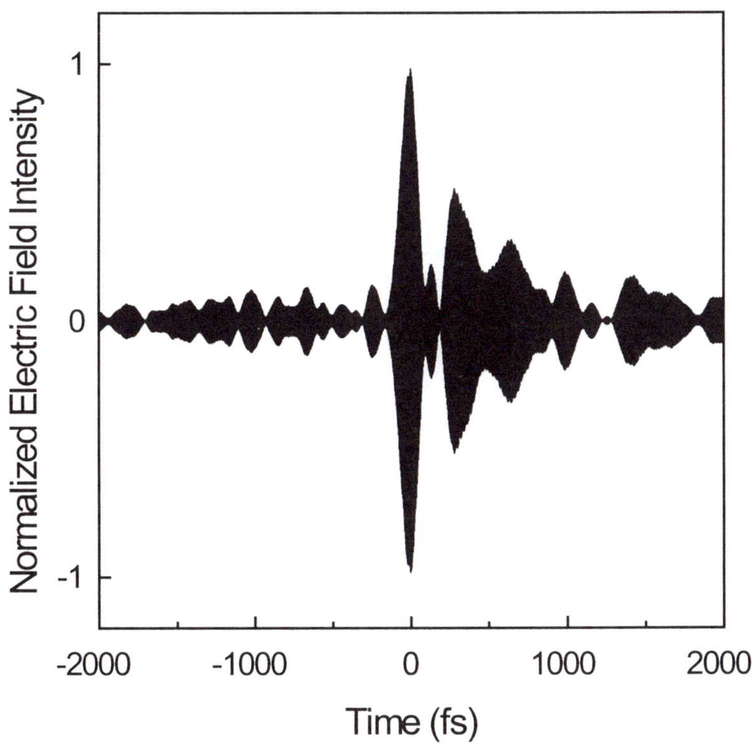

Figure 3. The electric field shape as a function of time for the best individual of the last generation following maximization of CH_2Br^+/CH_2Cl^+.

obtained with a bandwidth-limited pulse (Figure 1b). We observe a clear increase in the ratio of interest (seen as the strong increase in the CH_2Br^+ fragment height) as well as enhancement of smaller fragments such as CH_2^+, Cl^+, and Br^+, while at the same time formation of the parent ion. These features are a good indication that the pulse optimized for this task is in fact manipulating fragmentation dynamics in order to achieve its goal.

The results presented show that shaped laser pulses can be found which maximize the product ratio CH_2Br^+/CH_2Cl^+ with respect to that achieved with short laser pulse. This is done automatically and without prior knowledge of the molecular potential energy surfaces involved. Analysis of the mass spectra presented in Figure 1 indicates the CH_2Br^+/CH_2Cl^+ product ratio is 0.08 for the bandwidth-limited pulse and 0.16 for the shaped maximization pulse. Thus, a complex laser pulse has been discovered which doubles the ion ratio corresponding to a significant relative change in the breaking of the carbon-chlorine bond. Again we stress that this is the stronger of the two carbon-halogen bonds. We are confident that this effect is not simply a result of field intensity changes. Rather, it occurs because the phase-shaped pulse found by the adaptive optimization alters the fragmentation dynamics. The question that arises is what is the nature of these dynamics that are subjected to control in this system? Unfortunately this is not a simple question to answer as there are multiple mechanisms by which CH_2Br^+ and CH_2Cl^+ can be formed as photofragments following multi-photon excitation of CH_2BrCl (via dynamics on neutral surfaces, ionic surfaces, or ion-pair surfaces). These and possible mechanisms for control are discussed in detail elsewhere.(*26*) We can state, however, that even in the challenging CH_2BrCl system – where excited states are highly coupled – adaptive femtosecond pulse shaping is able to discover laser fields which facilitate bond-selective photochemistry.

Control of Charge-Transfer Excitation in the Solution Phase

The growing body of successful demonstrations suggests that adaptive feedback pulse shaping could develop into a general methodology for controlling the photochemistry of molecules. This is particularly true in the case of complex systems where knowledge of the excited-state Hamiltonian is difficult to obtain. From this, it is exciting to speculate that this technique might someday find its way into the toolbox of synthetic chemists as a means of manipulating the production of reactive intermediates or generating specific products. However, if adaptive pulse shaping can have promise as a general synthetic method, it must be shown to be viable in the solution phase medium where most preparative chemistry is done. There are substantial challenges facing experimentalists wishing to demonstrate this and at the forefront is the question of feasibility. In comparison to most gas-phase control problems, the introduction of a solvent environment increases the complexity of the time-dependent distribution of energy as well as in general, drastically decreasing the time-scale in which vibrational coherence remains in the system. Coherent control in the solution phase must work within these constraints. We believe that adaptive pulse shaping – because of its ability to explore electric field shapes within an extremely large parameter space – is currently the only feasible means of approaching this difficult physical problem.

Our starting place is a technique that requires a single laser beam (the shaped pulse) and relies on the detection of photophysical emission for feedback. We have chosen to consider an emissive charge-transfer chromophore $[Ru(dpb)_3](PF_6)_2$ (where dpb = 4,4'-diphenyl-2,2'-bipyridine)(Figure 4a) in methanol solution. Like other members of its class, $[Ru(dpb)_3]^{2+}$ has a strong visible metal-to-ligand charge transfer (MLCT) absorption band (Figure 4b). When excited, the metal center is formally oxidized (II → III) concomitant with reduction of a single ligand species.(29,30) Since we are using 800nm light, at least two-photon absorption is required to promote this electronic transition. Following excitation under typical solution phase conditions (i.e. room-temperature fluid solutions), emission is observed on a microsecond time scale as a broad spectrum red-shifted with respect to the MLCT excitation (Figure 4b).(30) The emission occurs from a ^{3}MLCT state efficiently accessed via ultrafast intersystem crossing and non-radiative relaxation.(31) Since the ^{3}MLCT emissive state is not directly accessed by the photoexcitation, the emission yield is only sensitive to the excited-state population. We can therefore use it as a feedback signal for controlling the excitation itself. The methodology is represented in Figure 4c where the ^{1}MLCT Frank-Condon state is excited by the shaped electric field and the emission used as feedback comes from the ^{3}MLCT state formed following ultrafast intersystem crossing and nonradiative relaxation.(31)

There are several examples in the literature where phase-shaped laser pulses have been used to influence the excited-state population of complex molecules in solution.(12,32,33) Included within these is the Wilson group's demonstration of automated feedback control using the fluorescent dye IR125 and a five parameter search space.(12) In that experiment, the shaped pulse is able to excite the chromophore under control with a one-photon absorption. The influence of the pulse shape arises because a second photon interaction can stimulate emission from the evolving wave-packet. For intense laser fields, a genetic search algorithm finds laser pulses with strong positive linear chirp consistent with previous one-parameter control experiments. As stated, our experiments require the absorption of at least two 800nm photons for excitation of the $[Ru(dpb)_3]^{2+}$ chromophore. We believe this fact, combined with the use of

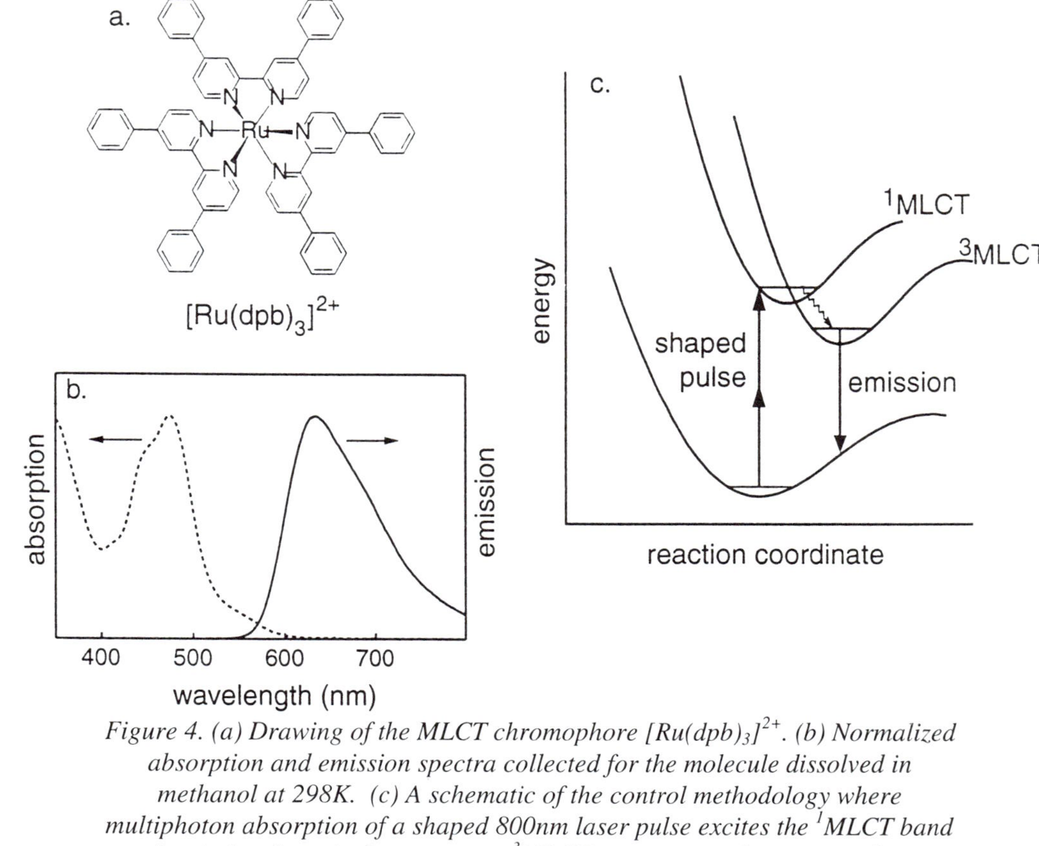

Figure 4. (a) Drawing of the MLCT chromophore [Ru(dpb)$_3$]$^{2+}$. (b) Normalized absorption and emission spectra collected for the molecule dissolved in methanol at 298K. (c) A schematic of the control methodology where multiphoton absorption of a shaped 800nm laser pulse excites the ^{1}MLCT band and emission from the lower energy ^{3}MLCT state accessed via non-radiative relaxation is used as a feedback signal.

a 128 parameter search space, allows us to consider the potential utility of more complex electric field shapes in controlling molecular excitation.

The setup for our control experiments in the liquid phase is described as follows. Laser pulses are actively manipulated in a pulse shaper and then directed into a flowing sample (~2 mm) of $[Ru(dpb)_3](PF_6)_2$ in a methanol solution. Emitted radiation (nominally phosphorescence) is collected at an approximate right angle, spectrally filtered at 620nm to remove scattered pump light, detected with a photomultiplier tube, and integrated on a microsecond time scale. During optimizations there are two different signals monitored and used in the optimization goals. The first is the molecular emission while the second is second harmonic generation (SHG) in BBO of the same laser beam used to excite the sample.

Figure 5a shows evolution curves for two initial optimization goals: maximization of molecular emission (filled circles) and maximization of SHG (open circles). What is plotted is the value of molecular emission during these two optimizations. We know that in the optimization of SHG, the end result is a transform limited pulse.(*34*) Therefore, the values of emission indicated by the open circles to the upper right of the diagram are a result of the most intense laser pulses available. When molecular emission itself is optimized we observe an evolution curve that is very similar indicating that emission alone is strongly intensity dependent. This result is entirely reasonable because a multi-photon process is required to excite the molecule in the first place.

With the experiment just described, we cannot expect to find pulse shapes that discriminate between MLCT excitation of a molecule in solution and SHG of the laser beam in a non-linear crystal because the control objective is not well defined. As such, we have explored two additional optimization goals; namely, maximization and minimization of the ratio emission/SHG. The idea behind this choice is to cancel the dominating effect of the non-linear excitation mechanism because of its presence in both the numerator and denominator of the ratio. If the mechanisms leading to the emission or SHG are identical then optimization of their ratio should be impossible. Optimization curves for these experiments are shown in Fig. 5b. What is seen is very clear evolution of these ratios from the initial generation of random-phased individuals to maximization as well as minimization conditions. This shows that the learning algorithm is able to find laser fields that distinguish between the two excitation processes involved. Also included within the figure is the value of the ratio throughout the optimization procedure that results from a short and intense laser pulse that has not been modulated by the pulse shaper. It is seen that for both optimization goals, intense laser pulses are not satisfactory for achieving an optimal condition.

In order to visualize the pulse shapes found by the learning algorithm, we plot the measured electric field in time and frequency space using the Husimi representation.(*35,36*) These are shown in Figure 6 for the best individual of the

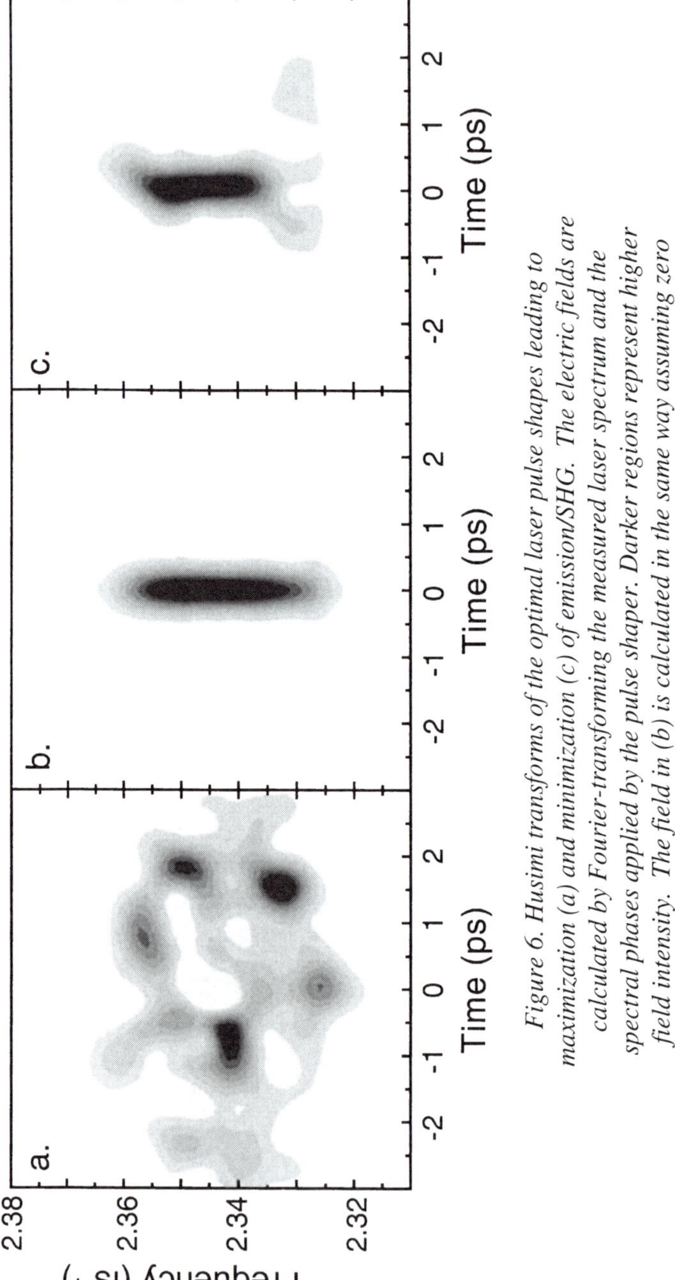

Figure 6. Husimi transforms of the optimal laser pulse shapes leading to maximization (a) and minimization (c) of emission/SHG. The electric fields are calculated by Fourier-transforming the measured laser spectrum and the spectral phases applied by the pulse shaper. Darker regions represent higher field intensity. The field in (b) is calculated in the same way assuming zero phase and thus represents a bandwidth-limited pulse.

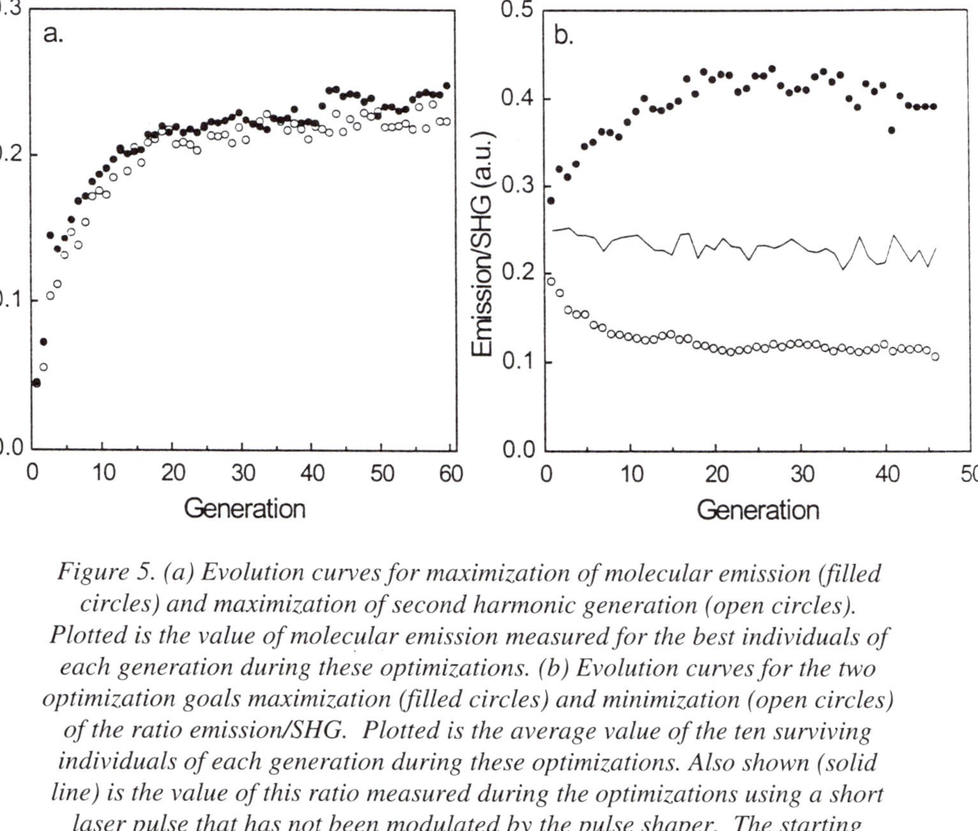

Figure 5. (a) Evolution curves for maximization of molecular emission (filled circles) and maximization of second harmonic generation (open circles). Plotted is the value of molecular emission measured for the best individuals of each generation during these optimizations. (b) Evolution curves for the two optimization goals maximization (filled circles) and minimization (open circles) of the ratio emission/SHG. Plotted is the average value of the ten surviving individuals of each generation during these optimizations. Also shown (solid line) is the value of this ratio measured during the optimizations using a short laser pulse that has not been modulated by the pulse shaper. The starting population for each optimization shown consists of sixty individuals with randomly chosen phases.

last generation of the maximization (a) and minimization (c) optimizations. The electric field for the shortest possible pulse (b) is calculated for comparative purposes using a flat spectral phase and the measured laser spectrum. What is observed following the optimizations are complex shapes – most notably in the maximization case – as well as marked differences between the fields suitable for the two particular goals. At this time, an interpretation of why these favor their respective optimization goals is difficult due to the electronic and physical complexity of the solution phase charge-transfer system itself. Nonetheless, certain features deserve a brief comment. In the maximization experiment it is intriguing that the pulse energy becomes discretized into smaller pulse-like features in both time and frequency space. It is of course reasonable that these are necessary for achieving two-photon excitation of the molecule. However, it is also possible that the discretization in frequency space indicates a sensitivity of the field to the vibrational structure of the charge-transfer molecule or to vibrational dynamics occurring concomitant with the electronic excitation. In support of this, it is noted that the maximization and minimization fields exhibit certain complimentary frequency characteristics. This can be seen by first considering Figure 6c, which shows a field with considerable temporal intensity following minimization of emission/SHG. Presumably, this is important in order to increase the amount of SHG in the denominator. However, short laser pulses are not as successful (such as the monitor signal in Figure 5b) and the algorithm finds a field with little intensity in the frequency region around 2.33 fs^{-1}. On the other hand, the field found following maximization of emission/SHG (6a), shows its most intense feature in this same frequency region. By considering the bandwidth-limited electric field of Figure 6b, it is clear that such frequency components are available with considerable intensity. Thus, the possibility exists that this frequency component ($2.33 fs^{-1}$) plays a specific role in the excitation of the charge-transfer species. Although not shown here, other pulses with comparable fitness found during both optimizations (for example, the best pulses of generations 20 and 30) show similar intensity features and this same complimentary behavior. The final point we make regarding pulse features concerns the apparent importance of the large (128) parameter search space for finding optimal electric fields. It is difficult to imagine simple control schemes manipulating one or even several variables that would be able to find comparable solutions. With a large parameter space, electric fields can be found which exploit the differences in the physical mechanisms leading to charge-transfer excitation versus SHG.

We have demonstrated feedback optimized control of the electronic excitation of a complex MLCT chromophore in solution. This has been done using a large (128 parameter) search space and multi-photon excitation of the molecule. Maximization and minimization of the ratio emission/SHG has shown that electric field shapes can be found which exploit the differences in the mechanisms of excitation leading to these observables. It is clear from these fields that successful optimization exploits the complexity afforded by the large parameter space. It appears that the optimized pulse shapes are sensitive to and thus have adapted to the electronic structure of the charge-transfer chromophore in solution. The result is promising because it implies that selectivity of two non-interacting photophysical processes may be achieved for a given laser pulse bandwidth. Under these conditions, the opportunities for control may be improved by adaptive shaping of input laser pulses with larger bandwidth. If, as we have argued, selectivity arises because of unique time and frequency domain requirements of the molecule undergoing multi-photon excitation,*(37)* the technique would provide a new method for finding and using complex pulse shapes to selectively excite one type of molecule among several non-interacting species in a single solution.*(38)*

The authors would like to thank T. Brixner, C. Dietl, B. Kiefer, G. Krampert, and S. H. Lee for their contributions to the work presented. N. D. would like to thank the Alexander von Humboldt Foundation for generous support of a postdoctoral position.

References

1. Rice, S. A.; Zhao, M. *Optical control of molecular dynamics.* Wiley: New York., 2000.
2. Gordon, R. J.; Rice, S. A. *Annu. Rev. Phys. Chem.* **1997**, *48*, 601.
3. Warren, W. S.; Rabitz, H.; Dahleh, M. *Science* **1993**, *259*, 1581.
4. Park, S. M.; Lu, S. P.; Gordon, R. J. *J. Chem. Phys.* **1991**, *94*, 8622.
5. Brumer, P.; Shapiro, M. *Chem. Phys. Lett.* **1986**, *126*, 541.
6. Baumert, T.; Grosser, M.; Thalweiser, R.; Gerber, G. *Phys. Rev. Lett.* **1991**, *67*, 3753.
7. Tannor, D. J.; Kosloff, R.; Rice, S. A. *J. Chem. Phys.* **1986**, *85*, 5805.
8. Rabitz, H.; Zhu, W. *Acc. Chem. Res.* **2000**, *33*, 572 and references therein.
9. Heritage, J. P.; Weiner, A. M.; Thurston, R. N. *Opt. Lett.* **1985**, *10*, 609.
10. Weiner, A. M.; Leaird, D. E.; Patel, J. S.; Wullert, J. R. *Opt. Lett.* **1990**, *15*, 326.
11. Judson, R. S.; Rabitz, H. *Phys. Rev. Lett.* **1992**, *68*, 1500.
12. Bardeen, C. J.; Yakovlev, V. V.; Wilson, K. R.; Carpenter, S. D.; Weber, P. M.; Warren, W. S. *Chem. Phys. Lett.* **1997**, *280*, 151.
13. Assion, A.; Baumert, T.; Bergt, M.; Brixner, T.; Kiefer, B.; Seyfried, V.; Strehle, M.; Gerber, G. *Science* **1998**, *282*, 919.
14. Weinacht, T. C.; White, J.; Bucksbaum, P. H. *J. Phys. Chem. A* **1999**, *103*, 10166.

15. Hornung, T.; Meier, R.; Zeidler, D.; Kompa, K. L.; Proch, D.; Motzkus, M. *Appl. Phys. B* **2000**, *71*, 277.

16. Butler, L. J.; Hintsa, E. J.; Shane, S. F.; Lee, Y. T. *J. Chem. Phys.* **1987**, *86*, 2051.

17. Tzeng, W. B.; Lee, Y. R.; Lin, S. M. *Chem. Phys. Lett.* **1994**, *227*, 467.

18. This is consistent with CH2BrI and was confirmed at 234nm by Lee et al. (c.f. 19) using a molecular beam imaging system and at 248nm and 268nm by McGivern et al. using a REMPI technique (c.f. 20).

19. Lee, S. H.; Jung, Y. J.; Jung, K. H. *Chem. Phys.* **2000**, *260*, 143.

20. McGivern, W. S.; Li, R.; Zou, P.; North, S. W. *J. Chem. Phys.* **1999**, *111*, 5771.

21. Baum, G.; Huber, J. R. *Chem. Phys. Lett.* **1993**, *213*, 427.

22. Lee, Y. R.; Tzeng, W. B.; Yang, Y. J.; Lin, Y. Y.; Lin, S. M. *Chem. Phys. Lett.* **1994**, *222*, 141.

23. Lee, Y. R.; Yang, Y. J.; Lin, Y. Y.; Lin, S. M. *J. Chem. Phys.* **1995**, *103*, 6966.

24. Takayanagi, T.; Yokoyama, A. *Bull. Chem. Soc. Jpn.* **1995**, *68*, 2225.

25. Bergt, M.; Brixner, T.; Kiefer, B.; M., S.; Gerber, G. *J. Phys. Chem. A* **1999**, *103*, 10381.

26. Damrauer, N. H.; Dietl, C.; Krampert, G.; Lee, S. H.; Jung, K. H.; Gerber, G. **2001**, in preparation.

27. Modifications to the phase structure of the pulse are then applied throughout the optimization by the operators of the evolutionary algorithm (c.f. 28).

28. Baumert, T.; Brixner, T.; Seyfried, V.; Strehle, M.; Gerber, G. *Appl. Phys. B* **1997**, *65*, 779.

29. Juris, A.; Balzani, V.; Barigelletti, F.; Campagna, S.; Belser, P.; Von Zelewsky, A. *Coord. Chem. Rev.* **1988**, *84*, 85.

30. Damrauer, N. H.; Boussie, T. R.; Devenney, M.; McCusker, J. K. *J. Am. Chem. Soc.* **1997**, *119*, 8253.

31. Damrauer, N. H.; McCusker, J. K. *J. Phys. Chem. A* **1999**, *103*, 8440.

32. Cerullo, G.; Bardeen, C. J.; Wang, Q.; Shank, C. V. *Chem. Phys. Lett.* **1996**, *262*, 362.

33. Bardeen, C. J.; Yakovlev, V. V.; Squier, J. A.; Wilson, K. R. *J. Am. Chem. Soc.* **1998**, *120*, 13023.

34. Brixner, T.; Strehle, M.; Gerber, G. *Appl. Phys. B* **1999**, *68*, 281.

35. Husimi, K. *Proc. Phys. Math. Soc. Jpn.* **1940**, *22*, 264.

36. Lalovic, D.; Davidovic, D. M.; Bijedic, N. *Phys. Rev. A* **1992**, *46*, 1206.

37. Brixner, T.; Damrauer, N.H.; Kiefer, B.; Gerber, G. **2001**, submitted for publication.

38. Brixner, T.; Damrauer, N.H.; Niklaus, P.; Gerber, G. **2001**, submitted for publication.

Chapter 14

The Mechanisms of Strong-Field Control of Chemical Reactivity Using Tailored Laser Pulses

Noel P. Moore, Getahum M. Menkir, Alexei N. Markevitch, Paul Graham, and Robert J. Levis

Department of Chemistry, Wayne State University, Detroit, MI 48202

Control of chemical photoionization, dissociation, and rearrangement is demonstrated using tailored intense (10^{13} W cm^{-2}) laser pulses. The laser pulses are created using a closed-loop learning algorithm with the product distribution guiding the shape of the laser pulses. Control is demonstrated for the production of the acetone parent ion (($(CH_3)_2CO^+$), the competition of the products CH_3CO^+ vs. CH_3^+ during CH_3COCF_3 laser induced dissociation, and the rearrangement acetophenone ($C_6H_5COCH_3$) to produce $C_6H_5CH_3$ vs. C_6H_5 + CH_3CO. The generic nature of the technique is demonstrated and the fundamental underlying principles of such strong-field control experiments are considered.

Controlling the electronic and nuclear dynamics of large polyatomic molecules requires, at the present time, tailored strong-field laser pulses (with intensity of approximately 10^{13} W·cm^{-2}). Strong-field in this context implies the use of laser pulses where the maximum amplitude of the electric field (~VÅ^{-1}) rivals the fields binding valence electrons to nuclei (1). Tailored implies that the time-dependent electric field of the laser has been shaped (as described subsequently) to optimize a desired reaction channel (2). Controlling the outcome of matter-radiation coupling is desirable from both theoretical and practical points of view. From the theoretical perspective there are many new phenomena present in the interaction of intense laser pulses with molecules and furthermore it may be possible to unravel the details of a molecule's Hamiltonian from a family of control experiments (3) that would otherwise remain inaccessible using conventional methods. From the practical standpoint, the control of chemical reactivity using strong-field optical pulses opens the twin horizons of combinatorial photochemistry and the development of an optical scalpel. The latter may now complement the methods of optical tweezers, cooling, and molasses. Computing a tailored optical field to control a chemical reaction from first principles is intractable for all but the simplest model systems because the molecular Hamiltonian is largely unknown. All is not lost for controlling chemical reactivity, as optimum solutions maybe converged upon using closed-loop learning algorithms to selectively enhance one product channel at the expense of others (2). In this sense, the chemical control problem is like many other non-deterministic problems, such as those found in geometry optimization or in neural net design.

In the closed-loop paradigm for controlling chemical reactivity, one specifies a desired outcome for the interaction of the laser pulse and the molecule of interest. The laser pulse is then manipulated to produce a well-defined time-dependent electric field. This time-dependent electric field interacts with the molecule resulting in a series of photochemical/physical products. Measurement of the appropriateness of the product distribution with respect to a desired distribution allows the design of new pulses using an optimization algorithm. Closed-loop control with strong-field laser pulses is at the heart of all recent control experiments on complex systems. Such schemes have been used to control dissociation in inorganic (4) van der Waals (5), and now, organic molecules (6); chemical rearrangement channels (6); stimulated Raman emission from molecules (7); and high harmonic generation from atoms (8).

From a practical point of view, teaching lasers to control reaction dynamics (2) is now possible because of the convergence of three optical technologies. The first is the production of ultrashort laser pulses where a large number of discrete frequencies are coherently superimposed to produce a transform-limited pulse. Pulses on the order of 40 fs at 800 nm have tens of nanometers of frequency components that may be controlled to produce the desired laser pulse. The second technology is the use of spatial masking techniques (9) to selectively alter the phase and amplitude of the component frequencies of the transform limited pulse, thus producing a new laser pulse with a distinct time-dependent electric field. The third is the use of pulse amplification to increase the energy of the tailored laser to an intensity regime where chemical reactions may be induced (1). The control aspect then involves using a feedback-loop managed by a learning algorithm to converge upon the optimal mask to shape the time-dependent electric field to produce a desired reaction channel. The degree of control rests in part with the number of component frequencies (the bandwidth) available for tailoring the pulse. Even with the increased bandwidth of modern ultrafast laser systems, there are an insufficient number of frequencies to control complex chemical reactions, particularly if there are no resonant states within the range of the laser. This limitation may be circumvented by using strong-field laser pulses which serve to "artificially" increase the excitation bandwidth as described (10).

We demonstrate control of photoionization, dissociation and rearrangement using a strong-field, ultrafast closed-loop excitation scheme. We show that control of the photoionization/dissociation distribution is possible both by altering the intensity and/or pulse duration of the near transform limited pulse, as well as by controlling the detailed temporal profile of the intensity of the pulse. Examples of each of these control methods are presented. Finally the underlying mechanisms of multiphoton excitation, transient shifting of intermediate states and lifetime broadening of eigenstates are discussed to explain this novel control paradigm.

Experimental

The oscillator pulse is an 86 MHz pulse train of 20 fs pulses centered at 800 nm. The pulses are stretched to 100 ps using two gratings. The stretched pulses are then amplified to 3 mJ in a 10 Hz regenerative amplifier and are compressed to 80 fs using a dual grating system. Pulse shaping to produce the time-dependent electric fields is achieved by means of a spatial light modulator (SLM) placed at the Fourier plane of the pulse stretcher, as shown in Figure 1. The SLM consists of two registered linear arrays containing 128 pixels (liquid crystals) each. In the present work, groups of 16 pixels are linked together to

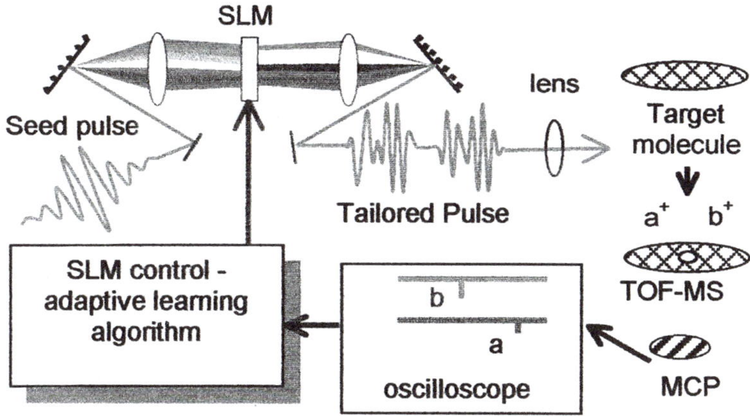

Figure 1. A schematic of the optical layout used to
produce the tailored, strong-field laser pulses.

create a genome with 2×8 variable regions or genes. Each variable region is biased to a voltage that controls the relative phase and amplitude of the frequency components travelling through each region. The tailored laser pulses are focused to a 100 μm diameter spot in the analysis chamber using a 20 cm lens. Spatial inhomogeneities in the laser focal volume present no significant problem as the genetic algorithm can cope with this instability (10). Target molecules are thermalized in vacuum to a pressure of $\sim 10^{-6}$ torr. The outcome of the interaction of the tailored pulses with the target molecule is monitored by means of time-of-flight mass spectroscopy.

A genetic algorithm (11) is employed to iteratively converge the evolving laser pulses to the optimal pulse for controlling the reaction dynamics as desired. Concepts of evolutionary biology are applicable in development of efficient genetic algorithms. Individual pulses can be evaluated in terms of their 'fitness,' which is a quantitative measure of how close an outcome is to the desired target. This may be written as:

$$J = (O - O^*)^2 \qquad (1)$$

where J is the cost, O is the observable and O^* is the desired target for the observable. The concept is to minimize the cost by finding the time-dependent electric field, $\varepsilon(t)$, that optimally matches O to O^*. A scheme that allows only the fittest members (laser pulse shapes or $\varepsilon(t)$) of a population to survive and create new pulses will eventually converge on an optimal control genome. The details of the algorithm are specific to each objective posed. In our experiments, a population of 40 distinct members (laser fields) was propagated to find the optimal pulse. Despite the large search space, an optimum solution can be converged upon rapidly by judicious choice of optimization algorithms (12). In subsequent generations new members are generated after proportional selection by random crossover and a mutation rate of 6% per individual, per generation.

In an effort to determine the viability of the method as a tool for controlling photoprocesses, reference experiments were performed. The target molecules were exposed to the unshaped reference pulse (80 fs) and the mass spectra were recorded as a function of laser intensity. The laser intensity was adjusted by placing glass slides (200 μm thick) with known transmission characteristics into the beam path. The use of such slides was not found to affect the group velocity dispersion of the pulse in any significant manner, but reduced pulse energy by approximately 7% per slide. The target molecules were also studied as a function of laser pulse duration by linearly chirping the pulses. The reference experiments serve to give some knowledge of most likely reaction pathways for a given molecule under the influence of the intense 800 nm ±10 nm laser pulses.

Results

An example of pulse energy and pulse duration control of the photoionization and dissociation proceses is shown in Figure 2 for the molecule p-nitroaniline (PNA). The pulse intensity was varied from 2.5-15 X 10^{13} W cm^{-2} in Figure 2A by varying the laser from 0.1 to 0.6 mJ per pulse with pulse duration held constant at 80 fs. An alternative method to alter the laser intensity was to maintain constant pulse energy while increasing the laser pulse duration. The duration can be increased in our apparatus by altering the grating separation in the compressor. The intensity is varied in Figure 2B from 0.3-15 X 10^{13} W cm^{-2} by altering the pulse duration while holding the energy constant at 0.6 mJ per pulse.

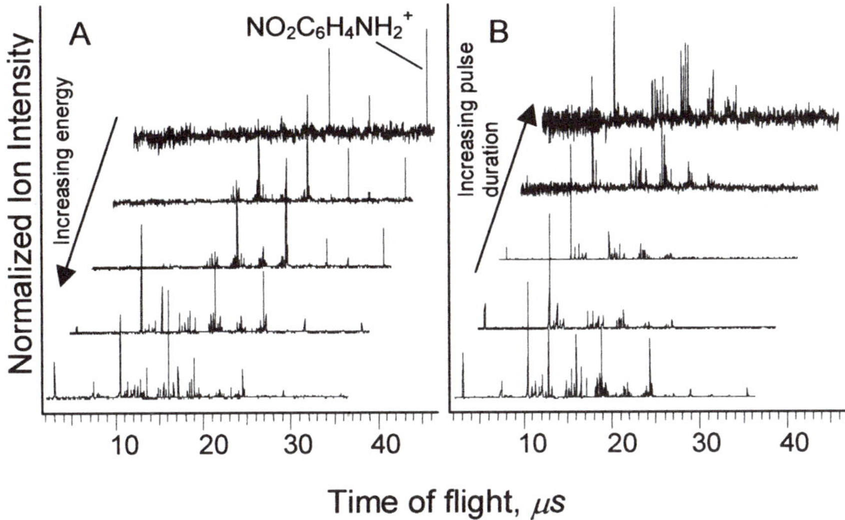

Figure 2. Time-of-flight ion spectra of p-nitroaniline subjected to 790 nm laser pulses. In Figure 2A the pulse energy was varied from 0.60 to 0.10 mJ/pulse, the pulse duration was 80 fs. In Figure 2B the pulse duration was varied from 100 fs to 5 ps, with 0.60 mJ/pulse.

The major trend observed in Figure 2A is the increase in the degree of fragmentation with increasing laser pulse energy. The molecular ion and large ion fragments are the major species observed at low laser intensity. If the laser pulse duration was increased while keeping the pulse energy constant, a significant degree of fragmentation was observed at all pulse durations, including at the laser intensity corresponding to the ion appearance threshold. Thus at the same laser intensity, as shown in the top two traces in Figure 2, we observe limited fragmentation for the transform limited pulse and considerable fragmentation for the stretched pulse. This experiment demonstrates that a degree of control over various fragmentation pathways can be achieved by simply varying laser pulse energy and/or duration.

Pulse shaping was investigated to attempt more sophisticated control over the photoionization and fragmentation distribution of the molecule acetone, $(CH_3)_2CO$. Figure 3 displays the results for the case where optimization of the parent peak in the mass spectrum of acetone was the specified goal in the control algorithm. Figure 3A shows two of the mass spectra taken as a function of generation number during the optimization. These spectra are representative of the 40 members of a given generation. The control algorithm was able to substantially increase molecular ion peak from a minimal level at the start of the experiment to being an easily detectable ion signal within a few generations. Subsequent increase in the parent signal was not found after the 13th generation despite the action of the genetic operators. Fluctuations in the relative abundance

212

of the parent peak after the 13th generation are expected due to the algorithm's attempts to improve upon the trial field by crossover and mutation. These fluctuations are local and do not affect the globally converged solution. The relative signal of parent peak as a function of generation number is shown in Figure 3B. In accord with the results of the survey experiments on acetone, the algorithm converged on a near transform limited pulse with which to maximize the parent signal. This simple experiment shows that it is possible to use shaped pulses to alter the appearance of the mass spectral distribution not only in terms of overall ion intensity, but also in terms of the fragmentation distribution.

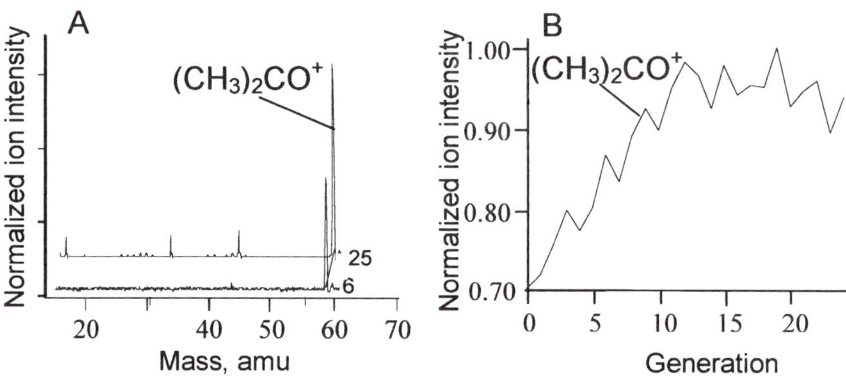

Figure 3. A) The mass spectra at two generations during optimization; B) The optimized acetone ion signal as a function of generation.

The survey experiments for acetone reveal that the optimum pulse for parent production is one that increases the laser intensity with a transform limited pulse. This optimal pulse is not surprising because increasing the laser pulse duration typically leads to increased dissociative ionization. Thus, the control exhibited in this case can be described as *trivial*, as there is no specificity to the molecular Hamiltonian incorporated into the optimum solution. *Non-trivial* control schemes depend on the evolution of the molecular Hamiltonian under the influence of the radiation and are hence molecule-specific.

Comparing the yields of two different ions from the simple intensity and pulse duration reference experiments to the yields from the genetic algorithm experiment may reveal whether the control exerted is trivial or non-trivial. Figure 4 presents the results from a closed-loop optimization on the molecule trifluoroacetone, CH_3COCF_3, where maximization of the ratio CH_3^+/CH_3CO^+ was specified. In these experiments the ion signals and ratios are normalized to unity at generation zero as the experiment starts with a randomly generated optical field. At the start of each optimization experiment, the absolute yields of the CH_3^+ and CH_3CO^+ ions were within a factor of two of each other. In this experiment the largest ratio between the two ions was 1.8. The increase in the ratio was mainly due to a significant increase in CH_3^+ yield, however both ions increased in overall intensity. Reference experiments reveal no trend in the CH_3^+/CH_3CO^+ ratio as a function of laser energy. Furthermore, the ratio decreases with increasing pulse duration, opposite to that shown in Figure 4. The trends in the reference experiments suggest that that non-trivial control is exerted in this case.

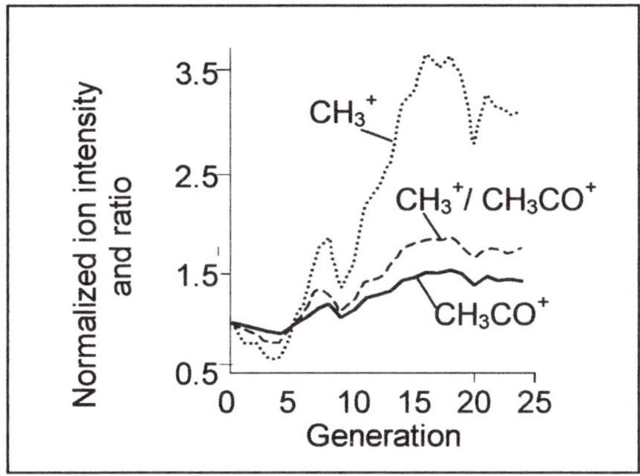

Figure 4. The relative intensity for CH_3^+, CH_3CO^+, ion formation from CH_3COCF_3, and the ratio of ions as a function of generation in the optimization algorithm. All curves are normalized to unity at the initial generation for this plot.

To eliminate the possibility that the non-trivial control observed was influenced by the choice of functional groups in triflouroacetone, we attempted control on the photochemical reaction dynamics of the molecule acetophenone, $C_6H_5COCH_3$. The mass spectrum for acetophenone exposed to the transform limited pulse is presented in Figure 5A. Observable are peaks corresponding to laser-induced cleavage of the methyl and phenyl species from the parent molecule. A previous investigation (6) detailed control of the branching ratios for cleavage of the phenyl and methyl species from the acetone precursor. Control over the branching ratios of a factor of four was measured. The corresponding reference experiments suggested that the control fell into the nontrivial class, similar control could not be achieved by varying either the laser pulse duration or the laser energy.

In addition to expected peaks in the mass spectrum, a peak exists at m/z = 92 amu. This peak corresponds to toluene, $C_6H_5CH_3$, as confirmed by complementary investigations on the dueterated acetophenone $C_6H_5COCD_3$. The toluene is a product of field-induced molecular rearrangement of the precursor acetophenone molecule. To test whether control is possible we first attempted to minimize the toluene yield by maximizing the ratio $C_6H_5^+/C_6H_5CH_3^+$. This optimization goal seeks to enhance phenyl group cleavage representing dissociation in comparison to the rearrangement channel. This ratio of the averages is plotted as a function of generation number in Figure 5B. The yield reaches a maximum of a factor of 1.9 at generation 15. We next attempted to maximize the rearrangement of the toluene product in preference to the phenyl dissociation product, thereby maximizing the ratio $C_6H_5CH_3^+/C_6H_5^+$. The

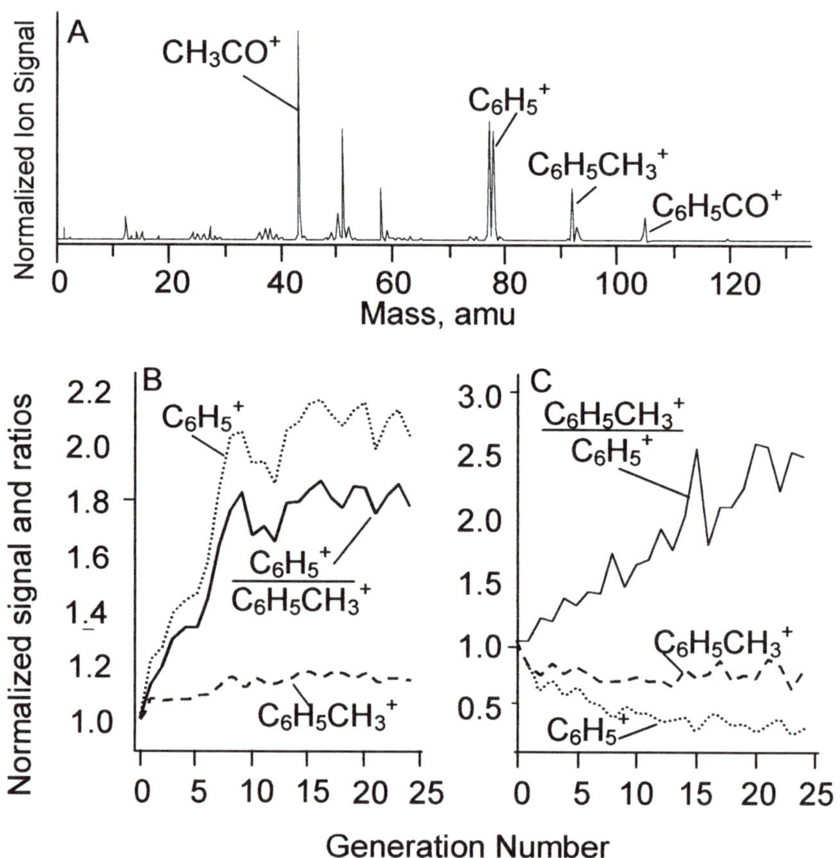

Figure 5. A, The strong-field mass spectrum for acetophenone; B, the learning curve and ion yields when minimization of the toluene:phenyl yield is specified; C, the learning curve when maximization of the toluene:phenyl yield is specified.

method manages to increase the ratio to 2.5 by holding the $C_6H_5CH_3^+$ constant and suppressing the $C_6H_5^+$ peak. This optimization as a function of generation number is shown in Figure 5C. In both control optimizations, the toluene ($C_6H_5CH_3^+$) signal remains approximately constant while the algorithm adjusts the phenyl ion ($C_6H_5^+$) yield. Clearly both pathways compete with each other starting from an excited parent molecule. However, the genetic algorithm can control the relative abundance of each pathway. The tailored pulses used in

Figure 5B and C can change the $C_6H_5^+/C_6H_5CH_3^+$ ratio over a dynamic range of ~5. In the case of trivial control, it is unlikely that one ion yield would remain constant while another would be free to fluctuate in intensity. Finally, absolute ion yield is seen to increase monotonically with decreasing pulse intensity or increasing pulse duration. Such correlation is not evident in the results of the control experiment shown in Figure 5.

Discussion

The results shown in Figure 2 demonstrate that control of the ion intensities in the mass spectrum of paranitroanaline can be accomplished using intensity and pulse duration variation in a transform limited pulse of high intensity ($\sim 10^{13}$ Wcm^{-2}). The degree of control is substantial and is indicative of photoinduced nuclear excitation of the molecule under the given excitation conditions. The pulse shaping experiments described in this work (see Figures 3 4 and 5) and in previous investigations (4-5) strongly suggest that control over the photo-induced dissociation, and now rearrangement of molecules, can be accomplished using tailored laser pulses in the strong-field regime. All of the control systems studied previously (from simple liquids, to inorganics to and array of organics) were excited using short duration laser pulses of with wavelengths centered in the region of 800nm. Since none of the systems have even vaguely similar absorption profiles, one may ask the question as to what the control mechanism is? The solution to this question, we propose, is a combination of strong-field excitation processes. In the strong-field regime multiphoton absorption of up to 50 photons may occur. In addition, field-induced shifts of molecular eigenstates (13) allows a greater degree of control over reaction dynamics. Finally, with the intense excitation scheme there is also a lifetime broadening mechanism that increases the "bandwidth" of the overal excitation scheme. The wide range of bond selective chemistry displayed using intense, near-infrared radiation of femtosecond duration with the rather narrow bandwidth (~0.1eV) is not surprising in light of the highly non-linear multiphoton excitation coupled with the marked effects of the electric field of the radiation on the energy levels of the molecule.

To understand the nature of strong-field excitation for control purposes it is useful to realize that field-induced electronic motion during the pulse essentially determines the nuclear motion after the pulse. This is because the laser pulse has short duration, on the order of a picosecond or less, while nuclear resulting in dissociation or rearrangement of large molecules typically requires longer time scales. All of the electronic states available in a polyatomic molecule form the basis set in which an arbitrary 3-dimensional motion of electrons inside the potential energy surface can be expressed (14). To be able to shape the 3-dimensional motion of the electrons inside electrostatic surface during a laser pulse in a specific way, a large number of electronically excited states must be excited and combined with a definite phase relationship.

A significant amount of experimental data (16-20) suggests that polyatomic molecules subjected to a dynamic strong-field can undergo multiple electronic excitations. The strong-field multi-electron excitations are the initial stage of highly inelastic (diabatic) laser/molecule coupling which may result in prolific nuclear motion, extensive fragmentation and large kinetic energy release of the fragment ions. It is encouraging that a large number of coupling channels can be accessed simultaneously in the strong-field regime as shown, as shown for instance in Figures 2 and 5.

Studying the coupling of molecules with unshaped strong-field laser pulses may reveal some rules of thumb for control experiments in the strong-field regime. Of particular interest to control experiments is the total amount of energy deposited during the laser/molecule interaction and the subsequent partitioning of the deposited energy among product channels. Recent data on kinetic energy release of H^+ fragment ions resulting from dissociative ionization of polyaromatic hydrocarbons demonstrate that H^+ ions with energies up to 60 eV are formed (17). This implies that up to 40 photons may be deposited in a single dissociative coupling channel. The large number of photons available for excitation is presumably responsible for the fact that a wide variety of molecules have been excited, ionized and detected using femtosecond duration radiation centered at 800 nm. Such highly nonlinear excitation must also play an enabling role in the mechanism of strong-field control. Unlike linear excitation schemes, strong-field excitation may employ a wide array of excited states to construct the most appropriate wave function for the desired final product states. Thus, the fact that a series molecules have (or do not have) similar absorption spectra has little to do with the degree of intense laser-molecule coupling.

Further evidence that a variety of coupling channels may be accessed simultaneously in the strong-field regime is found in photoelectron kinetic energy measurements of polyatomic molecules (15). The dynamics that determine the final product distribution are controlled by the mechanisms of energy coupling and subsequent partitioning. Despite the wealth of experimental (16, 17, 18, 19, 20) and computational studies (21, 22, 23, 24) probing strong-field excitation of molecules, the topic is still far from well-understood.

Insight into the interaction of intense lasers on atomic and molecular energy levels has been gained by investigating electron dynamics, both theoretically and experimentally (25). The strong-field photoelectron spectroscopy of acetylene (18), for instance, revealed the presence of above threshold ionization. This process involves signatures in the photoelectron spectrum resulting from, in this case, up to eight photons being absorbed in addition to that required to overcome the ionization potential. Such measurements demonstrate that even for a small molecules such as acetylene, at least 15 photons may be involved in the strong-field excitation event. The assignment of the features observed in these measurements revealed that the excited states of the molecule shifted up to 5 eV in the presence of the field. The myriad of strong-field phenomena revealed in photoelectron spectra underpin the wide range of coupling mechanisms which are central to exerting non-trivial control over chemical reactivity.

Further effects of the strong radiation fields on the molecular Hamiltonian are highlighted in a study of the photoelectron spectroscopy (*26*) of the molecules benzene (C_6H_6), naphthalene ($C_{10}H_8$) and anthracene ($C_{14}H_{10}$) at the single laser intensity, 4.0×10^{13} W·cm^{-2}. The measurements revealed an apparent decrease in the resolution of the photoelectron peaks with increasing molecular size. The smallest molecule, benzene, displayed a discrete spectrum of photoelectron peaks. Naphthalene displayed a series of discrete photoelectron peaks superimposed upon a broad, featureless distribution of electron kinetic energies. Anthracene, the largest molecule displayed an unstructured photoelectron kinetic energy spectrum. The modal and maximum kinetic energies recorded were seen to increase with molecular size. The data was interpreted in light of a *structure-based* model (16) that considers aspects of the electrostatic potential energy surface and was used to infer an evolution from a regime dominated by multiphoton ionization (MPI) for benzene to a regime dominated by field-ionization for anthracene. Such results clearly demonstrate that the degree of coupling increases in the strong-field limit as the characteristic length of the molecule increases.

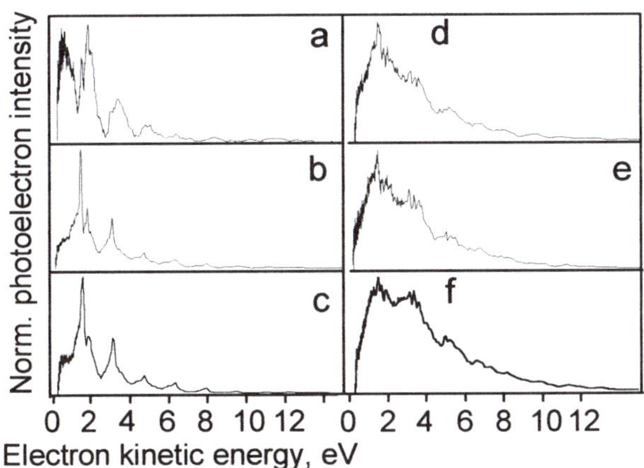

Figure 6. The photo- electron spectra for acetylene excited using 130 fs duration pulses of 800nm radiation. The intensity of the radiation is: a, 5.7 $\times 10^{13}$ W cm^{-2} ;b, 6.9 $\times 10^{13}$ W cm^{-2}; c, 8.6$\times 10^{13}$ W cm^{-2}; d, 10^{13} W cm^{-2}; e, 1.1$\times 10^{13}$ W cm^{-2}; f, 1.4 $\times 10^{13}$ W cm^{-2}.

The loss of discrete features for larger characteristic lengths in the benzene, naphthalene, anthracene series is proposed to be due to a lifetime broadening mechanism. To test this hypothesis the photoelectron spectra for naphthalene were measured for a range of laser intensities (13). In the context of the structure-based, tunnel ionization model as the intensity is increased the uncertainty in state lifetime should decrease. The measured spectra evolved from a series of assignable peaks at the lower intensities to a broad, featureless distribution of electron kinetic energies at higher intensity. The modal and maximum kinetic energies recorded were noted to increase as the laser intensity increased. Similar trends have been observed for benzene (15) and anthracene suggesting the existence of a generic field-induced mechanism for broadening of molecular eigenstates.

Figure 6 presents the strong-field photoelectron spectra of acetylene, C_2H_2 (panels a-f) and benzene. The trend observed is qualitatively similar to that of the naphthalene data reported previously (13). The spectra evolve from a structured photoelectron spectrum at the lower intensities toward a broader, less-featured distribution of kinetic energies as the laser intensity is increased. This suggests that even a small molecule, such as acetylene may undergo the lifetime broadening mechanism.

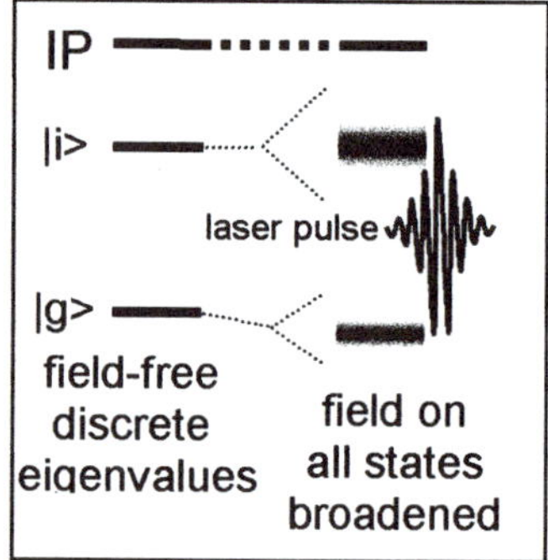

Figure 7. A schematic of the field-induced broadening that contributes to the effective bandwidth of the laser. Also shown is the field-induced shifting of the ground state to lower energy as a result of the intense laser pulse.

The structure-based model supports the inherently intuitive picture that field-ionization becomes easier as the laser intensity increases. The increasing rate of field ionization is key to understanding the loss of structure in the photoelectron spectra as the laser intensity is increased. If field-ionization is enhanced at higher laser intensity, eigenstate population is likely to decrease accordingly. Thus field ionization and any other events that cause eigenstate population decay will shorten the lifetime of such eigenstates. As the lifetime of a state is shortened, the uncertainty in the lifetime is also shortened. From the Heisenberg uncertainty principle ($\Delta t \cdot \Delta E > 1$, in atomic units), it is clear that the eigenvalue uncertainty is equal to or greater than the inverse of the uncertainty in the eigenstate lifetime. If we assume that a state can survive no longer than the time required to field ionize, we can equate the uncertainty in state lifetime with the field-ionization lifetime. This is because the state lifetime should have an upper bound of the field-ionization lifetime. Thus a broadening of the photoelectron peaks with increasing laser intensity is anticipated because the ionization probability increases with laser intensity. The corresponding broadening of intermediate states is illustrated in Figure 7. Calculation of the ionization probability for the molecule naphthalene at an intensity near 5×10^{13} W cm^{-2} suggests that the ionization rates will result in lifetime broadening on the order of 5 eV. The broadening calculated increased with the laser intensity and was on the same order of magnitude to that measured experimentally.

The field-induced broadening mechanism has been demonstrated in various molecules and seems to be a generic phenomenon in strong-field excitation process. While, field-induced eigenstate broadening mechanism might be viewed as a limitation for spectroscopic investigations, for strong-field optical control, the broadening mechanism might be viewed favorably as adding additional bandwidth to the nominally narrow bandwidth femtosecond laser. Such eigenstate broadening in combination with nonlinear (multiphoton) absorption in the strong-field regime should work in tandem to provide diverse coupling schemes. Non-trivial strong-field control can only be exerted if the full dynamic capabilities of a molecule are employed. The strong-field, time-dependent laser pulse appears to have the ability to create resonances as needed, and therefore circumvents the spectral limitations present in virtually all weak field excitation schemes.

Conclusions

We have shown control of reaction dynamics for p-nitrotoluene using the variables of intensity and pulse duration and in a series of ketones using tailored strong-field pulses under closed-loop control. We have presented data that shows control in both trivial and non-trivial schemes. The observations are discussed within the context of a multiphoton picture that includes transient shifting of intermediate states as well as lifetime broadening. Within these strong-field processes the range of coupling and partitioning channels is large and it is anticipated that laser-induced control of nuclear motion in the strong-field regime will be a viable tool for guiding chemical reactivity.

Acknowledgements

The authors would like to acknowledge the support of the Office of Naval Research and the National Science Foundation for this work.

References

1. DeWitt, M. J.; Levis, R.J. *J. Chem. Phys.* **1995**, *102,* 8670.

2. Judson, J.S.; Rabitz, H. R. *Phys. Rev. Lett.* **1992**, *68,* 1500.

3. Rabitz, H.R. *et. al. Science.* **2000**, *288,* 824.

4. Assion, A, *et. al. Science* **1998**, *282,* 919.

5. Daniel, C. *et. al. Chem. Phys. 2001 in press.*

6. Levis, R. J.; Menkir, G. M.; Rabitz, H. R. *Science* **2001**, *292,* 709.

7. Weinacht, T. C.; Ahn, J.; Bucksbaum, P. H. *Nature,* **1999**, *397,* 233.

8. Bartels, R.; Backus, S.; Zeek, E.; Murnane, M. M.; Kaptyen, H. C. *Nature* **2000**, *406,* 164.

9. Weiner A. M.; Leaird, D. E.; Patel, J. S.; Wullert, J. R. *IEEE J. Quant. Electron.* **1992**, *28,* 908.

10. Sunderman, E.; Rabitz, H. R.; De Vivie-Riedle, R. *Phys. Rev. A.* **2000**, 6201

11. *Genetic Algorithms in Search optimization and Machine Learning,* Goldberg, D. Addison Wesley, Reading, Mass, 1989.

12. Demiralp, M.; Rabitz, H. R.; *Phys. Rev. A.* **1993**, *47,* 809.

13. Moore, N. P., Levis, R. J., *J. Chem. Phys. Submitted.*

14. Mukamel S.; Tretiak S.; Wagersreiter T.; Chernyak V. *Science* **1997**, *277,* 781.

15. Moore, N. P.; Markevitch, A. N.; Levis, R. J. *in prep.*

16. DeWitt, M. J.; Levis, R. J., *J. Phys .Chem. A* **1999**, *103,* 6493.

17. Markevitch, A. N.; Moore, N. P.; Levis, R. J. *Chem. Phys.***2001**, *267,* 131.

18. Moore, N. P.; Levis, R. J. *J. Chem. Phys.* **2000**, *112,* 1316.

19. Moore, N. P.; Levis, R. J. *J. Chem. Phys. submitted*

20. Lezius, M.; Blanchet, V.; Rayner, D. M.; Villeneuve, D. M.; Stolow, A.; Ivanov, M. Yu. *Phys. Rev. Lett.* **2000**, *86,* 51.

21. Corkum, P. B.; Burnett, N. H.; Brunel, F. *Phys. Rev. Lett.* **1989**, *62,* 1259.

22. Corkum, P. B. *Phys. Rev. Lett.* **1993**, *71,* 1993.

23. Walker, B.; Sheehy, B.; Kulander, K. C.; DiMauro, L. F. *Phys. Rev. Lett.* **1996**, *77,* 5031.

24. Hankin, S. M.; Villeneuve, D. M.; Corkum, P. B.; Rayner, D. M. *Phys. Rev. Lett.* **2000**, *84,* 5082.

25. Blanchet, V. *et al. Faraday Disscussions,* **2000**, *115,* 33.

26. DeWitt, M.J. and Levis, R. J., Phys. Rev. Lett. **1998**, *81,* 5101.

Chapter 15

Laser-Phase Control of Dissociative Ionization of Molecules: Exact Non-Born–Oppenheimer Simulations for $H_2{}^+$

A. D. Bandrauk[1], J. Levesque, and S. Chelkowski

Laboratoire de Chimie Théorique, Faculté des Sciences, Université de Sherbrooke, Québec J1K 2R1, Canada
[1]Visiting professor: Department of Physics, University of California, Santa Barbara, CA 93106

Exact non-Born-Oppenheimer numerical solutions of the time-dependent Schroedinger equation for a 1D H_2^+ molecule in an intense two-color $(\omega + 2\omega)$ laser field have been obtained in order to clarify and identify the imporrtant mechanisms for pulse control of electron-nuclear dynamics. This benchmark simulation allows for a study of unexpected asymmetries in electron ionization and proton dissociation. A quantum regime is identified where electron quasistatic tunnelling is shown to be the dominant contribution to asymmetric and non-classical ionization in competition with proton dissociation. It is further shown that the quasistatic model of electron tunneling ionization and above barrier proton dissociation is a very useful concept in understanding intense field dissociative ionization and the concomitant counterintuitive (i.e. non-classical) behavior of electrons and nuclei in intense short laser pulses.

Laser control of nuclear motion in molecules is a growing area of research due to the ever improving laser technology making available laser pulses with variable and controlable amplitude, intensity and phase (1). Thus various laser coherent control schemes have been proposed to control photochemical products in chemical reactions by coherent superpositions of electronic or (and) nuclear states (2). We have focused on using the simple superposition of a field of frequency ω and its second harmonic 2ω (i.e. $\omega + 2\omega$) coherent superposition of laser fields. Such a superposition creates a periodic but nonsymmetric electromagnetic field (Fig. 1(a)). Previous control scenarios, such as $\omega + 3\omega$ (2) or $2\omega + 4\omega$ (3) always create symmetric fields which allow for laser control of molecular processes by interference of symmetry conserving quantum pathways. The $\omega + 2\omega$ superposition on the other hand leads to interference between nonsymmetry conserving quantum pathways. This leads to control of angular distributions of atomic ionization processes (4-6), of molecular photodissociation (3), the directional control of photocurrents in quantum wells (7) and in semiconductors (8). In the molecular case, we have shown previously that by using $\omega + 3\omega$ or $\omega + 2\omega$ coherent superposition one can control ionization (9) and also enhance high order harmonic generation (10) in the simplest one-electron molecules H_2^+, H_3^{++} and recently the two-electron molecule H_2 (11). These results were obtained from exact 3-D or 1-D numerical solutions of the time-dependent Schroedinger equation, TDSE for static nuclei (i.e. in the Born-Oppenheimer approximation). Recently we have performed the first non-Born-Oppenheimer simulation of dissociative ionization for a 1-D model of H_2^+ (13-14) and have applied this numerical methodology for the TDSE to propose a new method of imaging nuclear wave functions (13-14) and to study laser phase directional control of dissociative ionization of H_2^+ (15-16).

Previous experimental attempts to control molecular dissociative ionization using a $\omega + 2\omega$ coherent superposition scheme were reported for H_2 and HD molecules (17-18). Strong asymmetries or anisotropies of positively charged nuclear fragments and ionized electrons were found. Unexpectedly, both positively charged nuclear fragments and negatively charged electrons were found to be preferentially emitted in the same direction (i.e. in a "counterintuitive", non-classical direction). Thus for the case of the most asymmetric combined electric field (see inset in Fig. 1(a), phase $\phi = 0$ or π), one would expect the positively charged nuclei to be preferentially ejected in the direction of maximum positive electric field, following precepts of classical mechanics, whereas negatively charged electrons would be accelerated in the opposite direction (i.e. downfield), contrary to what is observed experimentally. (16-18). In the experimental interpretation of this unusual asymmetry, it was assumed that electrons behave " normally" and the anomalous angular distribution was attributed to complex nuclear effects (i.e. to "abnormal" protons). We have succeeded in explaining these unusual experimental results in two ways (15-16). First by solving the complete dynamic, non-Born-Oppenheimer TDSE for a 1-D H_2^+, we

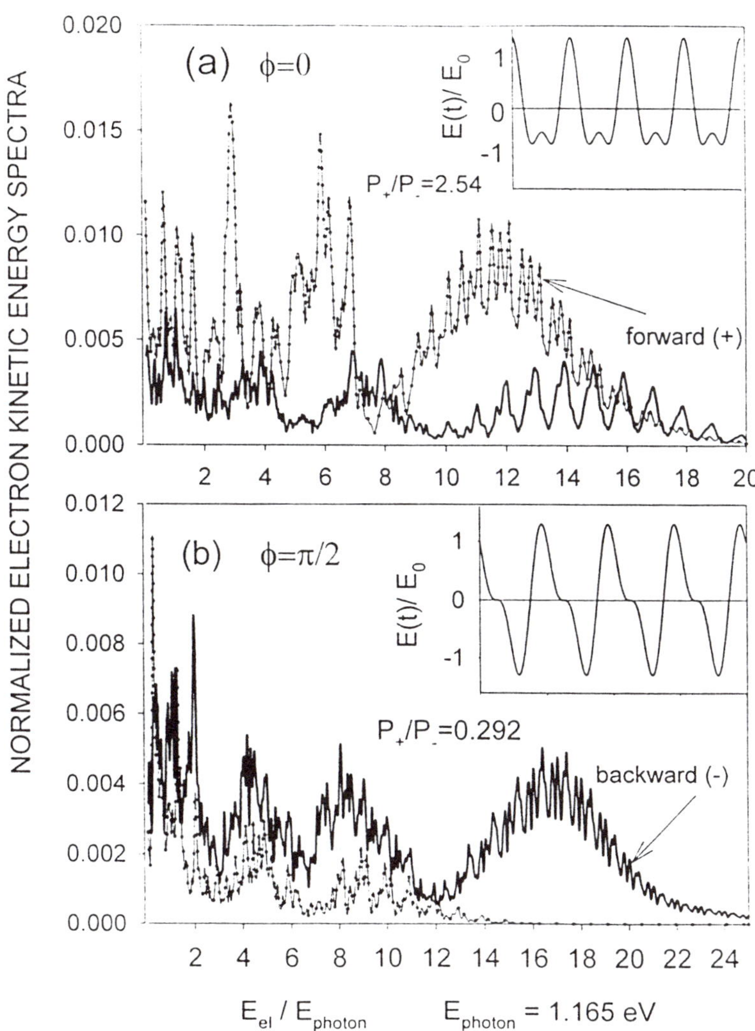

Figure 1: Electron kinetic energy, ATI, spectrum for H_2^+ in a laser field with frequencies ω ($\lambda = 1064nm$) and 2ω ($\lambda = 532nm$), relative amplitude f=0.5 and two different phases: a)$\phi = 0$, b)$\phi = \frac{\pi}{2}$ in equation (1). Energies are in photon numbers.

can obtaim the exact kinetic energy spectra of the electrons, called ATI (Above Treshold Ionization (*19*)) spectra and proton dissociation spectra, called ATD (Above Treshold Dissociation (*20*)) spectra. We have thus obtained the same kind of electron-proton correlated asymmetries as seen in experiment (i.e. we obtain from the numerical TDSE solutions of the exact electron-proton time-dependent wave-functions a preferential emission of electrons and protons in the same direction). Thus for a coherent laser field superposition described by

$$E(t) = E_0(t) \bullet (\cos(\omega t) + f \cos(2\omega t + \phi)) \quad , \tag{1}$$

where $E_0(t)$ is the field envelope, f the relative amplitude and ϕ the relative phase between the ω and 2ω fields, with $f = 0.5$ and $\phi = 0$ corresponding to the largest field asymmetry (see inset in Fig. 1(a)), we obtain the counterintuitive emission of electrons in H_2^+ (i.e. more electrons are ejected in the same direction of the protons, towards the positive field maximum). Secondly, as shown below, a quasistatic tunneling ionization model, successfully applied previously to explain intense field ionization and high-order harmonic generation in atoms (*21*)) and molecules (*22*)) can be used to interpret the electron ionization asymmetries.

The current non-Born-Oppenheimer numerical methodology, applied to dissociative ionization of H_2^+ allows us also to calculate the proton kinetic energy spectra. These are reported in Fig. 2. The integrated dissociation probabilities, P_+ for forward, exhibit intuitive (i.e. classical) asymmetries in the $\phi = 0$ case (Fig. 2(a)). For the phase difference $\phi = \frac{\pi}{2}$ between ω and 2ω fields where the net field is periodic symetric (see inset in Fig. 1(b)), the numerical results show now a *backward* asymmetry in both electron ATI (Fig. 1b) and proton ATD (Fig. 2b) kinetic energy spectra. Such anisotropy behaviour for a periodic symmetric electromagnetic field is clearly *counterintuitive* for both electrons and protons at high intensities. In the present chapter, we present detailed calculations and simple quasistatic models for both electron ionization and proton dissociation to elucidate these counterintuitive processes at high intensities. Any attemps to achieve simultaneous control of electron ionization and nuclear dissociation with intense short laser pulses (*23-24*) must confront the above unexpected results: counterintuitive (i.e. non-classical) behaviour of electrons and nuclei in such intense $(10^{14} \mathrm{W/cm^2} < I < 10^{15} \mathrm{W/cm^2})$, short $(t < 25\mathrm{fs})$ laser pulses. The present calculations are an attempt to elucidate these counterintuitive, non-classical effects in order to advance the new science of control and manipulation of molecules with intense short laser pulses.

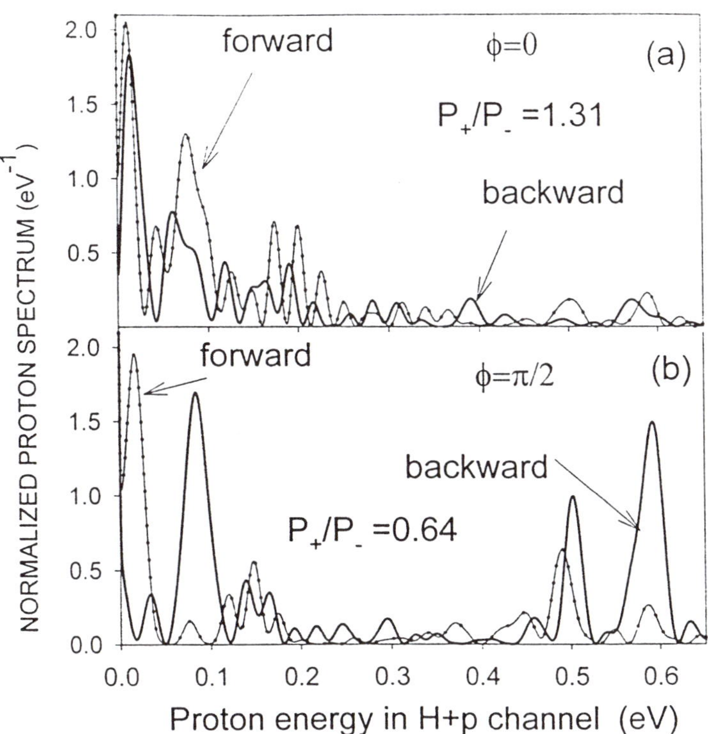

Figure 2: Corresponding proton kinetic energy, ATD, spectrum for Fig 1.

Theory and Model

We have solved previously the complete three body H_2^+ TDSE in 3-D (*25*), using absorbing boundary conditions for the high energy ionizing electrons whose kinetic energy is proportional to the ponderomotive energy U_p (1a.u. = 27.2eV) and intensity I(W/cm^2) at a given wavelenght λ(nm),

$$I = \frac{cE_0^2}{8\pi} \tag{2}$$

$$U_p(\text{a.u.}) = \frac{e^2 E_0^2}{4m\omega^2} = 3.4 \times 10^{-21} \; I(\text{W/cm}^2)\lambda^2(\text{nm}) \quad , \tag{3}$$

and the corresponding ponderomotive radius (1a.u. = 0.053nm)

$$\alpha_0(\text{a.u.}) = \frac{eE_0}{m\omega^2} = 2.4 \times 10^{-12} \; I^{\frac{1}{2}}(\text{W/cm}^2)\lambda^2(\text{nm}) \quad . \tag{4}$$

Thus at an intensity $I = 10^{14}$W/cm^2, $\lambda = 1064$nm, these field induced electronic parameters are $U_p = 0.4$a.u. (10.5eV) and $\alpha_0 = 27$a.u. (1.4nm). These high field parameters necessitate large numerical grids of size $\sim$ 1000a.u.. In order to study completely the ATI (electron) and ATD (proton) kinetic energy spectra we restrict ourselves to an exact non-Born-Oppenheimer 1-D model of H_2 which allows us to obtain complete ATI and ATD spectra with reasonable computational effort as described in (*12*). We thus solve the three-body TDSE for a 1-D H_2^+ molecule in a strong linearly polarised field along the electronic and internuclear, z and R axes respectively

$$\frac{i \; \partial\psi(z,R,t)}{\partial t} = H(z,R,t) \; \psi(z,R,t) \tag{5}$$

for simultaneous electron(z) and proton (R) motion. The numerical solution of the 1-D TDSE, equation (5), gives the exact non-Born-Oppenheimer electron-proton wavefunction $\psi(z,R,t)$. Projection onto exact solutions of a free electron in a laser field, called Volkov states (*12*), allows us to compute the complete electron ATI spectra (see Fig. 1). The asymmetric proton ATD spectra are calculated by integrating out the electron coordinate z after projecting on the atomic $1s$ orbitals localized at $z = \pm\frac{R}{2}$, thus giving two R-dependent functions

$$\psi_\pm(R,t) = \int_{-\infty}^{+\infty} \psi_{1s}(z \pm \frac{R}{2})\psi(z,R,t)\,\mathrm{d}z \quad . \tag{6}$$

The Fourier transform of the latter gives the total proton kinetic energy spectra, which at low energies (Fig. 2) corresponds to the ATD spectra whereas at higher energies one obtains CE (Coulomb Explosion) spectra due to CREI (Charge Resonance Enhanced Ionization) (*11,26*).

a) ATI spectra

As discussed in the introduction, perusal of the ATI spectra during the dissociative ionizaton of H_2^+ in a two-color (1064nm + 532nm) laser pulse with $I_0 = \frac{cE_0^2}{8\pi} = 4.4 \times 10^{13} \mathrm{W/cm^2}$ and relative amplitude $f = 0.5$ (equation 3) shows surprising forward at $\phi = 0$ and backward at $\phi = \frac{\pi}{2}$ anisotropies or asymmetries (Fig. 1). The intensity $I_0 = 4.4 \times 10^{+13} \mathrm{W/cm^2}$ was chosen so that the peak maximum intensity of the two overlapping color fields at phase $\phi = 0$ becomes $I = I_0(1 + f)^2 = 10^{14} \mathrm{W/cm^2}$.

We provide next a simple explanation of these anomalies using the quasistatic atomic tunnelling model illustrated in Fig. 3(a). At first the electron ionizes with a tunneling rate $\Gamma(\mathrm{s}^{-1})$ at time t_0 that depends on the instantaneous field $E(t)$ (*5,21*).

$$\Gamma(t_0) = \alpha^{\frac{5}{2}} \frac{4}{E(t_0)} \exp\left[-\frac{2}{3}\frac{\alpha^{\frac{3}{2}}}{E(t_0)}\right] \quad , \quad \alpha = \frac{I_{H_2^+}}{I_H} \quad , \tag{7}$$

where the I's in (7) are ionization potentials. In the second step the electron moves in the laser field (without the Coulomb attraction $V(z)$) as a classical particle starting at rest at the tunneling time t_0 and position $z = z_0$ (Fig. 3(a)) which is determined by the initial condition $V(z, t_0) = -I_p$, in the 1-D potential

$$V(z, t) = -\frac{1}{(z^2 + 1)^{\frac{1}{2}}} + zE(t) \quad . \tag{8}$$

We next solve the Newton equations of motion in the two-color laser field only, Eq. (1), with initial conditions appropriate for tunneling

$$v(t_0) = 0 \quad , \quad z = z(t_0) = z_0 \quad . \tag{9}$$

We thus get the following solutions for the electron position $z(t)$

$$z(t) = z_0 + z_{osc} + v_\alpha(t - t_0) \quad , \tag{10}$$

$$z_{osc} = \frac{E_0}{\omega^2}\left[\cos(\omega t) + \frac{f}{4}\cos(2\omega t + \phi) - \cos(\omega t_0) - \frac{f}{4}\cos(2\omega t_0 + \phi)\right] \quad . \tag{11}$$

The instantaneous electron velocity $v(t)$ is readily found to be (by integrating once the equations of motion),

$$v(t) = v_d(t_0) - \frac{E_0}{\omega}\left[\sin(\omega t) + \frac{f}{2}\sin(2\omega t + \phi)\right] \quad , \tag{12}$$

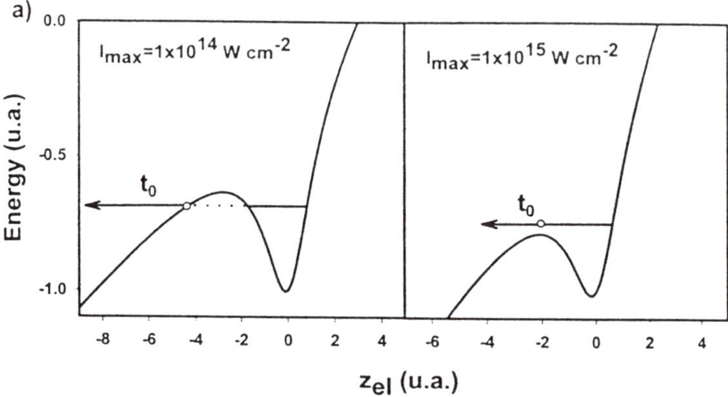

Figure 3(a): Quasistatic potential (1a.u.=27.2eV) for a 1D H atom at intensity $I = 10^{14} W cm^{-2}$ where underbarrier *ionization occurs at t_0 by tunneling and at intensity $I = 10^{15} W cm^{-2}$ where* overbarrier *ionization occurs at t_0.*

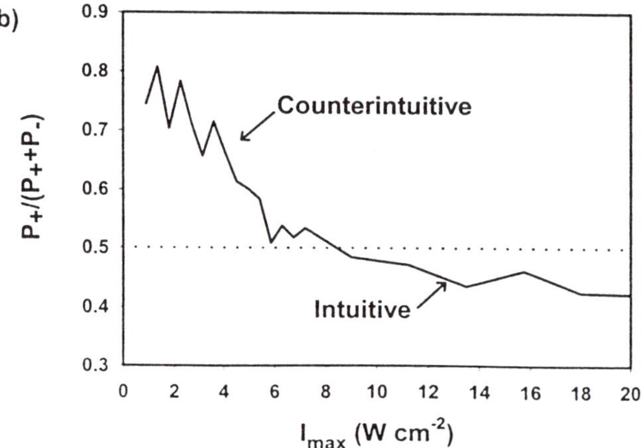

Figure 3(b): Forward (P_+) and backward (P_-) ionization probabilities and relative probability ($P_+/(P_+ + P_-)$), called asymmetry ratio, for the hydrogen atom in the same field as in Fig. 1(a). Counterintuitive corresponds to underbarrier tunneling whereas intuitive is the overbarrier classical regime. I_{max} corresponds to the maximum field intensity reached by the two-color field (for $\phi = 0$ and $f = 0.5$, $I_{max} = 2.25I_0$).

$$v_d(t_0) = \frac{E_0}{\omega}\left[\sin(\omega t_0) + \frac{f}{2}\sin(2\omega t_0 + \phi)\right] \quad . \tag{13}$$

$v_d(t_0)$ is called the "drift velocity" and is the result of applying the initial condition, equation (9), after the tunneling time t_0, (5,21). After the turn-off of the pulse, the oscillatory terms in equation (12) average out to zero so that the final electron velocity, as measured by experiment, coincides with the drift velocity $v_d(t_0)$. Similarly, neglecting oscillatory terms in equation (10) for long times, one obtains the measured electron position

$$z(t) = v_d(t_0)(t - t_0) + z_0 \quad . \tag{14}$$

Thus the two results, equations (13) and (14) determine in this simple one-electron model the measurable forward/backward electron asymmetry. In particular, equation (13) predicts a clear asymmetry for $\phi = \frac{\pi}{2}$, since in this case the last term in (13) becomes an even cosine function, so that $v_d(t_0) \neq v_d(-t_0)$. The symmetry about the field maximum (minimum) (Fig. 1(b)) for $\phi = \frac{\pi}{2}$ is thus broken. This can be attributed to the fact that the net field, equation (1), is not symmetric about each field maxima (minima) as seen in the inset of Fig. 1(b). As an example, around each field maxima, there is a sharp rise of the field to the left, and a slow descent into a kink. Since ionization is expected to occur mainly at the field maxima (minima) according to the tunneling expression (7), the asymmetry in the ionization can thus be attributed to the local asymmetry of the electric field around its maxima (minima), without assistance from the permanent Coulomb fields of the nuclei (protons).

The case of the phase $\phi = 0$ is more problematic. The electron field for relative amplitude rate $f = 0.5$ is twice larger in the positive field direction (see inset, Fig. 1(a)). Therefore one expects tunneling to occur, from equation (8), predominantly at one side of the nucleus, i.e. at negative z_0 (see Fig. 3(a)). However, taking field maxima to occur at $t_0 = 0$, one now notes from equation (13) that the drift velocity is antisymmetric with time, $v_d(t_0) = -v_d(-t_0)$. Hence, in this $\phi = 0$ case, neglecting the Coulomb attraction of protons should result theoretically in symmetric electron ionization distributions. The above result indicates that in this case Coulomb refocusing is essential to explain the experimentally observed (17-18) and the exactly calculated asymmetry (Fig. 1(a)).

A simple explanation of this asymmetry at $\phi = 0$ can be further gleaned from Fig. 3(a) where we show the effective static potentials in an H atom for two field intensities: i) $I < 4 \times 10^{14}\text{W}/\text{cm}^2$ called the *underbarrier* tunneling regime, and ii) $I > 4 \times 10^{14}\text{W}/\text{cm}^2$, the *overbarrier* ionization regime. In the first, weak field case, the tunneling electron has zero velocity as it leaves the atom, $v_d(t_0) = 0$ so that it will be under maximum influence of the atom Coulomb field and the electron field E which will decrease with time and eventually bring the electron back after half a cycle (Fig. 3(a)). In the strong field overbarrier limit, the electron has sufficient energy to completely overcome the concerted refocusing effect of the Coulomb and electric fields, equation (8), Fig. 3(a). The standard tunneling model in fact predicts a decrease of phase sensitivity and response of tunneling ionization with increasing laser intensity (17). This is corroborated by the calculated ATI asymmetry for the H-atom illustrated in Fig. 3(b). Thus for $I < 4 \times 10^{14} \text{W/cm}^2$, the asymmetry as measured by the rates of the integrated forward, P_+, and backward P_- probabilities, is large (i.e. counterintuitive) around $I \simeq 10^{14} \text{W/cm}^2$, and becomes intuitive (classical) for $I > 4 \times 10^{14} \text{W/cm}^2$. The separation between the counterintuitive (*underbarrier*) tunneling regime and the intuitive (*overbarrier*) classical regime occurs at the critical intensity I_0 where the ionization process occurs when the initial atomic $1s$ state becomes degenerate with the maximum of the total potential, equation(8). This gives the result that $I_c = \frac{(I_p)^4}{16}$ and since the ionization potential $I_p = 0.67 \text{a.u.}$ for a 1-D H atom, one obtains readily $I_c = 0.0125 \text{a.u.} = 4 \times 10^{14} \text{W/cm}^2$, in agreement with Fig. 3(b).

b)ATD spectra

Finally, we comment on the proton kinetic energy, or ATD spectra illustrated in Fig. 2. The asymmetry in the $\phi = 0$ case is as expected mainly in the forward direction, showing that protons behave classically contrary to previous interpretations (17-18). The intuitive asymmetry is nevertheless less for the protons (Fig. 4(a)) than the counterintuitive electronic result, Fig. 1(a). This is evidently related to the greater quantum nature of the electron. However for the case $\phi = \frac{\pi}{2}$, Fig. 4(b), one finds an unexpected asymmetry for the proton dissociation. Since from the field structure (inset in Fig. 1(b)) one has equal maxima-minima field strenghts, one would have expected a symmetric proton dissociation. It is to be observed that in both $\phi = 0$ and $\frac{\pi}{2}$ cases, the electron and proton asymmetries are in the same direction, with the electron asymmetry about twice larger than the proton asymmetry. Two factors must be operative: a)Coulomb interaction between electrons and protons, or (and) b) the local asymmetry of the net field amplitude around each maximum-minimum of these fields as discussed above for electrons.

In order to understand more the asymmetry in the calculated ATD spectra in presence of ionization, we illustrate in Fig. 4, the calculated ATD spectra for pure ATD, i.e. a two surface simulation, with the $^2\Sigma_g$ coupled radiatively to the $^2\Sigma_u$ state, *without* ionization. Comparing figures 4 and 2, one observes that the asymmetries calculated with ionization (Fig. 2) correspond better with the $v = 5$ results in the two surface calculations (Fig. 4). This last figure shows that strong asymmetries occur

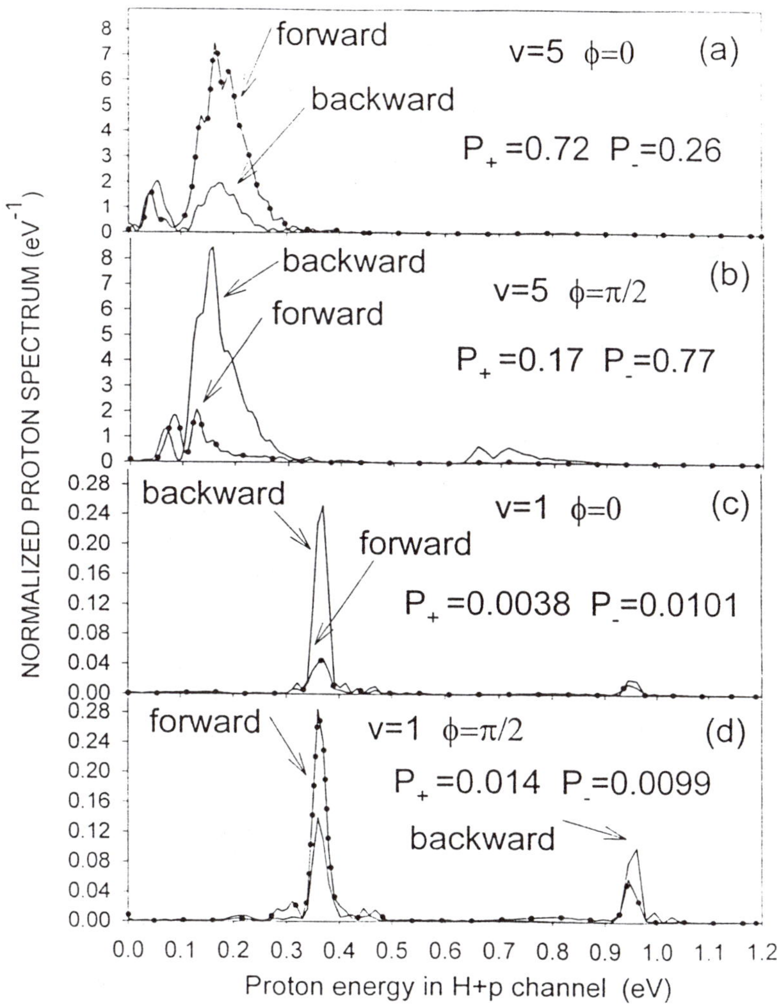

Figure 4: Proton kinetic energy (ATD) spectra in the dissociative H+p channel calculated using the two-surface model. the same laser parameters as in Figs. 2,3 ($\lambda = 1064 + 532$nm) were used but with different wave-packet initializations: (a),(b) initialization from v=5; (c),(d) initialization from v=1.

for the higher vibrational levels. This can be readily understood from the two-surface dressed staste representation of the $\omega + 2\omega$ photodissociation processes illustrated in Fig. 5. Thus the $v = 1$ state can dissociate to the $^2\Sigma_u$ potential after 3 photon absorption to the $|u, -3, 0\rangle$ repulsive state and interfere with the $^2\Sigma_g$ potential via the corresponding $|g, -1, -1\rangle$ dressed state (we use the notation $|\psi, n_1, n_2\rangle$ for the electronic ψ state with n_1 photons ω_1 (1064nm) and n_2 photons ω_2 (532nm)). The $v = 5$ state dissociation interferes via two isoenergy pathways with final states $|g, -2, 0\rangle$ and $|u, 0, -1\rangle$ corresponding to two 1064nm photons ($n_1 = 2, n_2 = 0$) and one 532nm photon ($n_1 = 0, n_2 = 1$) Furthermore since the latter one photon (532nm) dissociation of the $v = 5$ state is about 70 times larger than its two photon (1064nm) dissociation, the preponderance of the higher v levels to ATD is thus responsible for the enhanced backward-forward asymmetry at $\phi = 0$ (Fig. 2(a)). The same reasoning applies at $\phi = \pi/2$ since in Fig. 2(a), $P_+/P_- \simeq 0.2$ for $v = 5$ and 1.4 for $v = 1$, Fig. 4, without ionization.

Summarizing the comparison of Figs. 2 and 4 at $\phi = 0$ we conclude that the $v = 5$ ATD with its dominant forward asymmetry follows the classical intuitive model, whereas the $v = 1$ level ATD is counterintuitive, with the backward asymmetry dominant. Both $v = 5$ and $v = 1$ ATD asymmetries for H_2^+ are reduced by a factor of two in the presence of ionization which was simulated assuming a Franck-Condon distribution from an H_2, $v = 0$ initial molecule. Thus the dissociative ionization result, Fig. 2, represents the average over vibrational levels of H_2^+ excited from H_2. This interpretation over an average vibrational level distribution of H_2^+ also explains the results for the $\phi = \pi/2$ case. Thus in the dissociative-ionization case, Fig. 2(b), $P_+/P_- \simeq 0.6$ whereas it is 0.2 for $v = 5$ and 1 for $v = 1$ in the purely dissociative (no ionization) two-state calculation, Fig. 4.

For the $\phi = 0$ case, Posthumus et al. (27), have proposed previously that at the high intensities used here, the electron is always localized at the down-field nucleus, i.e. in the opposite direction of the field. For the $\phi = 0$ case, since the negative field is on for a longer time than the large positive field (see Fig 1(a)) the electron will be therefore localized in the opposite direction of the negative field most of the time thus forcing the proton to be ejected in the negative field direction, i.e. in the counterintuitive direction, in agreement with the purely dissociative $v = 1$ level (Fig. 4(c)), but not in agreement with the $v = 5$ level (Fig. 4(a)) nor the exact dissociative ionization result, Fig. 2(a). One can surmise that the $v = 1$ case is closer to a quasistatic picture since a larger number of photons are involved than the $v = 5$ level dissociation, so that the $v = 1$ behaviour follows the Posthumus quasistatic model whereas the $v = 5$ level falls more within the multiphoton regime (3).

The $\phi = \frac{\pi}{2}$ case is more problematic since the quasistatic model predicts

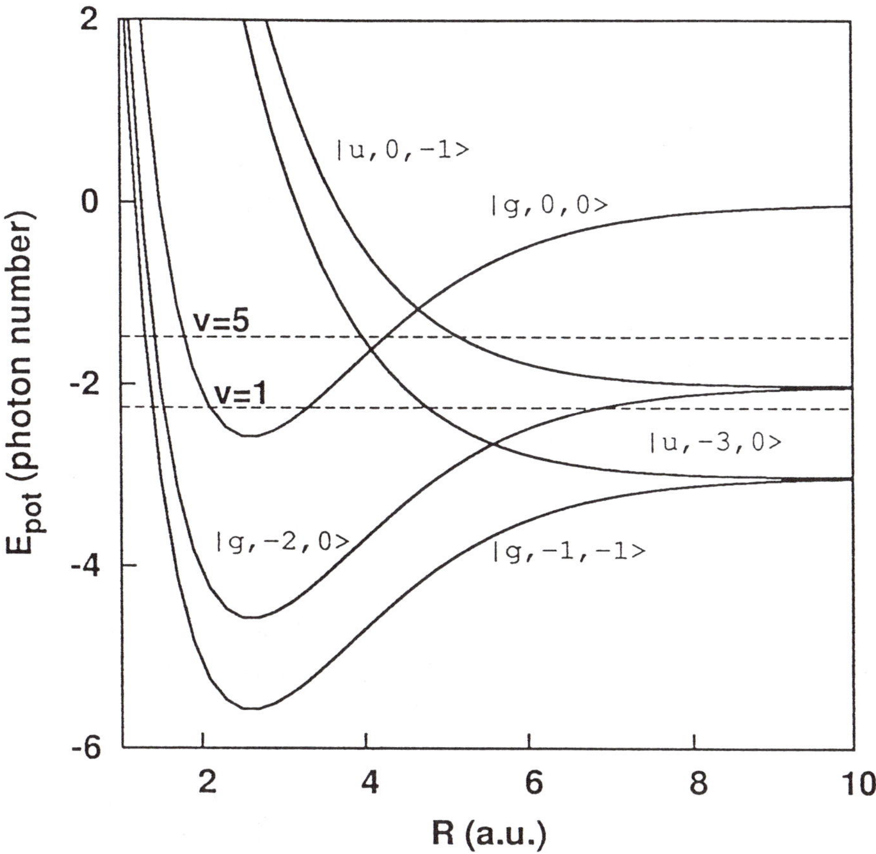

Figure 5: Dressed states for $\omega(1064\text{nm}) + 2\omega(532\text{nm})$ dissociation of $v = 1$ and $v = 5$ vibrational levels. $v = 1$ dissociates to two isoenergy states: $|g, -1, -1\rangle$ and $|u, -3, 0\rangle$ whereas $v = 5$ dissociates to the two isoenergy states $|u, 0, -1\rangle$ and $|g, -2, 0\rangle$.

a complete symmetric dissociation. The $v = 1$ level results (Fig.4(d)) agree with the quasistatic model but not the $v = 5$ case, (Fig. 4(b)), which shows a very large backward (5:1) asymmetry. The dissociative-ionization comparative result, Fig. 2(b), is a backward/forward asymmetry of 1.6, reflecting the average over low and high vibrational levels. One tempting explanation is the local asymmetry of the net field, illustrated in the inset of Fig. 1(b). The sharp rise from negative to positive amplitude of the electric field to the left of each maximum will inhibit an adiabatic following of the proton motion from the positive to negative amplitude to the right of field maxima. Thus protons will remain always longer under the influence of the negative field amplitude, thus being ejected predominantly backwards in this symmetric field, as obtained in the simulation, Fig. 2(b) and 4(b).

In conclusion, we have shown that at high intensities ($I > 1 \times 10^{14} \text{W/cm}^2$) where radiative processes are no longer perturbative the quasistatic model offers a simple explanation for the calculated asymetries in both electron and proton emission in $\omega + 2\omega$ fields.

Aknowledgments

We thank NSERC (Natural Sciences and Engineering Research of Canada) and CIPI (Canadian Institute for Photonic Innovations) for financial support of this research on laser control and manipulation of molecules.

References

1. T. Brabec,F. Krausz. *Rev. Mod. Phys.*, 72, 545, 2000.

2. M. Shapiro, P. Brumer. *Acc. Chem. Res.*, 22, 407, 1989.

3. E. Aubanel, A.D. Bandrauk. *Chem. Phys. Lett.*, 229, 169, 1994; *Chem. Phys.* , 194, 159, 1995.

4. N.B. Baranova, B.Y. Zeldovitch. *J. Opt. Soc. Am. B*, 8, 27, 1991.

5. D.W.Schumacher, P.H. Bucksbaum. *Phys. Rev. A*, 54, 4271, 1996.

6. K.J. Schafer, K.C. Kulander. *Phys. Rev. A*, 45, 8026, 1992.

7. E. Dupont, P.B. Corkum, H.C. Liu, M. Buchanan, Z.R. Wasilewski. *Phys. Rev. Lett.*, 74, 3596, 1995.

8. R. Atanasov, J.E. Sipe, H. Van Driel *Phys. Rev. Lett.*, 76, 1703, 1996.

9. T. Zuo, A.D. Bandrauk. *Phys. Rev. A*, <u>54</u>, 3254, 1996.

10. A.D. Bandrauk, S. Chelkowski, H. Yu, E. Constant. *Phys. Rev. A*, <u>56</u>, R2357, 1997.

11. A.D. Bandrauk, H. Yu. *Int. J. Mass Spectrom.*, <u>192</u>, 379, 1999.

12. S. Chelkowski, C. Foisy, A.D. Bandrauk. *Phys. Rev. A*, <u>57</u>, 1176, 1998.

13. S. Chelkowski, P.B. Corkum, A.D. Bandrauk *Phys. Rev. Lett.*, <u>82</u>, 3416, 1999.

14. A.D. Bandrauk, S. Chelkowski. *Chem. Phys. Lett.*, 2001.

15. A.D. Bandrauk, S. Chelkowski. *Phys. Rev. Lett.*, <u>84</u>, 3562, 2000.

16. S. Chelkowski, M. Zamojski, A.D. Bandrauk. *Phys. Rev. A*, <u>63</u>, 023409, 2001.

17. B.Sheehy, B. Walker, L.F. Di Mauro. *Phys. Rev. Lett.*, <u>74</u>, 4799, 1995.

18. M.R. Thompson, J. Posthumus, L.J. Frasinski *J. Phys. B*, <u>30</u>, 5755, 1997.

19. *Atoms in Intense Laser Fields.* M. Gavrila, ed., Academic Press, New-York, 1992.

20. *Molecules in Laser Fields.* A. D. Bandrauk, ed., M. Dekker Publ., New-York, 1994.

21. P.B. Corkum. *Phys. Rev. Lett.*, <u>71</u>, 1994, 1993.

22. M. Ivanov, P.B. Corkum, T.Zuo, A.D. Bandrauk. *Phys. Rev. Lett.*, <u>74</u>, 2933, 1995.

23. See contribution of G. Gerber *et al* to this volume.

24. See contribution of R. Lewis *et al* to this volume.

25. A. D. Bandrauk in. *The Physics of Electronic and Atomic Collisions.* Y. Itikawa *et al*, ed., AIP Conf. Proc., vol. 500, AIP, New-York, 2000, p.1020.

26. P.B. Corkum, P. Dietrich. *Comments At. Mol. Phys.*, <u>28</u>, 357, 1993.

27. J.H. Posthumus, M.R. Thompson, A.J. Giles, K. Codling. *Phys.Rev.*, <u>A54</u>, 955, 1996.

Chapter 16

Molecular Restructuring in Intense IR Laser Fields

T. Tung Nguyen-Dang[1], H. Abou-Rachid[1], N. A Nguyen[1],
N. Mireault[1], and O. Atabek[2]

[1]Départment de Chimie, Université Laval, Québec, Québec G1K 7P4, Canada
[2]Laboratoire de Photophysique Moléculaire (CNRS), Bât. 210,
Campus d'Orsay, Université de Paris-Sud, Orsay, Cedex, France

In an Infra-red (IR) laser field, synchronizing molecular wavepacket motions and the field oscillations can give interesting dynamical effects, suggesting new control schemes. We demonstrate that these effects are intrinsically due to a local time-asymmetry in the time-dependent molecular structure and force field impressed by the laser field on the molecule. We then show how this time-dependent molecular restructuring can be described generally by a simple LCAO scheme involving field-shifted atomic orbitals and show that a laser-induced time-asymmetric force-field is obtained for the water molecule prealigned and strongly driven by an intense IR laser pulse.

In an IR laser field, the nuclear dynamics within a molecule can be synchronized with the laser oscillations to produce effects that are reminiscent of a control scenario. This observation has been illustrated in a series of papers[1],[2],[3] where we showed how the dissociation of diatomic molecules, ranging from the simplest one, the H_2^+ molecular ion to a heavy, many electron molecule, such as HCl^+, can be either quenched or facilitated by synchronizing properly the birth of the initial nuclear wavepacket with the start of the laser's electric field oscillations. In the review to be found in the next section, we will show that this dynamical dissociation quenching, DDQ, effect is a reflection of a local time-asymmetry in the laser-induced time-dependent molecular force-field. This in turn reflects the molecular restructuring in the IR field, a result of the strong radiative interactions that also operates in complex molecules. In this sense, it is expected that effects similar to the DDQ effect in diatomic should also exist in polyatomic molecules and could be exploited in time-resolved control schemes involving energy flows among their many internal degrees of freedom.

Molecular restructuring, embodied in the changes in the MO structure induced by a IR laser field, has so far been analyzed in terms of laser-induced interactions between the field-free MOs[4]. A complementary and equivalent view of the molecular orbital restructuring process exists and may prove useful and more insightful when applied to polyatomic molecules. In this view, one considers that the MOs formed in the presence of the field result from the interaction of atomic orbitals (AO) of the constituent atoms already conditioned and shifted by the field. We will outline the principle of the new LCAO scheme, called LCfiAO scheme, using these field-shifted AOs in section 3, and show how the MO structure of H_2O is predicted to change under the action of a IR laser field polarized linearly along the C_2 symmetry axis of the molecule.

Time-asymmetric force field for molecular motions in an IR laser pulse.

The laser-induced Dynamical Dissociation Quenching (DDQ) effect described in our previous work[1]-[3] hinges on how the wavepacket in a diatomic molecular ion, such as H_2^+, HD^+, HCl^+, present themselves with respect to the time-dependent potential barrier to dissociation that the laser field is impressing upon the molecular force field. Let us concentrate for the moment on the H_2^+ molecule. We considered[1] wavepackets launched, on the ground state PES of the molecular ion, at time $t = 0$ of the field oscillations described by an electric-field function of the form $\vec{E}(t) = \varepsilon(t).\cos(\omega_L t - \delta)$, where $\varepsilon(t)$ denotes a quasi-rectangular pulse shape of amplitude $E_0(\propto \sqrt{I})$, so that the situation with $\delta = 0$ corresponds to the initial wavepacket being prepared, by vertical promotion from the parent neutral molecule ground state for instance, while the field is at its maximum intensity. In contrast, for $\delta = \pi/2$, the initial state preparation occurs at the start of an optical cycle, i.e. at zero field intensity. The dynamical difference between these two situations resides basically in the timing of the arrival of the wavepacket in the so-called gap region, where the adiabatic potentials W_+ and W_-, defined by diagonalizing the instantaneous potential matrix $\underline{\hat{W}}(R,t)$:

$$W_{ij}(R,t) = V_i(R)\delta_{ij} - \mu_{ij}(R).\varepsilon(t).\cos(\omega_L t - \delta) \tag{1}$$

i.e.

$$\underline{C}(R,,t)\underline{\hat{W}}(R,t)\underline{C}^{-1}(R,t) = \begin{pmatrix} W_-(R,t) & 0 \\ 0 & W_+(R,t) \end{pmatrix} \tag{2}$$

are closest to each other and where a time-dependent barrier is exhibited on the lower potential W_-, Figure 1(a). Arrival at a receding barrier, i.e. closing gap, tends to prevent escape toward the asymptotic region; this leads to the dynamical dissociation quenching (DDQ) effect. Wavepacket arrival at an incoming barrier, or opening gap, permits the escape of an important part of this wavepacket towards the asymptotic, dissociative limit. This is the barrier lowering (and/or suppression) mechanism. In the more general case of the heteronuclear ion HD^+, the presence of a permanent dipole moment introduces an asymmetry in the adiabatic potentials $W_\pm$'s temporal dependence[4]. Thanks to this temporal asymmetry, the DDQ effect can, in this case, distinguish even the $\delta = 0$ and π cases, corresponding to different initial orientations of this molecule [2]. Figure 1(b) illustrates the dynamical effect of this asymmetry through a simple classical 'ballistic' model in which we imagine a particle launched at time $t = 0$ towards the gap region, and would exit this region only if its energy exceeds the barrier height encountered there, otherwise the particle is reflected back towards the repulsive wall from where it started. The figure shows the position of the potential barrier as a function of time along with the trajectory of the ballistic particle. The distinction between the cases $\delta = \pi$ and $\delta = 0$ is clearly seen in this simple picture. Figure 1(c) shows how the wavepacket simulation results support this interpretation. On the left panel, we reproduce the curves representing the time evolution of the total bound state population $P_{bound}(t) = \sum_v | < v|\psi(t) > |^2$ as obtained for HD^+ prepared at time $t = 0$ in a field of frequency $\omega_L = 1680\ cm^{-1}$, of intensity $I = 5 \times 10^{13}\ W/cm^2$ and of varying absolute phase δ. This is to be compared with the results for H_2^+ obtained in the same conditions (but for $\omega_L = 943\ cm^{-1}$).

That the DDQ effect in diatomic is in fact an effect of the time-asymmetry in the time-dependent force field can further be verified by considering the dynamics of the homonuclear HD^+ molecule in a two-color $(\omega + 2\omega)$ field of the form[5]

$$E(t) = \varepsilon(t).[\cos(\omega_L t - \delta) + 0.5\cos(2\omega_L t - 2\delta + \phi)] \tag{3}$$

Figure 2(a) illustrates how the combined two-color electric field oscillates in time for various values of the relative phase ϕ between the two waves,

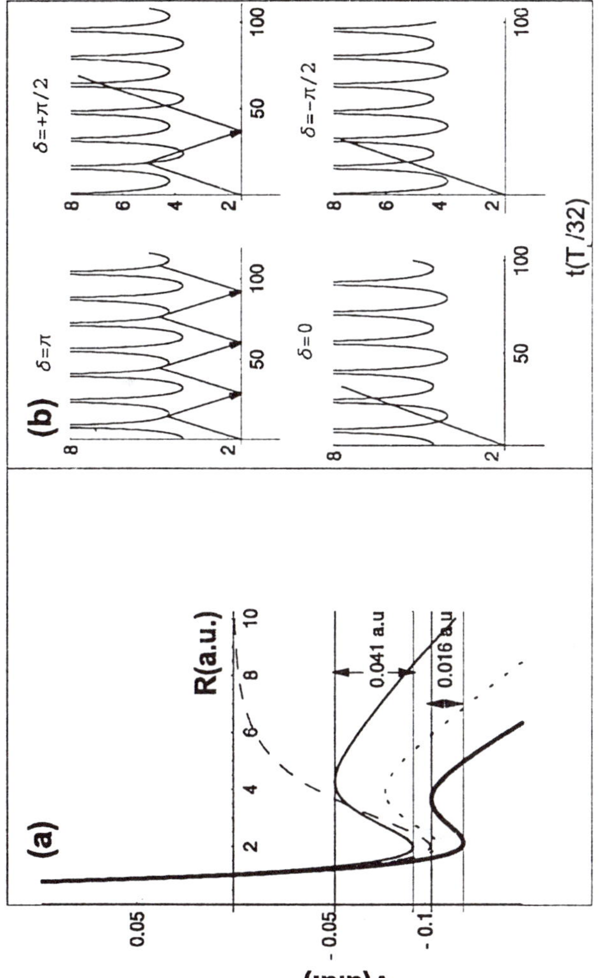

Figure 1(a): Time-dependent ground state potential energy curve W_- for HD^+ when its dipole moment is parallel to $\vec{E}$ (thin solid line), antiparallel to $\vec{E}$ (thick solid line) as compared to the situation in H_2^+ (dotted line) and to the field free potential (dashed line). The field intensity is $I = 5 \times 10^{13} \ W/cm^2$.

Figure 1(b): Straight-line trajectory of a ballistic particle (modeling HD^+), launched at time $t = 0$, in a field $\vec{E}(t) = E_0 cos(\omega_L t - \delta)$, $\delta = 0, \pi, +\pi/2, -\pi/2$, towards the moving barrier.

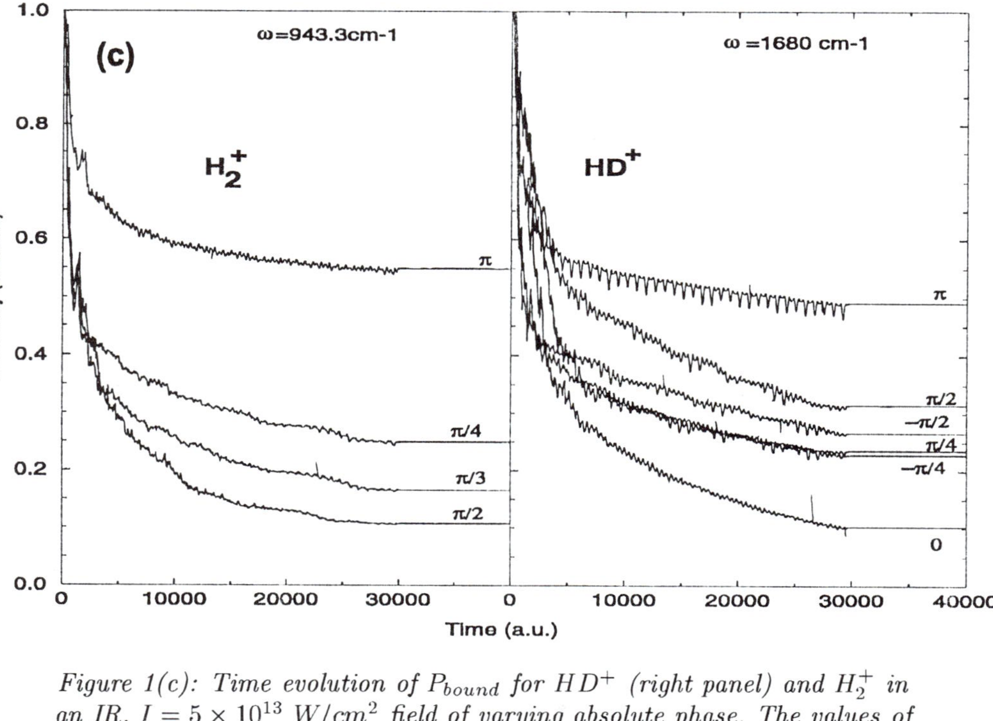

Figure 1(c): Time evolution of P_{bound} for HD^+ (right panel) and H_2^+ in an IR, $I = 5 \times 10^{13}$ W/cm^2 field of varying absolute phase, The values of the field frequency ω_L and absolute phase are indicated explicitly.

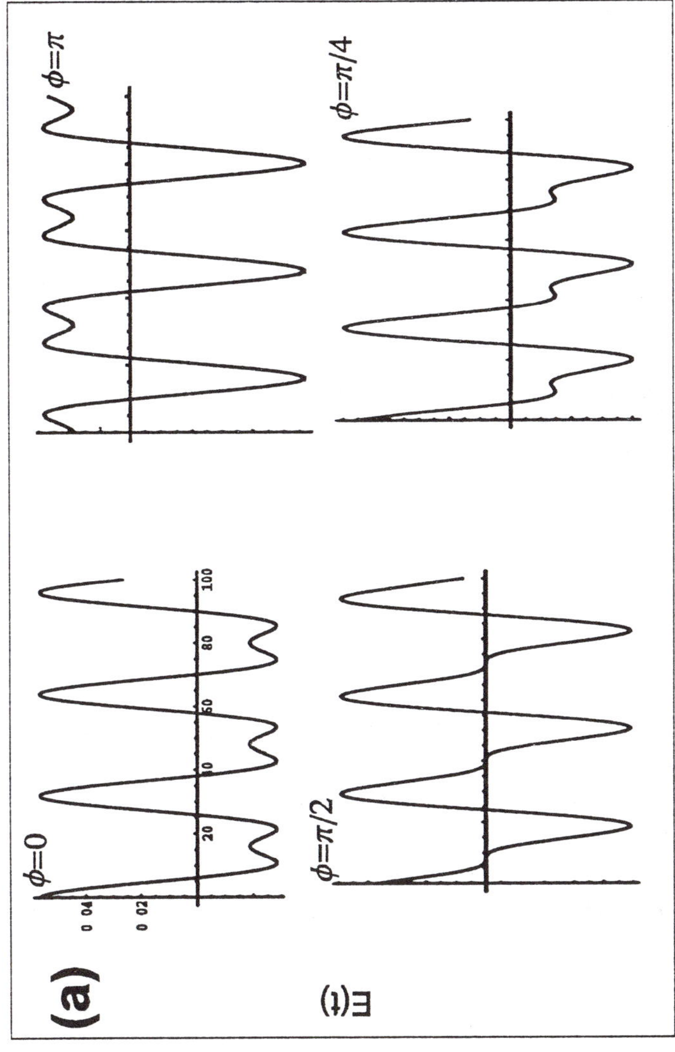

Figure 2(a): Time-variation of a field of the form given in eq.(3), with $\delta = 0$ and $\phi = 0, \pi/4, \pi/2, \pi$.

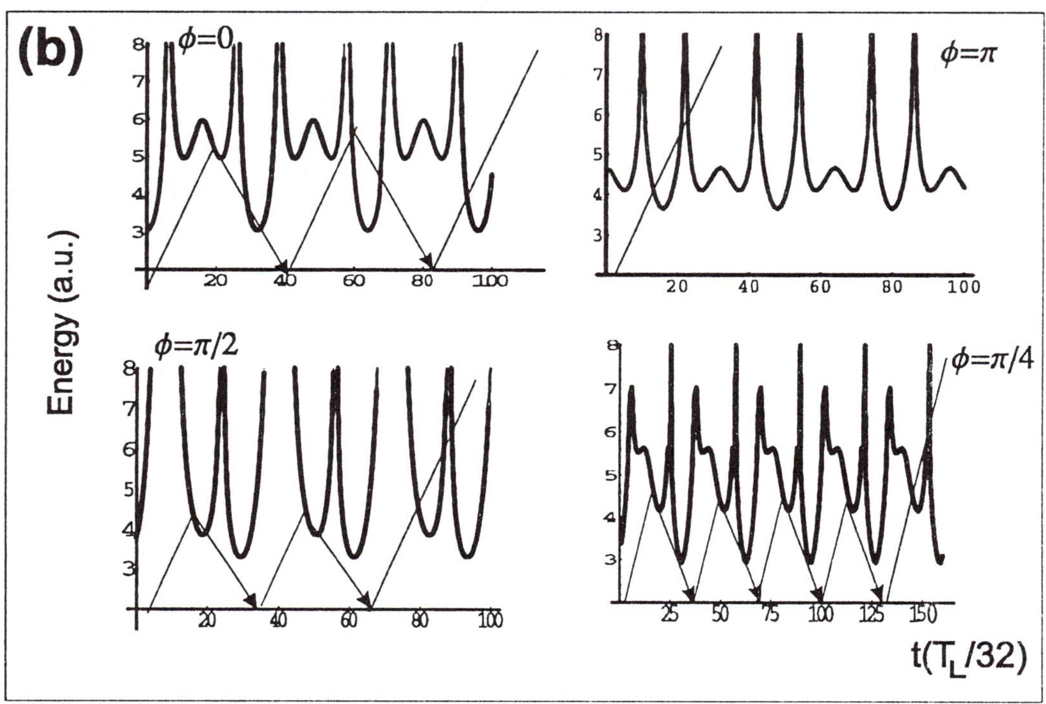

Figure 2(b): Straight-line trajectory of a ballistic particle, modeling HD^+ in the same field as in Figure 2(a), launched at time $t = 0$ towards the moving barrier the position of which is traced by thick solid lines.

and for a fixed absolute phase $\delta = 0$. In Figure 2(b), we show, as a function of time, for HD^+, (treated in the usual two-channel model with but parallel coupling, induced by the field of panel (a), between the lowest two σ orbitals), the position of the potential energy barrier found on the ground-state adiabatic surface $W_-(R,t)$. Like in Figure 1(c), we also show, in the various panels of Figure 2(b), the trajectory of a ballistic particle representing the wavepacket launched on the $W_-(R,t)$ at $t = 0$.

One dimensional wavepacket calculations have been performed for HD^+ under the two-color field with the relative phase ϕ taking on the values illustrated in Figure 2(a). The same methodology and initial conditions as described in our previous work are employed. The results of these calculations are shown in Figure 3(a), for a field frequency $\omega_L = 1680 \ cm^{-1}$, and peak intensity $I = 5 \times 10^{13} \ W/cm^2$. These are to be compared with the results shown in Figure 1(c) for H_2^+ and HD^+ under a one-color excitation. Here, the control parameter is not the field's absolute phase, but the dephasing between the second harmonic and the fundamental waves. We also give, in Figure 3(b), the results obtained for the same two-color fields as in panel (a), but with an absolute phase $\delta = \pi$. Our results demonstrate clearly that roughly the same selectivity (about 50 % of difference) is achieved between $\phi = 0$ and $\phi = \pi$, in this two-color time-asymmetric excitation of HD^+, as found previously between $\delta = 0$ and $\delta = \pi$ in the one-color excitation of the same molecule. This is true for the two choices of the absolute phase illustrated in Figure 3(a) and (b), although in the fine details of the time-resolved dynamics, significant differences exist between these two cases.

The selectivity of the DDQ effect is up to this point limited to about 50 %. It can certainly be improved if we search to optimize the initial stage of the scheme by a tracking control procedure[6], properly adapted to the specific objective we have in mind (initial drop of $P_{bound}(t)$ for instance), and restricted to the type of field such as the two-color field above. Works in this direction are in progress in our laboratory.

Molecular dynamical restructuring in an IR laser field

Principle

In current views of laser-induced molecular restructuring, the instantaneous electronic dipole interaction potential

$$V_{int}^{EF}(\{\vec{r}_i\}) = -e \sum_i \vec{r}_i \cdot \vec{E}(t) \tag{4}$$

is thought to cause mixing of field-free MOs, $\{\phi_k\}$ themselves constituted of field-free AOs, $\{\chi_j\}$, i.e. $\phi_k(1) = \sum_j c_{kj}\chi_j(1)$, yielding new, field-induced,

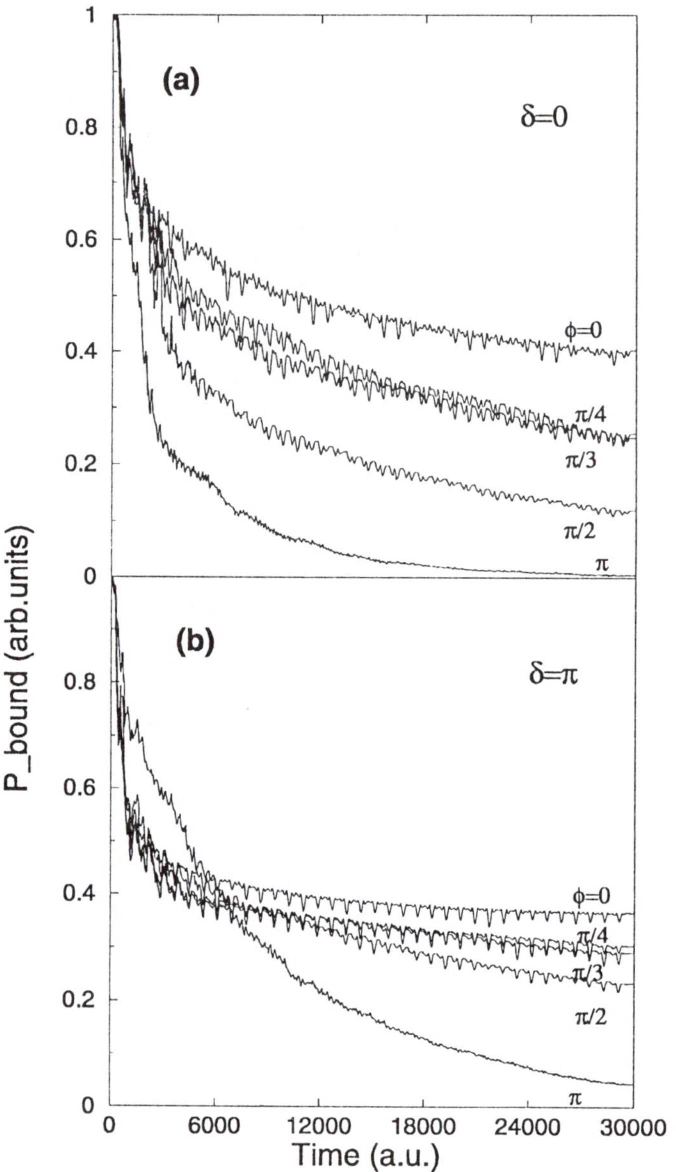

Figure 3: Time evolution of P_{bound} for HD^+ in an IR, $I = 5 \times 10^{13}$ W/cm² field of the form given in eq.(3) and of varying relative phase and with (a), $\delta = 0$, (b), $\delta = \pi$.

time-parameterized MOs, $\{\tilde{\phi}_k\}$, (henceforth, tilded symbols will always denote field-induced orbitals and orbital properties). Alternatively, these field-induced MOs can also be regarded as resulting from the interactions of AOs $\tilde{\chi}_j(1,t)$ that are already shifted and modified by the intense electric field, i.e. satisfying

$$\left[\hat{h}_j^0(c[j]) - e\vec{r}_1.\vec{E}(t)\right]\tilde{\chi}_j(1,t) = \tilde{\epsilon}_j(t)\tilde{\chi}_j(1,t) \tag{5}$$

where $c[j]$ denotes the atomic center which supports the considered AO, and $\hat{h}_j^0(c[j])$ denotes the one-electron effective hamiltonian defining the field-free atomic orbitals centered on $c[j]$. Among the effects that the linear potential $-e\vec{r}_1.\vec{E}(t)$ has on the AO $\tilde{\chi}_j(1,t)$ and its energy $\tilde{\epsilon}_j(t)$, there always is a geometry-dependent energy shifts $\delta_j = -e\vec{R}_{c[j]}.\vec{E}(t)$ due to the displacement of the nuclear center $c[j]$ from the center of mass, which always is the center with respect to which is defined the electronic dipole moment of the molecule. Often, and in most interesting cases, where $\delta_j \neq 0$, as in the examples of the H_2^+, HD^+ molecules considered above, this shift dominates over the intrinsic atomic Stark shift, so that one expects the changes in the LCAO-MO scheme, as one goes from the field-free to the strong field cases, to be due mainly to this (geometry-dependent) shift. To examine how the shifts δ_j of AOs induce changes in the LCAO-MO structure, through a correlation diagram highlighting the interaction of field-shifted AOs, defines what we call the LCfiAO scheme.

As an example, consider the H_2^+ molecule described in the minimal $\{1s_A, 1s_B\}$ basis and aligned along the polarization direction of the field. In the LCfiAO approach, rather than starting by coupling the field-free $\sigma_{g(u)}$ MOs together, we would consider the interaction between the field shifted AOs $\widetilde{1s}_A(t), \widetilde{1s}_B(t)$[7] of energy (in atomic units) $\tilde{\epsilon}_{A(B)} = -1/2 \pm RE(t)/2$. Diagonalizing the hamiltonian matrix, $\tilde{H}(R,t)$ in this basis, we obtain the same (time-)adiabatic eigenvalues, $(W_\pm(R,t))$, as obtained in eq.(2).

Molecular restructuring of H_2O

We now apply the LCfiAO principle to a H_2O molecule subjected to an IR laser field linearly polarized along the C_2 symmetry axis of the molecule. This amounts to assuming that the molecule has been prealigned by the field through its interaction with the permanent-dipole[8]. Figure 4 shows, for a C_{2V} geometry close to the field-free equilibrium geometry, the typical correlation diagrams depicting the laser-induced MO formation from the field-shifted AOs, (I) in the case $\vec{\mu}.\vec{E} < 0$, corresponding to an antiparallel orientation of the molecule's permanent dipole moment $\vec{\mu}$ with respect to the instantaneous field (right panel), and (II) in the case $\vec{\mu}.\vec{E} > 0$ corre-

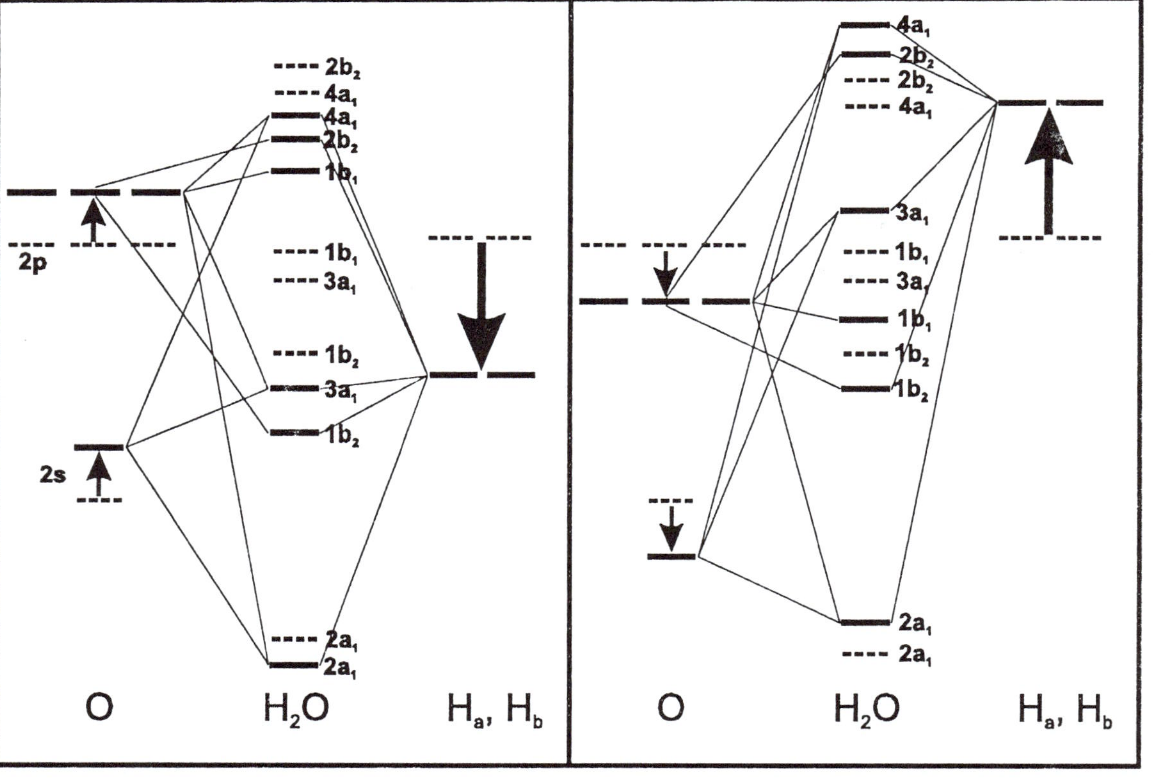

Figure 4: Correlation diagram illustrating the formation, by the LCfiAO principle, of molecular orbitals in H_2O in a static electric field with polarization vector oriented anti-parallel to the molecule's permanent dipole moment (left panel), and parallel to that moment (right panel).

sponding to a parallel configuration of $\vec{\mu}$ with respect to the instantaneous electric field (left panel).

Case I: Antiparallel orientation

In the case I, (right panel of Figure 4), the hydrogen $1s$ levels are strongly lowered while the oxygen $2s$ and $2p$ levels are shifted upward, by an amount less than the shift of the H levels, as the oxygen atom lies closer to the center of mass of the molecule. As a consequence of these shifts, all the bonding orbitals, the $2a_1$ and the $1b_2$ MOs, see their hydrogen $1s$ content increases, and are stabilized while their antibonding counterparts, the $4a_1$ and $2b_2$ MOs, should see their LCAO composition more dominated by the oxygen's $2p$ orbitals of appropriate symmetry. The $3a_1$ MO is expected to be lowered and to contain a more appreciable contribution of the hydrogen $1s$ a_1 symmetry-adapted combination. Overall, these changes in the MOs' compositions and in their energies result in (i) a lowering of the total energy, (ii) a net polarization of the electronic charge density from the oxygen atom to the hydrogens, (even one of the 'lone pairs', the one occupying the $3a_1$ MO, is now shifted towards the hydrogens). In the limit of a very intense field, and/or at a large OH internuclear distance R, which makes the geometry-dependent shifts δ_i become larger than the shifts due to the AO interaction in field-free condition, the charge transfer from the oxygen atom to the hydrogens would tend to be more and more complete, leading asymptotically to ionic fragments $O_2^{2+} + 2H^-$, and the total energy decreases linearly with E_0 or R. This asymptotic region is expected to be preceded by a region, found at a lower R (and/or E_0), where the intrinsic AO interactions continue to dominate over the geometric shift effect and give rise to bound MOs and a potential well which is separated from the asymptotic region by a barrier for dissociation. Thus one expects that, for a given value of the HOH bond angle θ, the laser-induced adiabatic potential energy curve would look like the W_- curve of the previous section. Moreover, for this antiparallel orientation of the molecule's permanent dipole, the geometric shift effects (charge polarization and energy lowering) are stronger for smaller bond angles, i.e we expect the potential well to be deeper for smaller values of θ.

All these predictions are well borne out by results of *ab-initio* calculations carried out with the option '*field*' of the GAUSSIAN 94 package[9] which allows for the inclusion of a finite field in the form of an external one-electron potential added to the *ab-initio* Fock operator. In so far as no information on ionization rate is required, these calculations are straightforward, and can be made at any level. Figure 5 shows results of calculations made at the Hartree-Fock Self-consistent-field (HF-SCF) level using a $6-31G^{**}$ basis, in the form of two-dimensional potential energy surfaces for a H_2O molecule in a static electric field. The upper panels (indicated

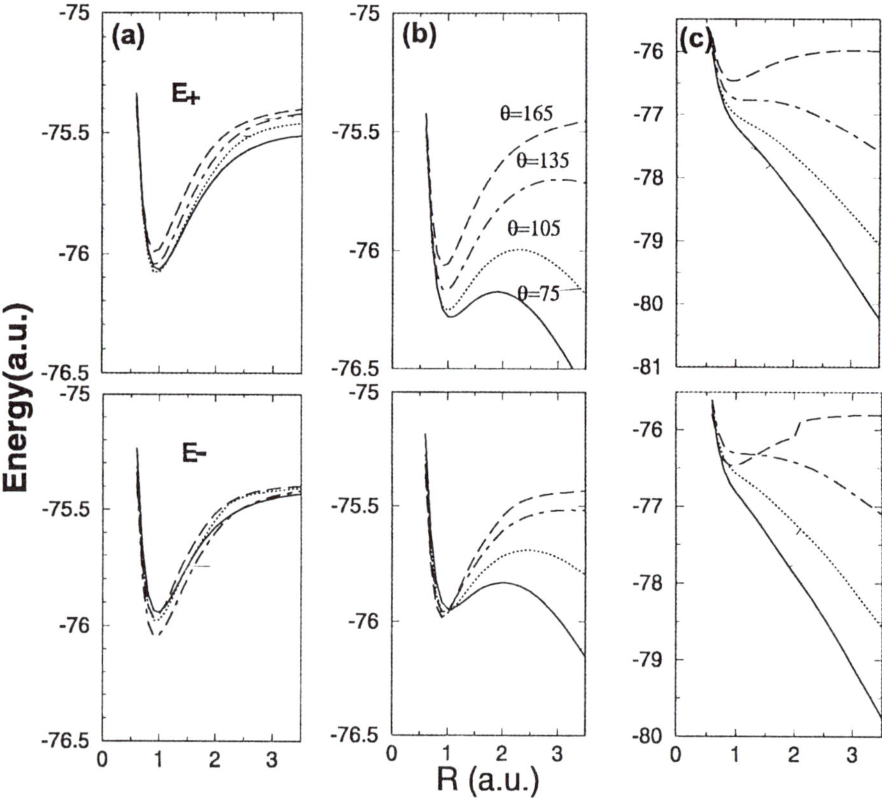

Figure 5: Two-dimensional PESs for a H_2O molecule in a static electric field with polarization vector oriented anti-parallel to the molecule's permanent dipole moment (upper panels), and parallel to that moment (lower panels); (a) $I = 5 \times 10^{13}$ W/cm^2, (b) $I = 4 \times 10^{14}$ W/cm^2, (c) $I = 2 \times 10^{15}$ W/cm^2. The values chosen for θ(deg) are given in (b).

$E+$) refer to the antiparallel configuration of the permanent dipole moment with respect to an instantaneous field of intensity (a) $I = 5 \times 10^{13} \; W/cm^2$, (b) $I = 4 \times 10^{14} \; W/cm^2$, (c) $I = 2 \times 10^{15} \; W/cm^2$. The PES clearly exhibit the features anticipated above.

Case II: Parallel orientation

In the case II of a parallel orientation of the molecule's permanent dipole moment, (left panel in Figure 4), the hydrogen $1s$ levels are strongly shifted upward while the oxygen $2s$ and $2p$ levels are lowered. This time, it is towards the O atom that electronic charge density would flow, resulting in an dissociative asymptotic charge distribution, $O^{2-} + 2H^+$, that is exactly the opposite of what was found for the antiparallel orientation case. What was said above about the overall, generic shape of potential energy curves at a fixed HOH angle remains true, however we would expect deeper potential wells at larger rather than at smaller angles. Also, the strong upward shifts of the hydrogen $1s$ levels would move them somewhat out of 'resonance' with the oxygen AOs, so that we expect the bonding character of all occupied MOs weakened, i.e the energy at a given geometry would tend to increase with respect to the field free situation, (it would certainly be larger than in the antiparallel case), a tendency which is smaller at larger values of θ and smaller R.

The lower panels of Figures 5(a)-(c) show the two-dimensional(2D) PES obtained in our ab-initio calculations for this case. At lower intensities, the predictions spelled out above seem to be verified to some extent: There indeed is a shift of the potential well toward a larger θ, although, in Figure 5(b), the 2D PES appears to exhibit rather an almost flat valley with respect to the varying bond-angle. At the level of the present ab-initio calculations,the potential wells in this configuration are about one eV higher than those obtained in case I above.

Laser-induced asymmetric force field

The results shown above, of the ab-initio calculations of molecular orbitals of H_2O in a IR laser field polarized along the molecular C_2 symmetry axis, and of the correlation diagram describing the LCfiAO-MO scheme for this laser-driven system demonstrate clearly that the first ingredient for the DDQ effect to operate in this case does indeed exists: The 2D PESs found above for the two configurations of the molecular permanent dipole moment with respect to the instantaneous electric field vector, corresponding to two different stages of the field oscillation, or two different times within an optical cycle, exhibit the precise time-asymmetry needed for a control scheme of the DDQ type to be possible. There are certainly alignment situations for which such a natural time asymmetry would not be possible, but given a polyatomic molecular system, there always exists a configuration (molecular alignment) which offers this opportunity to control the dynamics by a DDQ-like effect. In the case of H_2O, the alignment of the molecular permanent dipole moment along the field polarization is the necessary condition for a time-asymmetric laser-induced force field to exist and to give rise to a DDQ-like effect. Once this alignement is assured by the proper choice of laser intensity range, we can imagine the wavepacket dynamics being steered to favor one or the other dissociative pathway, using a two-color excitation for example.

Acknowledgements

Financial support of the research of TTND and his group by NSERC of Canada and the Fonds FCAR of Québec is gratefully acknowledged.

References

1. F. Châteauneuf, T. Tung Nguyen-Dang, N. Ouellet, and O. Atabek. *J. Chem. Phys.*, <u>108</u>, 3974, 1998.
2. H. Abou-Rachid, T.Nguyen-Dang, and O. Atabek. *J. Chem. Phys.*, <u>110</u>, 4737, 1999.
3. H. Abou-Rachid, T.Nguyen-Dang, and O. Atabek. *J. Chem. Phys.*, <u>114</u>, 2197, 2001.
4. A. D. Bandrauk, E. E. Aubanel, and J. M. Gauthier in. *Molecules in Laser Fields*. A. D. Bandrauk, ed., M. Dekker publ., New-York, 1994.
5. Such a field has also been used in a study of the directional dissociative ionization of H_2^+; see contribution of A. D. Bandrauk *et al* to this volume and also: A. D. Bandrauk and S. Chelkowski, *Phys. Rev. Lett.* <u>84</u>. 3562, (2000).
6. T. T. Nguyen-Dang, C. Chatelas, and D. Tanguay. *J. Chem. Phys.*, <u>102</u>, 1528, 1995.
7. I. Kawata, H. Kono, Y. Fujimura and A. D. Bandrauk, *Phys. Rev. A*<u>62</u>, 031401, (2000).
8. See contribution of D. Mathur *et al* to this volume.
9. M. J. Frisch, G. W. Trucks, H. B. Schlegel, P. M. W. Gill, B. G. Johnson, M. A. Robb, J. R. Cheeseman, T. Keith, G. A. Petersson, J. A. Montgomery, K. Raghavachari, M. A. Al-Laham, V. G. Zakrzewski, J. V. Ortiz, J. B. Foresman, C. Y. Peng, P. Y. Ayala, W. Chen, M. W. Wong, J. L. Andres, E. S. Replogle, R. Gomperts, R. L. Martin, D. J. Fox, J. S. Binkley, D. J. Defrees, J. Baker, J. P. Stewart, M. Head-Gordon, C. Gonzalez, and J. A. Pople. *Gaussian 94 and Revision B.3*. Gaussian and Inc., Pittsburgh PA, 1995.

Chapter 17

Ultrafast Dynamics of Molecules in Intense Laser-Light Fields: New Research Directions

Kaoru Yamanouchi

Department of Chemistry, School of Science, The University of Toyko, 7–3–1 Hongo, Bunkyo-ku, Toyko 113–0033, Japan

In intense laser light fields, molecules undergo a variety of characteristic processes such as alignment along laser polarization direction, dressed-state formation, ultrafast structural deformation, multiple ionization, and Coulomb explosion. The investigation of the dynamical behavior of molecules in intense laser fields has afforded us invaluable opportunities to understand fundamentals of interaction between molecules and light fields as well as to manipulate molecules using characteristics of laser light fields. In the present article, new directions of this rapidly growing interdisciplinary research fields are proposed by introducing our recent studies based on (i) a quasi-stationary Floquet approach, (ii) a pulsed gas electron diffraction measurement, and (iii) tandem TOF mass spectroscopy.

Introduction

When a molecule absorbs short wavelength light such as UV light, quite often its chemical bond is severed. Therefore, if we could break a specific chemical bond using light, control of chemical processes could be realized. However, due to a process called intramolecular vibrational energy redistribution, the energy deposited to a molecule by the light absorption is distributed quickly among the vibrational degrees of freedom in a complex way. This means that the breaking of a specific chemical bond could not be realized by simply irradiating a molecule with light

Owing to the development of short-pulsed laser technology, laser field intensity as high as $\sim 10^{16}$ W/cm^2 can be routinely generated. In such a strong laser intensity regime, electronic structure of atoms and molecules is perturbed strongly. This means that a potential energy surface (PES) of molecules whose shape is determined by its electronic configuration could be deformed largely in intense laser fields. Considering that the dynamics of a molecule is governed by its PES, if the shape of PES is varied in a desired way in intense laser fields, active control of the fate of a molecule by light would become realistic.

Such an idea of changing the shape of a PES by a light field is schematically shown in Fig.1. In a perturbative light intensity regime, light absorption occurs from the PES of an electronic ground state to the PES of an electronic excited state. If the electronic excited state is dissociative, breaking of a chemical bond proceeds along the slope of the PES. However, if two PESs are mixed through a strong coupling between two electronic states induced by the intense light fields, a new class of bound and dissociative PESs is formed, and a system evolving toward dissociation may be trapped in the newly formed bound part of the potential. It might be possible that its direction changes and evolves toward a new direction, resulting in the braking of another chemical bond.

It has been known that molecules behave in a characteristic way in intense laser fields (1-4). When the intensity of light fields is increased from a perturbative regime, a direction of a molecular beam can be deflected (5-6) and molecular axis are aligned along the laser polarization direction (7-10). In more intense fields as high as 10^{13} W/cm^2, light-dressed states (11) are formed through the coupling between electronic states of a molecule and the light field, and the nuclear dynamics of molecules can be described by the motion of the wave packet on the adiabatic light-dressed PESs. It was found for a number of molecules that the geometrical shape of a molecule is deformed within an ultrashort laser pulse duration, and this phenomenon was ascribed to the formation of such light dressed states in the intense laser fields (12-15).

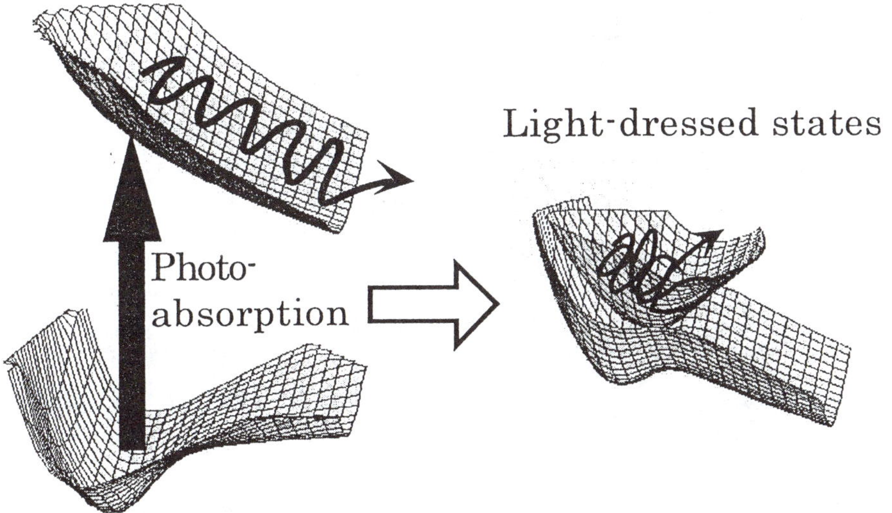

Fig. 1:

Light absorption by a molecule in a weak light field from its electronic ground bound potential energy surface (PES) to an electronically excited repulsive PES (left), and the formation of the dressed state PESs in intense laser light fields. The nuclear dynamics on the light-dressed PES could be controlled by the light fields.

When the light field intensity surpasses 10^{14} W/cm^2, a multiple ionization process tends to dominate, and multiply charged molecular ions begin to be formed. On the other hand, enhancement of the ionization process in a specific range of the bond length occurs, which is called enhanced ionization (16-19). If the light field intensity increases more, it reaches a classical tunneling regime, where electrons escape from a distorted Coulombic potential of a molecule. This multiple ionization results in the formation of multiply charged molecules, which then break into atomic fragment ions having large kinetic energy originating from a Coulombic repulsion force within a molecule. This type of fragmentation processes is called a Coulomb explosion.

In the present article, I introduce first our quasi-stationary Floquet treatment of the dissociation dynamics of H_2^+ in intense laser fields, since H_2^+ in intense laser fields is an ideal system from which we could learn how the dynamics of molecules in intense laser fields is characteristic and how it is interpreted. Then, I introduce our recent study on pulsed gas electron diffraction measurements which can probe an alignment process of molecules along the laser polarization direction in intense laser fields. Finally, by referring to our studies which revealed characteristic geometrical deformation of diatomic and triatomic molecules occurring before the Coulomb explosion process, I introduce a new trial of identifying an electronic state playing a central role in the geometrical deformation process by tandem type time-of-flight (TOF) mass spectroscopy

Quasi-stationary Floquet approach of nuclear dynamics of H_2^+ in intense laser light fields

In a Floquet (or dressed-state) picture, a dressed state is denoted as |a, N-n> using an electronic state of a molecule |a> and the number of photons in a photon field N before the interaction, where n stands for the number of photons involved in the coupling between a molecule and the light field. The PES of the |a,N-n> dressed state is described as that shifted by the energy corresponding with the number of photons. In a Floquet picture, we describe nuclear dynamics of the system by using such dressed states.

In Fig.2, the dressed states of H_2^+ generated by shifting an original set of two potential curves upward and downward are drawn with dotted lines. In the crossing regions among the original and dressed-state potential curves, an avoided crossing occurs to form a gap between a pair of the dressed state potentials. The magnitude of the gap increases as the laser field intensity increases, and the resultant dressed-state potentials become as drawn in a solid

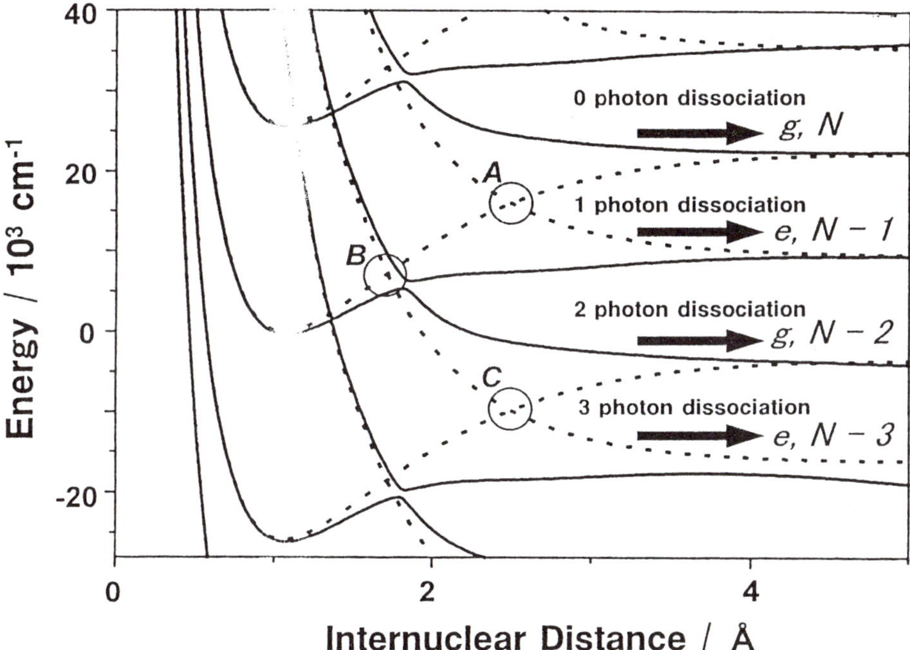

Fig. 2:

The dressed potential energy curves of H_2^+ in intense light fields. The dotted curves represent the bound $1s\sigma_g$ and repulsive $2p\sigma_u$ curves, and those shifted by an energy corresponding with the number of photons involved in the formation of the dressed states. When the laser light fields with the wavelength of 800 nm is 5.0 x 10^{13} W/cm^2, the dressed state potential curves becomes those shown in the solid curves. The energy gap at the one-photon crossing point (A) is significantly larger than that at the three-photon crossing (B). The three major dissociation pathways are zero-photon dissociation, one-photon dissociation, and two-photon dissociation.

line in Fig. 2. In this figure, "A" is called an one-photon crossing, and "B" is called a three-photon crossing.

It has been known from the previous studies (20-23) that there are three major pathways, i.e., (i) zero-photon dissociation proceeding without passing any crossing point, (ii) one-photon dissociation in which the dissociation proceeds through the one-photon crossing, leading to the dissociation along the one-photon dressed potential curve, and (iii) two-photon dissociation in which the dissociation proceeds through the three-photon crossing first, and then, changes its state into the two-photon dressed state at the one-photon crossing "C" between $|g, N-2>$ and $|e, N-3>$.

In the Floquet picture, it is considered that an interaction between a molecule and a light field is caused by a periodically varying oscillatory electric field with a constant amplitude. However, in recent experiments in which ultrashort laser pulses are employed, an assumption that the amplitude of the electric field is constant could not hold. Therefore, in order to describe the dynamics of molecules in such an ultrashort intense laser pulse within the framework of a Floquet-type treatment, it is necessary to introduce a temporary varying amplitude to the light fields. In this quasi-stationary treatment, the shape of a set of PESs of dressed states vary as a function of time depending on the temporal variation of the amplitude of the light fields.

In our recent study (24), we propagated a nuclear wave packet on the dressed state potential curves of H_2^+ whose shape changes as a function of time by adopting a multi-channel simplectic integrator method (25). Our treatment can be called a quasi-stationary Floquet method, in which the time dependence of the light-matter interaction is implemented within the framework of the Floquet approach.

By adopting this quasi-stationary Floquet approach and assuming that H_2^+ is prepared according to the Franck-Condon principle from the vibrational ground state in the electronic ground state of neutral H_2, the momentum distribution of H^+ produced through the dissociation in intense laser fields was calculated as shown in Fig.3. The momentum distribution in this figure was calculated when the shape of the light pulse is a Gaussian whose full-width-at-half-maximum is 100 fs and the light field intensity at the maximum is 5×10^{13} W/cm^2. It was found that this theoretically synthesized momentum distribution reproduced well the corresponding experimental results.

It is of course possible to propagate the wave packet without using the quasi-stationary Floquet approach and discriminate two types of dissociation pathways, i.e. *gerade* pathways and *ungerade* pathways, which are composed of the dissociation from the states with even or odd number photons, respectively. However, by a non-Floquet type approach, it is impossible to extract the contribution of a dissociation pathway dressed with a specific number of the coupling photons.

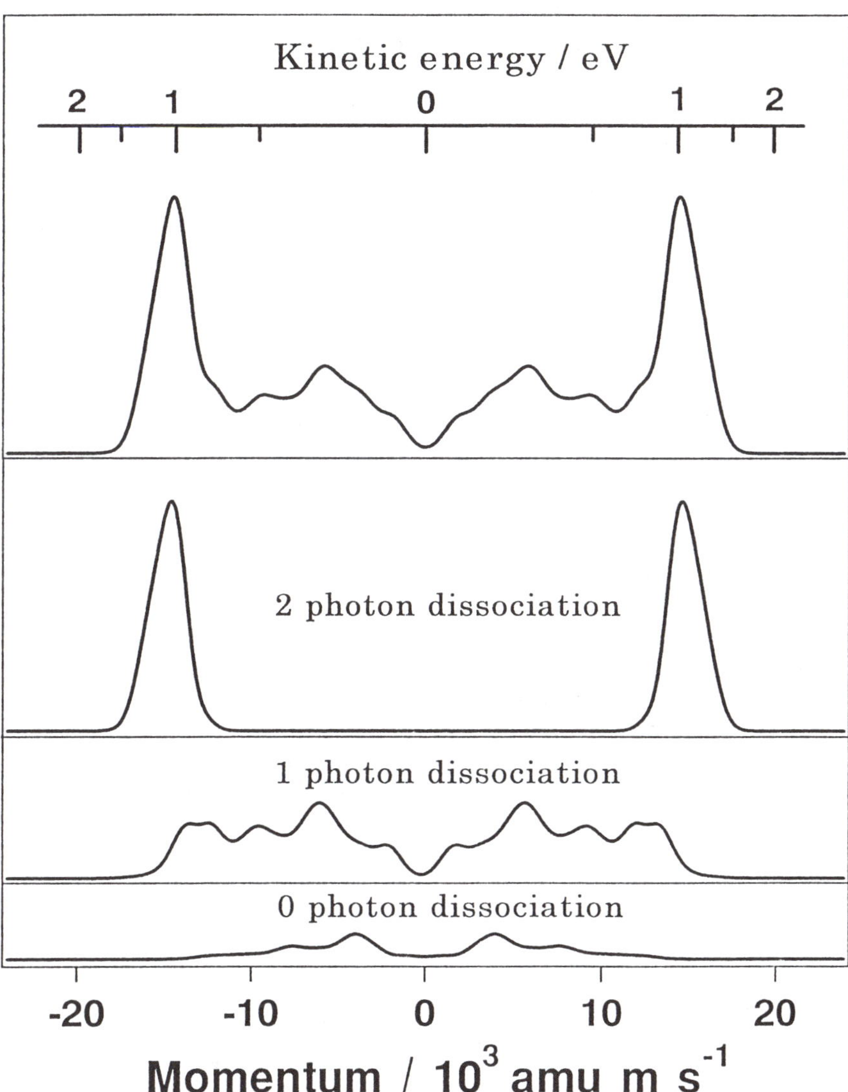

Fig. 3:

The kinetic energy and momentum distribution of H^+ ejected from H_2^+ in intense light fields obtained by the quasi-stationary Floquet approach. The light field conditions are the same as those in Fig.2.

Contrary, in the quasi-stationary Floquet approach, it is straightforward to discriminate the contribution from the zero-photon dissociation and the two-photon dissociation, both of which have the *gerade* symmetry as shown in Fig. 3. According to this figure, in the kinetic energy range below 0.4 eV the zero-photon dissociation dominantly contributes, while in the kinetic energy range above 0.4 eV, the two-photon dissociation dominates. The characteristic structure appearing in the energy range below 0.3 eV mainly comes from the one-photon dissociation, but the minor contribution from the zero-photon dissociation overlaps.

Within this quasi-Floquet framework, we can interpret the characteristic dynamics of H_2^+ in intense laser fields in terms of *bond-softening* and *bond-hardening* as follows. In the zero-photon dissociation, a wave packet trapped in the dressed state potential caused by the *bond-hardening* follows adiabatically a variation of the shape of the dressed state potential curve, and it is pushed up above the energy corresponding to the dissociation limit of the electronic ground state. The one-photon dissociation is the process in which a wave packet escapes through a gap at the one-photon crossing point "A" associated with the *bond-softening*. In the two-photon dissociation process, a wave packet transferred to the three-photon dressed state through the three-photon crossing point "B" emits one photon at the crossing "C" between the two-photon dressed state and the three-photon dressed state, and dissociates on the two-photon dressed potential curve.

In Fig. 3 the momentum peak assigned to the two-photon dissociation pathway is much narrower than the spread momentum distribution of the one-photon dissociation pathway. This observation could be ascribed to the narrow three-photon gap shown in Fig. 2 through which the two-photon dissociation occurs, because it could allow only a restricted range of the kinetic energy release.

As the light-field intensity increases, the contribution from the three photon dissociation increases, but the ionization to H_2^{2+} followed by the Coulomb explosion process begins to participate. Therefore, in the more intense light field range, it becomes inevitable to treat the ionization dynamics simultaneously with the nuclear dynamics on the time-dependent light-dressed potential curves.

Molecular alignment and structural deformation in intense laser fields probed by gas electron diffraction

In intense linearly polarized laser fields, it has been known that one of the principal axis of a molecule can be aligned along the laser polarization direction. This is due to the torque imposed on the molecule which is generated through the interaction between non-resonant light fields and an induced dipole moment of a molecule as interpreted by Friedrich and Hershbach (7).

One of the promising approaches to probe directly the extent of the alignment of molecules in intense laser fields may be a measurement of their electron diffraction pattern. From long ago, a gas electron diffraction method has been known as the method by which geometrical structure of molecules is determined with high precision. If a molecular axis is aligned along the laser polarization directions, such a process is expected to be reflected sensitively on the electron diffraction pattern. Recently, we have succeeded to record a gas electron diffraction pattern of CS_2 molecules whose molecular axes are aligned along the polarization direction of the intense laser fields (26). In the experiment using this apparatus, a pulsed electron beam is generated by irradiating a surface of a photo-cathode with a UV laser pulse through a photoelectric effect. The electron packet is then accelerated to collide with a sample gas introduced into a diffraction chamber through a pulsed nozzle. Because a pulse width of the electron packet is determined by a pulse width of the UV laser, it is possible to generate an ultrashort pulsed electron packet when adopting an ultrashort UV laser pulse. A simulation of the electron trajectory showed that the temporal packet of the electron beam could become as short as 1 ps, indicating that real time ultrashort dynamics of molecules could be probed as a series of snap shots of gas electron diffraction patterns.

When a diffraction pattern is observed for molecules aligned along the laser polarization direction, it could be significantly different from that observed for molecules randomly oriented in space. For example, in the case of CS_2, when the distribution of the molecular axis is isotropic, the diffraction halo pattern becomes concentric. On the other hand, if the molecular axis is aligned along the laser polarization direction, the resultant diffraction pattern becomes that compressed along the laser polarization direction. In our recent measurements performed under the laser field intensity of $\sim 10^{12}$ W/cm^2 showed that an effect of the alignment is clearly reflected in the observed diffraction pattern. Because a spacing between neighboring halos is sensitively dependent on the geometrical structure of molecules, this approach is also promising to probe a structural deformation process of molecules in intense laser fields.

On the theoretical side, Fujimura and coworkers proposed a new approach to design an optimal laser pulse shape for controlling orientation as well as alignment of molecules (27). The term "orientation" is defined for polar molecules in which a specific direction is defined along one of the molecular axes. For example, in the case of CO molecules, if they are aligned along the vertically polarized laser fields so that O atom is placed above C atom, it can be said that the orientation is achieved. This type of control of molecular

260

orientation would become an important factor to control chemical reaction on the solid state surface under the intense laser fields.

In the case of polyatomic molecules, there is a process of structural deformation which could compete with the alignment process. It has not been clear so far whether these two processes occur rather sequentially or almost simultaneously. Recently we investigated the alignment and structural deformation processes of CS_2 molecules by aligning them by a intense nano-second laser fields (~10^{12} W/cm^2) and by ionizing them with circularly polarized ultrashort laser fields, and successfully probed the alignment and deformation processes in real time within a nano-second laser pulse (10). It was found that the structural deformation from linear to bent occurs as the extent of the alignment proceeds until the laser pulse reaches a maximum, and after passing the maximum, the recovery of the geometrical structure form bent to linear occurs as the laser field intensity decreases, simultaneously with the decrease in the extent of the alignment. The structural deformation of CS_2 identified in this relatively weak laser field suggests that a pair of dressed states is efficiently formed with an electronic state having bent equilibrium geometry.

Coulomb explosion of molecules in intense laser fields: An approach by tandem type mass spectroscopy

When molecules are exposed to the strong short pulsed laser light whose intensity exceeds ~10^{14} W/cm^2, a Coulomb explosion process becomes evident. In the case of diatomic and triatomic molecules, momentum vector distributions of atomic fragment ions are sensitively dependent on the geometrical structure of molecules just before the Coulomb explosion. In our recent studies, such ultrafast structural deformation has been investigated for N_2 (12,28,29), NO (12,30), SO_2 (28), CO_2 (13), NO_2 (14), H_2O (15) by the method called mass-resolved momentum imaging (MRMI) (28-30), in which momentum distributions of atomic and molecular fragment ions are plotted in the form of the two dimensional or three dimensional plot.

From the analysis of the observed MRMI maps of diatomic and triatomic molecules, it was identified that the critical bond length at which the Coulomb explosion occurs increases commonly from ~1.4 r_e to ~2.7 r_e as the charge number of the parent ion increases, where r_e denotes an equilibrium internuclear distance. Furthermore, it was also identified that the increase of the critical bond length exhibits an even-odd alternation upon the increase of the total charge number of the parent molecular ions. The more striking finding was the existence of an ultrafast deformation process of molecular skeletal geometry of triatomic molecules.

In the case of CO_2, its linear structure is deformed towards bent structure in intense laser fields, resulting in a largely spread bond angle distribution centering at the linear configuration with a mean amplitude of as large as 40°. Similarly, in the case of NO_2 whose equilibrium structure is bent, it was found that the probability of taking linear configuration increases significantly, and in the case of originally bent H_2O its structure changes into that centered at linear configuration with a broad mean amplitude of ~60°. As demonstrated in our series of studies, the MRMI method is a powerful technique to elucidate such ultrafast structural deformation of molecules in intense laser fields.

The geometrical deformation may be interpreted by considering a set of dressed state potentials which are formed through the strong mixing of molecular electronic states with intense laser fields. As explained for H_2^+, a prepared wave packet would evolve on such deformed PESs, resulting in the deformation of skeletal geometry. For example, an electronic state whose equilibrium structure is linear may be coupled through light fields with an electronic state whose equilibrium structure is bent, and a wave packet located first at linear configuration would be spread along the bent direction through the formation of a pair of dressed states as the laser field intensity increases, resulting in an ultrafast structural deformation.

However, by measuring simply the momentum vector distributions of fragment ions, it would be difficult to identify whether the structural deformation occurs at the neutral stage, singly charged stage, or multiply charged stage, since the ionization process competes with the formation of the dressed states. In order to identify the ion stage at which crucial dynamics occurs through a major coupling among its electronic states induced by the light fields, it would be worthwhile to prepare a specific charge state of molecules by a mass separation technique and irradiate them with strong laser light. This type of study was performed recently by our group (31) using a tandem type time-of-flight (TOF) mass spectrometer shown in Fig. 4. By the first stage of the TOF mass separation, mass and charge selected parent ions are prepared at the interaction region, where they are irradiated with ultrashort pulsed strong laser light, and the momentum distribution of fragment ions are measured by the second stage of the TOF mass separation.

The measurements using the tandem TOF mass spectrometer were performed first for benzene cations. By comparing the ionization and fragmentation processes of neutral benzene with those of singly charged benzene cations, it was identified for the first time that the most dominant coupling between the electronic states are induced at the singly charged stage within the ultrashort intense laser fields (~50 fs) at ~400 nm. On the other hand, when the wavelength was ~800 nm, the dominant products were found to be doubly charged parent molecules, and no evidence was found for a specific coupling among its electronic states. This remarkable difference in the

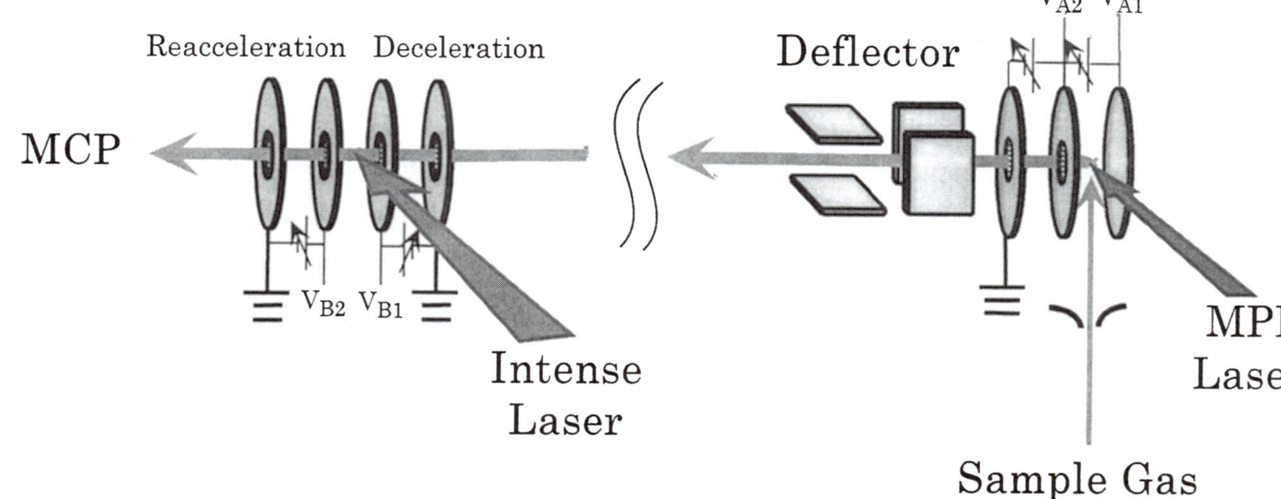

Fig. 4:

The schematic diagram of a tandem type time-of-flight mass spectrometer. In the first ionization region, state specific molecular ions are generated, and these ions are selectively introduced into the second ionization region, where the charge-and-state selected ions are exposed to intense laser light fields. The multiply charged parent ions as well as the fragment ions generated after the Coulomb explosion process are detected by a micro-channel plate detector (31).

response of molecules in the laser light fields at different wavelengths suggests a possibility of controlling of molecular dynamics in intense laser light field by treating the wavelength as one of the major control parameters.

On the theoretical side, in order to interpret the experimental findings of the ultrafast structural deformation of molecules in intense laser light fields, Kono, Koseki, and Fujimura (32) introduced a new idea in which molecular orbital calculations of PESs is combined with an electrostatic model of charge distribution within a molecule, and investigated the ultrafast dynamics of CO_2 in intense laser fields. They were able to identify that the most crucial stage of the dynamics occurs at the doubly charged CO_2^{2+} stage, and that the geometrical structure of CO_2 just before the Coulomb explosion determined from our experiment can be interpreted mostly by the coupling between the molecular states and the light fields occurring at the doubly charged stage.

Conclusion

As introduced in the present article, the systematic investigation of the behavior of molecules in intense laser fields was launched only very recently, and a wide range of progress of this new interdisciplinary research field is expected in the near future. For example, in our recent study of O_2^+, more than three electronic states were found to be coupled to form a sequence of dressed states (33). This study is the first example of the successful interpretation of the nuclear dynamics of multi-electron molecules in intense laser fields on the basis of the information from molecular spectroscopy. On the theoretical side, the dressed state picture was utilized to design an ideal control of molecular reaction processes (34,35) as well as to investigate ionization processes through overlapping resonances (36).

Another important aspect of intense laser fields is its capability of generating short-pulsed short wavelength light such as vacuum ultraviolet (VUV) (37) and extreme ultraviolet (XUV) (38-40) light. The intense laser field can also be used to generate short pulses of characteristic X rays, and interesting applications such as real-time probing of temporal variation of crystal structure have been reported (41). In our group, by focusing strong ultrashort laser pulses into a rare gas medium, short-pulsed high odd-order harmonics in the range from the 7th order to the 15th order were generated. In order to investigate a dissociative ionization process of molecules through the electronically highly excited states (37), they were irradiated with the harmonics with a specific order separated by a monochromator. The photon energy corresponding with this XUV wavelength region between 40~100 nm is in the same range of typical dissociation energy of chemical bonds in a molecule.   Therefore, a laser based pump-and-probe experiment in this wavelength range would be promising to gain new insight into photochemistry initiated by braking a specific chemical bond.   It would be noteworthy to mention that ionization and explosion dynamics of large clusters such as C_{60} is important in designing a new type of phenomena occurring only through collective dynamics of electrons and nuclei in large-size systems in intense laser fields (42).

Acknowledgements

After launching a five-year term research project called "Control of photochemical reaction in a femto-second time scale" in the fiscal year of 1996 supported as one of the CREST projects of JST (Japan Science and Technology Corporation), my group has conducted a series of studies related with dynamics of molecules in intense laser fields. The results of our studies introduced in the present article were obtained mainly through this CREST project. I would like to thank my colleagues, Dr. K. Someda, Dr. A. Hishikawa, Dr. K. Hoshina, Dr. A.Iwamae, Dr. M. Kono, Dr. S. Liu, Dr. T. Sako, Dr. Y. Fukuda, Dr. R. Itakura, and Dr. A. Iwasaki for their efforts in making the project fruitful. Finally, I thank Dr. H. Todokoro, Mr. T. Ohshima, and Mr. Y. Oze for their support in the construction of the gas electron diffraction apparatus.

References

1. *Molecules in Laser Fields*; Bandrauk, A. D., Ed.; M.Dekker, New York (1993).
2. Sheehy, B.; Dimauro, L. F. *Ann. Rev. Phys. Chem.* **1996**, 47, 463.
3. Codling, K.; Frasinski, L. J. *J. Phys. B: At. Mol. Opt.* **1993**, 26, 783.
4. Normand, D.; Lompre, L. A.; Cornaggia, C.; *J. Phys. B.: At. Mol. Opt.* **1992**, 25, 1497.
5. Sakai, H.; Tarasevitch, A.; Danilov, J.; Stapelfeldt, H.; Yip, R.W.; Ellert, C.; Constant, E.; Corkum, P. B.; *Phys. Rev. A* **1998**, 57, 2794.
6. Seideman, T.; *J. Chem. Phys.* **1997**, 106, 2881.
7. Friedrich, B.; Herschbach, D.; *Phys. Rev. Lett.* **1995**, 74, 4623.
8. Sakai, H.; Safvan, C. P.; Larsen, J. J.; Hilligsøe, K. M.; Hald, K.; Stapelfeldt, H.; *J. Chem. Phys.* **1999**, 110, 10235.
9. Larsen, J. J.; Sakai, H.; Safvan, C. P.; Larsen, I. W.; Stapelfeldt, H. *J. Chem. Phys.* **1999**, 111, 7774.
10. Iwasaki, A.; Hishikawa, A.; Yamanouchi, K. *Chem. Phys. Lett. in press*.

11. Giusti-Suzor, A.; Mies, F. H.; DiMauro, L. F.; Charron, E.; Yang, B. *J. Phys. B: At. Mol. Opt.* **1995**, 28, 309.

12. Yamanouchi, K.; Hishikawa, A.; Iwamae, A.; Liu, S. *Phys. Elec. At. Col.*, *AIP Conference Proceedings* **2000**, 500, 182.

13. Hishikawa, A.; Iwamae, A.; Yamanouchi, K. *Phys. Rev. Lett.* **1999**, 83, 1127.

14. Hishikawa, A.; Iwamae, A.; Yamanouchi, K. *J. Chem. Phys.* **1999**, 111, 8871.

15. Liu, S.; Hishikawa, A.; Iwamae, A.; Yamanouchi, K.; *Advances in Multiphoton Processes and Spectroscopy* 13, Y. Fujimura and R. J. Gordon Eds., World Scientific, **2000**, p.189.

16. Codling, K.; Frasinski, L. J. *J. Phys. B: At. Mol. Opt.* **1993**, 26, 783.

17. Posthums, J. H.; Frasinski, L. J.; Giles, A. J.; Codling, K. *J. Phys. B: At. Mol. Opt.* **1995**, 28, L349.

18. T.Seideman, T; Ivanov, M. Y.; Corkum, P. B. *Phys. Rev. Lett.* **1995**, 75, 2819.

19. Villeneuve, D. M.; Ivanov, M. Y.; Corkum, P. B. *Phys. Rev. A* **1996**, 54, 736.

20. Frasinski, L. J.; Posthumus, J. H. Plumridge, J.; Codling, K. *Phys. Rev. Lett.* **1999**, 83, 3625.

21. Talebpour, A.; Vijayalakshmi, K.; Bandrauk, A. D.; Nguyen-Dang, T. T.; Chin, S. L. *Phys. Rev. A* **2000**, 62, 042708.

22. Williams, I. D., McKenna, P.; Srigengan, B.; Johnston, I. M. G.; Bryan, W. A.; Sanderson, J. H.; El-Zein, A.; Goodworth, T. R. J.; Newell, W. R.; Taday, P. F.; Langley, A. J. *J. Phys. B: At. Mol. Opt.* **2000**, 33, 2743.

23. Sandig, K.; Figger, H.; Hansch, T. W. *Phys. Rev. Lett.* **2000**, 85, 4876.

24. Maruyama, I; Sako, T.; Yamanouchi, K. *in preparation*.

25. Takahashi, K.; and Ikeda, K. *J. Chem. Phys.* **1993**, 99, 8680.

26. Hoshina, K.; Yamanouchi, K.; Ohshima, T.; Ose, Y.; Todokoro, H. *in preparation*.

27. Hoki, K.; Fujimura, Y. *Chem. Phys.* **2001**, 267, 187.

28. Hishikawa, A.; Iwamae, A.; Hoshina, K.; Kono, M.; Yamanouchi, K. *Chem. Phys. Lett.* **1998**, 282, 283.

29. Hishikawa, A.; Iwamae, A.; Hoshina, K.; Kono, M.; Yamanouchi, K. *Chem. Phys.* **1998**, 231, 315.

30. Iwamae, A.; Hishikawa, A.; Yamanouchi, K. *J. Phys. B: At. Mol. Opt.* **2000**, 33, 223.

31. Itakura, R.; Watanabe, J.; Hishikawa, A.; Yamanouchi, K. *J. Chem. Phys.* **2001**, 114, 5598.

32. Kono, H.; Koseki, S.; Shirota, M.; Fujimura, Y. *J. Phys. Chem. A* **2001**, 105, 5627.

33. Hishikawa, A.; Liu, S.; Iwasaki, A.; Yamanouchi, K. *J. Chem. Phys.* **2001**, 114, 9856.
34. Teranishi, Y.; Nakamura, H. *Phys. Rev. Lett.* **1998**, 81, 2032.
35. Teranishi, Y.; Nakamura, H. *J. Chem. Phys.* **1999**, 111, 1415.
36. Hiyama, M.; Someda, K. *Phys. Rev. A* **2000**, 61, 023411.
37. Fukuda, Y.; Iwamae, A.; Hosaka, K.; Hoshina, K.; Hishikawa, A.; Yamanouchi, K. *Proceedings of the Second Symposium on Advanced Photon Research, Japan Atomic Energy Research Institute* **2001**, p.288.
38. Sekikawa, T.; Ohno, T.; Yamazaki, T.; Nabekawa, Y.; Watanabe, S. *Phys. Rev. Lett.* **1999**, 83, 2564.
39. Tamaki, Y.; Itatani, J.; Nagata, Y.; Obata, M.; Midorikawa, K. *Phys. Rev. Lett.* **1999**, 82, 1422.
40. Nakano, H.; Goto, Y.; Lu, P.; Nishiakwa, T.; Uesugi, N. *Appl. Phys. Lett.* **1999**, 75, 2350.
41. Hironaka, Y.; Yazaki, A.; Saito, F.; Nakamura, K. G.; Kondo, K.; Takenaka, H.; Yoshida, M. *Appl. Phys. Lett.* **2000**, 77, 1967.
42. Kou, J. K.; Zhakhovskii, V.; Sakabe, S.; Nishihara, K.; Shimizu, S.; Kawato, S.; Hashida, M.; Shimizu, K.; Bulanov, S.; Izawa, Y.; Kato, Y.; Nakashima, N. *J. Chem. Phys.* **2000**, 112, 5012.

Chapter 18

Intense-Laser-Induced Electron Transfer and Structure Deformation of Molecules

Hirohiko Kono[1] and Shiro Koseki[2]

[1]Department of Chemistry, Graduate School of Science,
Tohoku University, Sendai 980–8578, Japan
[2]Department of Material Science, College of Integrated Arts and
Sciences, Osaka Prefecture University, Sakai 599–8531, Japan

The electronic and nuclear dynamics of H_2^+ and H_2 in intense laser fields are examined by solving the time-dependent Schrödinger equations for the systems. We clarify the dynamics of bound electrons and the subsequent ionization process in terms of "field-following" adiabatic states. The adiabatic state analysis leads to an electrostatic model in which each atom in a molecule is charged by laser-induced electron transfer and ionization proceeds via the most unsatable atomic site. We apply this model to multi-electron molecules with the help of *ab initio* molecular orbital calculations; the structure deformations of CO_2 cations in an intense field are investigated. The control of electronic and nuclear dynamics is briefly discussed from the viewpoint of the dynamics on adiabatic states.

Interaction of atomic or molecular systems with high-power laser fields results in nonperturbative electronic dynamics such as above-threshold ionization (*1*) and tunnel ionization (*2, 3*) . In a high-intensity and electronically nonresonant low-frequency regime (intensity $I > 10^{13}$ W/cm^2 and wavelength $\lambda > 700$nm), a laser electric field significantly distorts the Coulombic potential that the electrons are placed in. The distorted potential forms a "quasistatic" barrier (or barriers) through which an electron or electrons can tunnel. This type of ionization is called tunnel ionization. For the case of atoms, the tunnel ionization regime can be distinguished by using the Keldysh parameter $\gamma = \omega\sqrt{2I_p}/f(t)$ (2), where I_p is the ionization potential of the atom, ω is the laser frequency, and $f(t)$ is the pulse field envelope at time t. As the electric field becomes stronger and its period becomes longer, an electron penetrates or goes beyond the barrier(s) more easily before the phase of the electric field changes. The quasi-static tunneling condition (4) is given by the inequality $\gamma < 1$, while the ordinary multiphoton ionization (MPI) regime is defined as $\gamma > 1$ (5). The straightforward application of the Keldysh parameter to diatomic and polyatomic molecules is questionable because a one-center (zero-range) potential is employed in Keldysh's treatment. Recently, DeWitt and Levise have proposed a "modified" Keldysh parameter that takes into account extensive electron delocalization in molecules (6).

For the case of molecules, the electronic wave function is so strongly deformed within a half optical cycle of an intense field that a large part of the electron density is transferred among nuclei. Such an intramolecular electronic motion induces nuclear motions as well as tunnel ionization; resultant laser-induced structure deformations in turn change the electronic response to the field, e.g., ionization rates. It is known that the resultant bond stretching enhances the rate of tunnel ionization (known as enhanced ionization). The observed kinetic energies of ionized fragments of a molecule are large ($>>$ a few eV) and are consistent with Coulomb explosions of multiply charged cations at a *specific internuclear distance* R_c in the range of $\sim 2R_e$ (7), where R_e is the equilibrium internuclear distance. Numerical simulations indicate that the tunnel ionization rates around R_c exceed those near R_e and those of dissociative fragments (8-10).

To quantitatively understand the electronic dynamics in intense fields (in a nonperturbative case), it is necessary to solve the time-dependent Schrödinger equation for the electronic degrees of freedom of a molecule. We have been developing an efficient grid point method, the dual transformation method, for accurate propagation of an electronic wave packet (11-13). In this method, both the wave function and the Hamiltonian are transformed consistently to overcome the numerical difficulties arising from the divergence of the Coulomb potentials. We have applied this method to small molecular systems such as H_2^+ and H_2. To the best of our knowledge, our treatment of H_2 is the first accurate evaluation of two-electron dynamics of a molecule. The vibrational degree of freedom is also incorporated in the calculation of H_2^+ without resorting to the Born-Oppenheimer (B-O) approximation (14). We have demonstarated that field-induced intramolecular electronic motion triggers tunnel ionization.

Intramolecular electronic dynamics is analyzed by means of "field-following" adiabatic states $\{|n(t)\rangle\}$ defined as eigenfunctions of the "instantaneous" electronic Hamiltonian $H_0(t)$ including the interaction with light. To obtain $\{|n\rangle\}$, we diagonalize $H_0(t)$ by using *bound* eigenstates of the B-O electronic Hamiltonian H_{el} at zero field as a basis set (15). In a high-intensity regime, field-following adiabatic potential surfaces can in general cross each other (*16,17*). Field-induced nonadiabatic transitions through avoided crossing points in time and internuclear coordinate space, as well as nuclear motion-induced ones that occur without external fields, govern the electronic and nuclear dynamics in intense fields. Tunnel ionization occurs from such an adiabtic state (or from adiabatic states) to Volkov states (quantum states of a free electron in a laser field) (*1,18*).

In this paper, we present the results of a theoretical investigation of the intense-field-induced electronic and nuclear dynamics of molecules, including the multi-electron triatomic molecule CO_2. First, the numerical results of electronic and nuclear dyanamics of H_2^+ and H_2 are summarized. A straightforward application of the dual transformation method to multi-electron systems is however impracticable. Instead, we first extract characteristic features of electronic dynamics in intense fields from the results of simulations of H_2^+ and H_2. We then present the electrostatic model that the tunnel ionization process of a molecule is governed by the excess charges due to intramolecular field-induced electron transfer. To calculate the charge distributions which can be used to estimate the probability of tunnel ionization, we propose to use *ab initio* molecular orbital (MO) methods. The nuclear dynamics prior to ionization can be evaluated by using field-following adiabatic potential surfaces calculated by an MO method. We apply this approach, which is applicable to multi-electron polyatomic molecules, to reveal the origin of the experimentally observed molecular structure deformations of CO_2 cations. We calculate adiabatic potential surfaces and charge distributions of CO_2 and its cations by MO methods and reveal the deformation stage of CO_2 in an intense laser field. This simple approach can serve as a theoretical basis for controlling the electronic and nuclear dynamics of molecules.

Electronic and nuclear dynamics of H_2^+

We here summarize the adiabatic state analysis of electronic and nuclear wave packet dynamics of H_2^+ (*12*). The electronic dynamics of H_2^+ prior to tunnel ionization is determined by the large transition element of the dipole interaction $z\mathcal{E}(t)$ between the lowest two electronic states, $1s\,\sigma_g$ and $2p\,\sigma_u$. Here, z is the electronic coordinate parallel to the molecular axis, and $\mathcal{E}(t)$ is the z component of the linearly polarized laser electric field. The dipole transition moment between the two states increases as $R/2$, where R is the internuclear distance. This large transition moment, which is characteristic of a charge resonance transition between a bonding and a corresponding antibonding molecular orbital (*19*), changes the potential surfaces E_g and E_u of $1s\,\sigma_g$ and $2p\,\sigma_u$ to field-following adiabatic surfaces, $E_{\mp}(R)$:

$$E_{\mp}(R,t) = \frac{1}{2}\left[E_g + E_u \mp \sqrt{\left(E_u - E_g\right)^2 + 4\left|\langle 1s\sigma_g |z|2p\sigma_u \rangle \mathcal{E}(t)\right|^2}\right],$$

$$\approx \left[E_g + E_u \mp R|\mathcal{E}(t)|\right]/2, \quad \text{for large } R \tag{1}$$

The eigenvalues $E_{\mp}(R)$ and corresponding eigenstates $|-\rangle$ and $|+\rangle$ of the instantaneous Hamiltonian $H_0(t)$ of H_2^+ are obtained by using the $1s\,\sigma_g$ and $2p\,\sigma_u$ states.

The instantaneous electrostatic potential for the electron has two wells around the nuclei, i.e., $z = \pm R/2$. For $\mathcal{E}(t) > 0$, the potential slants to the left because of the dipole interaction $z\mathcal{E}(t)$: the potential well formed around the right nucleus ascends by $\mathcal{E}(t)R/2$ and the well formed around the left nucleus descends. Then, there exist barriers between the two wells (inner barrier) and outside the descending well (outer barrier). At large R, E_+ and E_- change according to the energy shifts of the ascending and descending wells, respectively, as shown in eq. 1. $|+\rangle$ and $|-\rangle$ are localized near the ascending and descending wells, respectively. The adiabatic energies $E_{\mp}(R)$ at $\mathcal{E}(t) = 0.045$ a.u. are plotted in Figure 1 against R as well as the barrier heights. If $\mathcal{E}$ is assumed to be equal to the field envelope $f(t)$, the field strength $\mathcal{E}$ in atomic units corresponds to the intensity $I = 3.5 \times 10^{16} \mathcal{E}^2\,\text{W/cm}^2$. While E_- is usually below the barrier heights, E_+ can be higher than the barrier heights in the range $R_c = 7\text{-}8$ a.u. In this critical range of R, the upper adiabatic state $|+\rangle$ is easier to ionize than is $|-\rangle$.

The range of R_c values is consistent with the results of numerical simulations of ionization. In an intense and long wave length regime, after one-electron ionization from H_2, $|-\rangle$ of H_2^+ is prepared and the bond distance of H_2^+ then stretches on the E_- laser-induced dissociative potential (bond softening due to a laser field). At large R, nonadiabatic transitions occur between $|-\rangle$ and $|+\rangle$ when the field $\mathcal{E}(t)$ changes its sign, i.e., when the two adiabatic potential surfaces come closest to each other. A nonadiabatic transition between $|-\rangle$ and $|+\rangle$ corresponds to suppression of electron transfer between the nuclei. Around R_c, ionization proceeds via the $|+\rangle$ state nonadiabatically created from $|-\rangle$. This mechanism of enhanced ionization has been directly proved by monitoring the populations of adiabatic states such as $|+\rangle$ and $|-\rangle$ (*12,15*). As the field strength approaches a local maximum, a reduction in the population of $|+\rangle$ is clearly observed, whereas the population of $|-\rangle$ changes very little. An ionization process begins and ends within a half optical cycle (which is characteristic of tunnel ionization). In conclusion, enhanced ionization in H_2^+ is due to electron localization in the ascending well (*8-10*). The nonadiabatic transition probability can be increased as the intensity, frequency, and R are increased. The above analysis using the populations of *two* adiabatic states, $|+\rangle$ and $|-\rangle$, is validated by the fact that only two states, $|-\rangle$ and $|+\rangle$, are mainly populated before ionization. We have diagonalized $H_0(t)$ by using the lowest six σ-states (*15*). The total population of the resultant six adiabatic states is nearly equal to the sum of the populations of $|+\rangle$ and $|-\rangle$.

Electronic wave packet dynamics of two-electron molecule H_2

We have succeeded in accurately propagating the two-electron wave packet

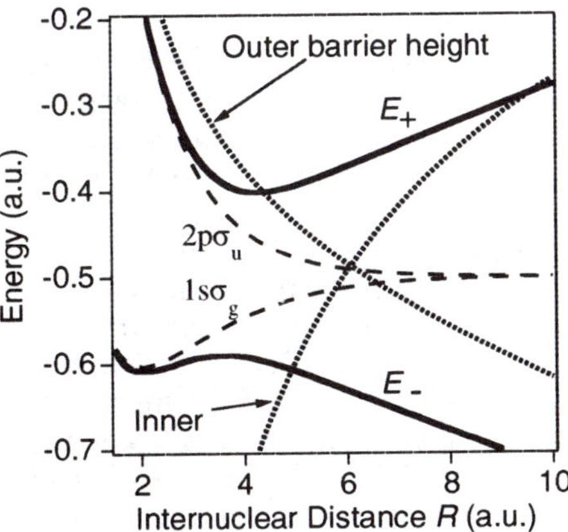

Figure 1. The potential energies of the lowest two field-following adiabatic states, E_- and E_+, of H_2^+ and the heights of the inner and outer barriers for tunnel ionization at $\mathcal{E}(t) = 0.045$ a.u. (dotted lines). The broken lines denote the Born-Oppenheimer potential surfaces of $1s\sigma_g$ and $2p\sigma_u$.

of H_2 in an intense field for the first time (*13*). In this subsection, we present the results of investigation of the electronic dynamics. The position of the jth electron can be designated by cylindrical coordinates (ρ_j, z_j and φ_j). The z_1 and z_2 axes are parallel to the molecular axis. The internuclear distance R is fixed. Here, we assume that the molecular axis is aligned, parallel to the polarization direction, by the applied field $\mathcal{E}(t)$. Electron transfer in H_2 is characterized by motion along the polarization direction z. To represent the wave packet, we therefore employ the reduced density $\bar{P}(z_1, z_2)$ obtained by integrating the square of the wave function, $|\psi(t)|^2$, over the degrees of freedom other than z_1 and z_2 (*20*). As an example, $\bar{P}(z_1, z_2)$ for the exact ground state at $R=4$ a.u. is drawn in Figure 2(a). The reduced density map clearly demonstarate that the covalent components ($H \cdot H$) around $z_1 = -z_2 = \pm R/2$ are dominant in the ground state at $R=4$ a.u.

The applied field $\mathcal{E}(t)$ is assumed to be $f(t)\sin\omega t$, where the envelope $f(t)$ is linearly ramped with t so that after one cycle $f(t)$ attains its maximum f_0. The field parameters used are $f_0=0.12$ a.u. and $\omega = 0.06$ a.u. ($\lambda = 760$ nm). The length of a quarter cycle is $\pi/2\omega = 26.2$ a.u.$=0.634$ fs. The ionic component around the left nucleus, of which the peak is located around $z_1 = z_2 = -R/2$, increases as the field approaches the first local maximum at $t = \pi/2\omega$, as shown in Figure 2(b). The field strength $\mathcal{E}(t)$ is 0.03 a.u. at $t = \pi/2\omega$. The laser field forces the two electrons to stay near a nucleus for a half cycle. At $t = \pi/2\omega$, the field returns to zero. The packet at this moment shown in Figure 2(c) is nearly identical with the initial one in Figure 2(a), indicating that the electronic response to the field is still adiabatic. A quarter cycle later, as shown in Figure 2(d), the density around $z_1 = z_2 = R/2$ is very high because of the stronger field $\mathcal{E}(t = 3\pi/2\omega) = -0.09$ a.u.

From an analysis of the 3D spatial configuration of the two electrons, the electronic state of the localized ionic structure is identified with the bound state of H^- at the nucleus in the descending potential well ($z\mathcal{E}(t) < 0$) (*13*). We define the localized ionic structures $|H^+H^-\rangle$ and $|H^-H^+\rangle$ as H^- ions located at $z = \pm R/2$, respectively. At $t=0$, $|\langle\psi|H^+H^-\rangle|^2=|\langle\psi|H^-H^+\rangle|^2 = 0.19$. The ionic character increases as the field strength increases; at $t = 3\pi/2\omega$, the ionic character is as large as $|\langle\psi|H^+H^-\rangle|^2= 0.54$. As indicated by the broken line in Figure 2(d), an electron is ejected from the localized ionic structure $|H^+H^-\rangle$. If the two nuclei are far away from each other, the ionization potential of the localized ionic structure is considered to be as low as $I_p(H^-)=0.75$ eV. The localized ionic structure is hence regarded as a doorway state to ionization. The direct ionization route from the covalent structure $H \cdot H$ is denoted by a dotted line in Figure 2(d), but the ionization current along the dotted line is relatively small. At $R=4$ a.u., the rate of ionization from a *pure* ionic state is at least ten times greater than that from a *pure* covalent state. As R increases, the rate of ionization from the ionic state increases owing to the less attractive force of the distant nucleus, while the rate of the covalent state is almost independent of R. On the other hand, the population of the H^-H^+ created

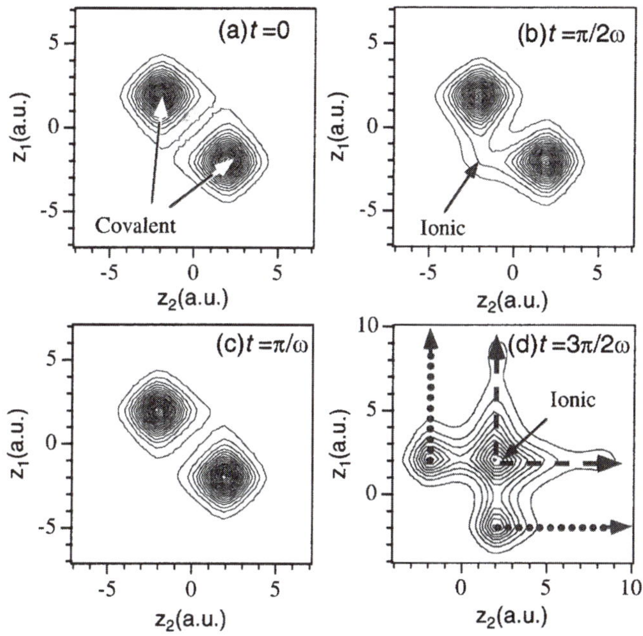

Figure 2. Snapshots of the electronic wave packet dynamics of H_2 at $R = 4$ a.u. in an intense field. The coordinates z_1 and z_2 for the two electrons are parallel to the molecular axis. The reduced density $\overline{P}(z_1, z_2)$ obtained by integrating the square of the wave function over the other degrees of freedom is drawn at quarter cycle intervals ($\omega = 0.06$ a.u.). The covalent components around $z_1 = -z_2 = \pm R/2$ are dominant in the initial ground state. The instantaneous field strengths at (a), (b), (c), and (d) are 0, 0.03, 0, -0.09 a.u., respectively. The contour intervals are the same for the four panels. An ionic component H^-H^+ or H^+H^- around $z_1 = z_2 = \mp R/2$ is created near the descending well. As indicated by the broken line in (d), an electron is ejected from the structure H^+H^-. Reproduced with permission from reference 13.

decreases with increasing R. As a result, ionization is enhanced at the critical distance R_c=4-6 a.u.

At relatively large R ($>R_e$), the field strength necessary for creating a localized ionic state H^-H^+, $\mathcal{E}_c$, is estimated as follows (*21*). The energy of the initial covalent-character-dominated state $H \cdot H$ is roughly estimated as (Atomic units are used for the equations.)

$$E(H \cdot H) \approx -2I_p(H). \tag{2}$$

The covalent-character-dominated initial state crosses the lowering excited ionic state H^-H^+ localized in the descending well. The energy of the localized ionic state H^-H^+ at the field $\mathcal{E}(t)$, $E(H^-H^+)$, is

$$E(H^-H^+) \approx -I_p(H) - I_p(H^-) - 1/R - |\mathcal{E}(t)|R, \tag{3}$$

where $-1/R$ is the energy of the Coulomb attraction between H^- and H^+, and $-|\mathcal{E}(t)|R$ is the dipole interaction energy of the two electrons in the descending well. A necessary condition for the formation of a localized ionic state is then given by $E(H^-H^+) < E(H \cdot H)$. We thus have the threshold intensity $\mathcal{E}_c$ from $E(H^-H^+) = E(H \cdot H)$. For $R = 4$ a.u., $\mathcal{E}_c \approx 0.06$ a.u. is obtained (*13*). If $|\mathcal{E}(t)|$ goes beyond the intensity required for the crossing, $\mathcal{E}_c$, the population of H^+H^- dramatically increases, as shown in Figure 2(d). The covalent-character-dominated state $H \cdot H$ is adiabatically connected with the localized ionic state in the descending well after the crossing (*16*). At large ($R > 8$ a.u.), the gap at the crossing between the two adiabatic states is very small, i.e., the case is diabatic: the main character is always covalent. As R increases, the electron density adiabatically transferred between the nuclei decreases.

Adiabaticity of vibrational motion with respect to the pulse envelope

We next describe the effect of field-induced electronic motion on the molecular vibration of H_2 (*22*). Since it is computationally demanding to treat even one vibrational mode quantum-mechanically in addition to two electrons, we have employed a 1D H_2 model in which the two electrons are allowed to move only along the molecular axis (The Coulomb potentials are softened.) (*23*). The potential surfaces of low-lying electronic states for the 1D model qualitatively agree with the experimental results, except that the 1D potentials are all shifted down. The transition moment between the 1D exact ground ($X\ ^1\Sigma_g^+$) and first excited ($B\ ^1\Sigma_u^+$) states increases as $R/\sqrt{2}$ up to $R \approx 3$ a.u [known as an electron transfer transition (*19*)]. This is consistent with accurate calculations for 3D H_2 (*24*). This 1D model reproduces essential features of the 3D H_2 such as the formation of localized ionic states in intense fields.

For excitation by Ti:Sapphire laser light, the following inequality in temporal or energy scale holds:

$$\omega_{\text{elec}} > \omega > \omega_{\text{vib}} > \dot{f}(t)/f(t) \sim 1/T_{\text{pulse}}, \tag{4}$$

where ω_{elec} represents the characteristic frequency for the electronic transitions, ω_{vib} is the vibrational frequency, $f(t)$ is the pulse envelope, and T_{pulse} is the pulse duration. For the 1D model, $\omega_{\text{elec}} \approx 0.36$a.u. for the $X \to B$ transition and $\omega_{\text{vib}} \approx 0.011$ a.u.$=2,400\,\text{cm}^{-1}$ in the X state. While the electronic motion adiabatically follows the field $\mathcal{E}(t)$, the condition $\omega > \omega_{\text{vib}}$ means that the change in $\mathcal{E}(t)$ is too fast for the vibrational motion to follow $\mathcal{E}(t)$ adiabatically. Our numerical results indicate that as long as the field-induced vibrational motion is a bound-type, it adiabatically follows the cycle-averaged pulse envelope $f(t)/\sqrt{2}$.

The adaiabatic following of the vibrational motion with $f(t)/\sqrt{2}$ is explained as follows. For $\omega_{\text{elec}} > \omega$, the vibrational motion in the jth adiabatic state is determined by the time-dependent potential $V_j(R, \mathcal{E}(t))$. In the case where the field-induced energy shift is dominated by the polarizability $\alpha(R)$, $V_j(R, \mathcal{E}(t))$ is constructed as follows:

$$V_j(R, \mathcal{E}(t)) = V_j(R, 0) - \alpha(R)\mathcal{E}^2(t)/2, \tag{5}$$

where $V_j(R, 0)$ is the adiabatic potential at zero field. Since $\omega > \omega_{\text{vib}}$ and $f(t)$ does not change in an optical period, i.e., $\omega > \dot{f}(t)/f(t)$, $\mathcal{E}^2(t)$ can be replaced with the cycle average $\left[f(t)/\sqrt{2}\right]^2$. The effective potential $\overline{V}_j(R, t)$ is then

$$\overline{V}_j(R, t) = V_j(R, 0) - \alpha(R)f^2(t)/4. \tag{6}$$

If the condition $\omega_{\text{vib}} > \dot{f}(t)/f(t) \sim 1/T_{\text{pulse}}$ is fulfilled, i.e., if the vibrational period is shorter than T_{pulse}, the "slow" vibrational motion can follow the effective pulse envelope $f(t)/\sqrt{2}$ adiabatically. If the field is not so strong that $\overline{V}_j(R, t)$ becomes a dissociative type, the vth vibrational state of $V_j(R, 0)$ is adiabatically connected with the vth vibrational state of $\overline{V}_j(R, t)$.

Electrostatic Model

On the basis of the characteristic features of electronic dynamics of H_2^+ and H_2 in intense fields, we present the simple electrostatic model in which tunnel ionization proceeds through the most unstable atomic site charged by field-induced electron transfer (or by suppression of electron transfer). The electronic wave function is deformed before ionization. For H_2, tunnel ionization proceeds via unstable localized ionic components H^-H^+ and H^+H^-. It should be pointed out that, besides the covalent-character-dominated state, only ionic-character-dominated adiabatic eigenstates are involved in bound-state electronic dynamics prior to ionization because of the large transition moments for charge transfer transitions such as $X \to B$. Only a limited number of adiabatic states participate in the dynamics of bound electrons and the subsequent ionization process. What are required to analyze the electronic dynamics prior to ionization are the populations of relevant adiabatic states and the charge distributions on individual atomic sites in a molecule. The field-induced nuclear motion is determined by the time-dependent adiabatic potential surfaces at instantaneous field strengths or by the cycle-averaged effective potentials.

We propose application of the above adiabatic state analysis to multi-electron molecules. It is virtually impossible to accurately solve the time-dependent Schrödinger equation for multi-electron systems. However, the proposed idea allows us to use a practical approach for multi-electron molecules, i.e., *ab initio* MO methods. The adiabatic potential surfaces calculated by an MO method can be used to predict whether or not the molecule is deformed; the charge distributions obtained by population analysis can be used to estimate the probability of tunnel ionization. The ionization rate from the most unstable (most negatively charged) atomic site increases as the distance between the opposite charges increases. In what follows, we apply this approach to reveal the mechanism of the experimentally observed structure deformations of CO_2 cations.

Structure Deformation of CO_2

Ultrafast field-induced deformation of molecular structure has been experimentally investigated for various triatomic molecules such as H_2O (*25*), NO_2 (*26*) and CO_2 (*27*). Hishikawa *et al.* (*28*) have experimentally determined the geometrical structure of CO_2^{3+} in a 1.1 PWcm^{-2}, 100 fs pulse ($\lambda = 795$ nm) just before undergoing Coulomb explosions, namely, that the C-O bond length is stretched to about 1.6 Å and the mean amplitude of bending is large (~40°). To investigate the deformation stage, the adiabatic potential surfaces and charge distributions of CO_2, CO_2^+, and CO_2^{2+} are calculated by using *ab initio* MO methods (*29*). The full-optimized reaction space multiconfiguration self-consistent-field (MCSCF) method (*30*) with the 6-311+G(d) basis set (*31*) is used for a zero-field case. For a nonzero field case, we employ the MCSCF or the full valence configuration interaction (CI) method where the MCSCF orbitals optimized for the zero field are used as a basis set. Although the MCSCF is more accurate than the CI, it is more laborious to go through the MCSCF procedure in the presence of an intense static field. All *ab initio* MO calculations have been performed using the GAMESS suite of program codes (*32*).

A linearly polarized laser field can align molecules (*33*). To achieve high alignment along the polarization direction by adiabatic following of the molecular rotation to the pulse envelope, the pulse duration must be longer than $2\pi/2B \approx$ 40 ps (*34*), where $B(\approx 0.39\,\text{cm}^{-1})$ is the rotational constant of CO_2. Since the pulse duaration used in the experiment (≈ 100 fs) is shorter than $2\pi/2B$, the alignment is incomplete. The ionization rate has, however, a maximumat the geometry where the molecular axis is parallel to the polarization direction (*35*). This parallel case should be studied as the main spatial configuration.

The adiabatic energy shift with respect to the antisymmetric stretching mode Q_a is given by $-\mu(Q_a)\mathcal{E}(t) - \alpha(Q_a)\mathcal{E}^2(t)/2$. The cycle average of the first term is zero and $\alpha(Q_a)$ is nearly independent of Q_a for CO_2 (a quadratic decreasing function of Q_a for CO_2^{2+}). Thus, the effective potential for the antisymmetric mode is just shifted down by a field (For CO_2^{2+}, the curvature along Q_a is increased.). Bond stretching along the antisymmetric coordinate is therefore hardly induced. The main stretching motion is either symmetric stretching under the condition $R_1 = R_2$ (at $Q_a = 0$) or asymmetric stretching under a condition such as $R_1 > R_2 = R_e$, where R_1 and R_2 are the two C-O bond distances and R_e is the equilibrium C-O bond distance. The dissociation energy at zero field is ca. 16 eV in the case of symmetric stretching and ca. 7 eV in the case of asymmetric stretching. On the other hand, while the polarizability α is nearly constant in the asymmetric case where $R_1 = R > R_2 = R_e$, α in the symmetric case is an increasing function of $R_1 = R_2 = R$. Therefore, the effective dissociation energy at $I \approx 10^{14} - 10^{15}$ W/cm^2 is smaller in the symmetric case than in the asymmetric case. We hence focus on the symmetric dissociation.

We now apply the above-mentioned electrostatic model to the CO_2 case. Consider a linear molecule $O^{P+} C^{Q+} O^{Z+}$ placed parallel to a field $\mathcal{E}$. As suggested by the following numerical calculations, we assume that the charge on C does not change. The electrostatic energy of $O^{(P-1)+} C^{Q+} O^{(Z+1)+}$ expected to be a doorway state to tunnel ionization is then given by (29)

$$E_I = I_p(O^{Z+}) - I_p(O^{(P-1)+}) - (Z + 1 - P)/2R - 2R|\mathcal{E}|. \tag{7}$$

The reference energy is that of the initial state $O^{P+} C^{Q+} O^{Z+}$. As in the H_2 case, we evaluate the intensity $\mathcal{E}_c$ required for the creation of the charge-transfer ionic state by using the crossing condition $E_I = 0$. An appreciable amount of charge is transferred between the O atoms when the field strength exceeds the value $\mathcal{E}_c$. The estimated value of $\mathcal{E}_c$ serves to predict the intensity required for tunnel ionization.

(i) Neutral CO_2 The charges on the three atoms in the *lowest adiabatic state* of linear CO_2 are plotted in Figure 3 for two field strengths: $\mathcal{E}(t) = 0$ a.u. and 0.1 a.u. The total charge is assigned to each atom by the Mulliken population analysis of the *ab initio* MO calculation. The molecular axis is parallel to the polarization direction. Near the equilibrium internuclear distance $R_e \approx 1.2$ Å, the charges of O and C at zero field are -0.22 and +0.45, respectively. The main charge distribution at zero field is expressed as $O^{0+} C^{0+} O^{0+}$. When a field is applied, an appreciable amount of negative charge is transferred from the O in the ascending well to the O in the descending one; on the other hand, the charge on C changes very little. The molecular axis component of the polarizability at R_e calculated by the MCSCF agrees with the experimentally observed value (4.1×10^{-24} cm^3) (36), while the CI value is about half of the experimental value. The inaccuracy of the CI method results in insufficient charge transfer as shown in Figure 3. The contribution of the hyperpolarizability (37) to the adiabatic energy shift is negligible (less than 10% up to $\mathcal{E}(t) = 0.2$ a.u.).

As R or $\mathcal{E}(t)$ increases, the difference in charge between the O atoms increases. For $\mathcal{E}(t) \geq 0.13$ a.u., a pure ionic structure $O^- CO^+$ is formed at R_e. The field strength necessary for the creation of $O^- CO^+$ from $O^{0+} C^{0+} O^{0+}$ is estimated by eq. 7 as $\mathcal{E}_c = \{[I_p(O) - I_p(O^-)] - 1/2R\}/2R$. Near R_e, $\mathcal{E}_c = 0.05$ a.u. As shown in the MCSCF case, the charge of the O atom in the descending

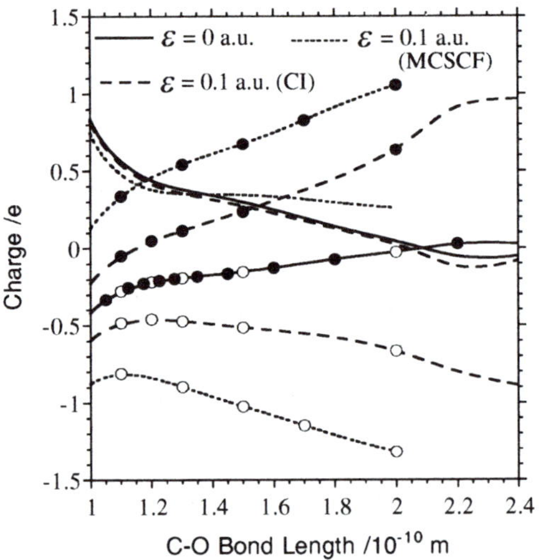

Figure 3. The charges of the three atoms of CO_2 for two field strengths ε. The charge distributions denoted by the broken and dotted lines are obtained by the CI and MCSCF methods, respectively. The lines with open circles denote the charge of the O atom in the descending well, the lines with closed circles denote the charge of the O atom in the ascending well, and the lines without marks denote the charge of C. Reproduced with permission from reference 29.

well is -0.5 at $\mathcal{E}_c$ and -0.83 at $\mathcal{E}(t) = 0.1$ a.u. $> \mathcal{E}_c$. The ionic component O^-CO^+ becomes dominant beyond $\mathcal{E}_c$.

The attractive force due to the positively charged atom exerted on the localized ionic component determines what portion is ionized out of the created ionic component. To estimate the effect of the positively charged atom on tunnel ionization, we refer to the H_2 at internuclear distance $R_{H-H} \approx 2.3$ Å. Since the value of R_{H-H} is equal to the distance between the positive and negative charges in O^-CO^+, the positively charged atom in the H_2 at $R_{H-H} \approx$ 2.3 Å exerts nearly the same attractive force on the localized ionic component as in the CO_2 case. This H_2 model has almost the same properties as CO_2 has (such as in $\mathcal{E}_c$ and ionization potential), though the total number of electrons of H_2 is greatly different from that of CO_2. According to the wave packet simulation for H_2 modeled after CO_2, the field intensity required for ionization is around 0.08 a.u. ($> \mathcal{E}_c = 0.06$ a.u. for the H_2 at $R_{H-H} \approx 2.3$ Å). We thus expect for CO_2 that tunnel ionization via the ionic structure occurs somewhere not far above its $\mathcal{E}_c = 0.05$ a.u. (say, ~0.08 a.u.).

We have calculated the potential surfaces of the lowest adiabatic state of CO_2 as a function of R and the bond angle θ. Although the dissociation energy is reduced from 16 eV to 8 eV at $\mathcal{E}(t) = 0.1$ a.u. ($> \mathcal{E}_c$) in the MCSCF calculation (10 eV in the CI calculation), the 2D potential is a bound-type. In the experiment, the condition of eq. 4 holds (The period of the symmetric stretching, 25 fs, is shorter than T_{pulse}.). If the same criterion for adiabatic following of the vibrational motion is applied to the CO_2 case, the vth vibrational state at zero field is adiabatically transferred to the vth state of the effective vibrational potential like eq. 6. We thus conclude that a stable linear structure around $R_e = 1.2$ Å exists even at field strengths $\mathcal{E}(t) \approx 0.1$ a.u. Large amplitude bending motion is not induced for $\mathcal{E}(t) < 0.1$ a.u. A field intensity > 0.15 a.u. is required for dissociation.

In the neutral stage, ionization occurs before the field intensity becomes high enough to deform the molecule. The same conclusion is drawn also for the CO_2^+ stage. For CO_2 and CO_2^+, in the ionization stage, R is small: only the lowest adiabatic state is expected to be populated and adiabatic electron transfer occurs therein (For CO_2^+, the lowest two states adiabatically connected to $^2B_2$ and $^2A_2$ at zero field may be involved.), as has been assumed above.

(ii) CO_2^{2+} In the subsequent stage of the ionization of CO_2^+, one must consider at least the lowest three adiabatic states of CO_2^{2+}, the ground triplet state connected with the $^3B_1$ state at zero field and the nearly degenerate lowest singlet states connected with $^1A_1$ and $^1B_1$ at zero field ($^1\Delta_g$ at zero field). The energy difference between $^1A_1$ and $^3B_1$ at zero field is as small as ~1.5 eV near the equilibrium geometry ($R_e \approx 1.2$ Å) and the potential surfaces of the lowest three states have nearly the same shape, irrespective of the field strength. Since the laser field under consideration is very strong, the difference of ~1.5 eV (~one photon energy $\hbar\omega$) is not decisive in electronic and nuclear dynamics. The lowest three states have nearly the same charge distribution. The discussion below applies to both singlet and triplet cases.

Two positive charges in CO_2^{2+} are nearly equally distributed among the three atoms as OC^+O^+, O^+C^+O, and O^+CO^+. An ionic structure favorable for tunnel ionization is $O^-C^+O^{2+}$ created from OC^+O^+. The field strength required for this crossing estimated by eq. 7 is $\mathcal{E}_c = 0.18$ a.u. at R_e, which is expected to be the minimum intensity for tunnel ionization. Near R_e, the charge of the O atom in the descending well is nearly zero at $\mathcal{E}(t) = 0.1$ a.u.; the main structure at $\mathcal{E}(t) = 0.1$ a.u. is OC^+O^+. In the intensity region up to $\mathcal{E}(t) \approx 0.1$ a.u., the electron transfer corresponds to the transition from O^+C^+O to OC^+O^+ (The energy of OC^+O^+ is always lower than that of O^+C^+O when $\mathcal{E}(t) > 0$.). Much higher field strengths (~ 0.2 a.u. $> \mathcal{E}_c$) are required to create an ionic structure favorable for tunnel ionization, $O^-C^+O^{2+}$, at R_e.

The potential surface of linear CO_2^{2+} in the lowest singlet state is presented in Figure 4. The dissociation energy is as small as ~ 2.2 eV at $\mathcal{E}(t) = 0.1$ a.u. as shown in the more reliable MCSCF case. The potential surface becomes dissociative for $\mathcal{E}(t) > 0.11$ a.u. The bond of CO_2^{2+} is stretched at field strengths 0.11-0.18 a.u., which do not cause tunnel ionization near R_e. In Figure 5, we show 2D potential surfaces of the lowest singlet adiabatic state of CO_2^{2+} at $\mathcal{E}(t) = 0$ and 0.1 a.u. The O-O axis is assumed to be parallel to the polarization direction. The adiabatic energies are calculated by the CI method. The potential curvature along θ is nearly the same as that in the MCSCF calculation.

The induced dipole of the lowest adiabatic state which shifts the energy down becomes smaller as the molecule becomes more bent. As a result, the curvature of the potential with θ is larger at nonzero fields than at zero field. If R is fixed, therefore, the electric field hardly induces large amplitude bending motion because it does not flatten the potential along θ. However, for field amplitudes $f(t) > 0.11$ a.u., the bond of CO_2^{2+} stretches. The kinetic energy gained when $|\mathcal{E}(t)|$ is large can be converted to bending motion. Along isoenergy contour lines starting from the equilibrium structure, the nuclear wave packet spreads. A typical case is indicated in Figure 5(b) by the stream of arrows that determines the maximum amplitude for the instantaneous potential. Another important factor that increases the bending amplitude as R increases comes from the fact that the potential at $\mathcal{E}(t) = 0$ is very flat against θ for $R > 1.5$ Å as shown in Figure 5(a). The cycle-averaged potential indicates that the wave packet in the large R region spreads along the bending coordinate more easily. We thus conclude that field-induced bond stretching accompanied by a large amplitude bending motion occurs before ionization. This is responsible for the experimentally observed structure deformation of CO_2^{3+} just before Coulomb explosions. As R increases, $\mathcal{E}_c$ decreases and the attractive force due to the distant nuclei C^+ and O^{2+} in the ionic structure $O^-C^+O^{2+}$ becomes weaker against an electron in O^-. Hence, after bond stretching in CO_2^{2+}, ionization to CO_2^{3+} occurs even at field strengths that do not ionize CO_2^{2+} at R_e.

Concluding Remarks

We have investigated the electronic and nuclear dynamics of H_2^+ and H_2 in intense laser fields by solving the time-dependent Schrödinger equation of the system. The mechanisms of intense-field-induced phenomena such as intramolecular electron transfer and enhanced ionization can be clarified by

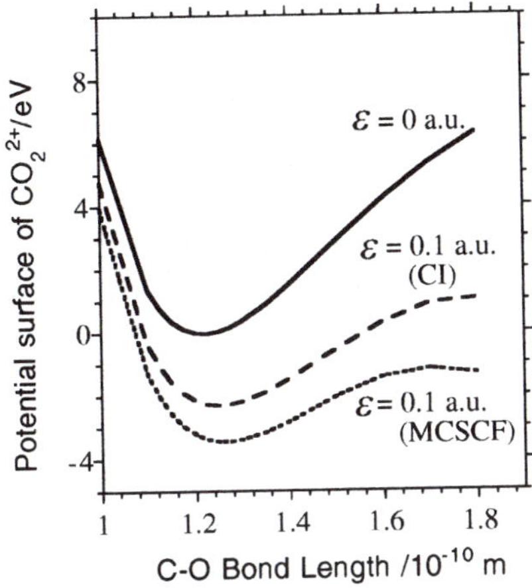

Figure 4. Potential surfaces of the lowest singlet state of linear CO_2^{2+} (under symmetric stretching): $\varepsilon(t) = 0$ a.u. (solid line), CI (broken line) and MCSCF (dotted line) calculations at $\varepsilon(t) = 0.1$ a.u.

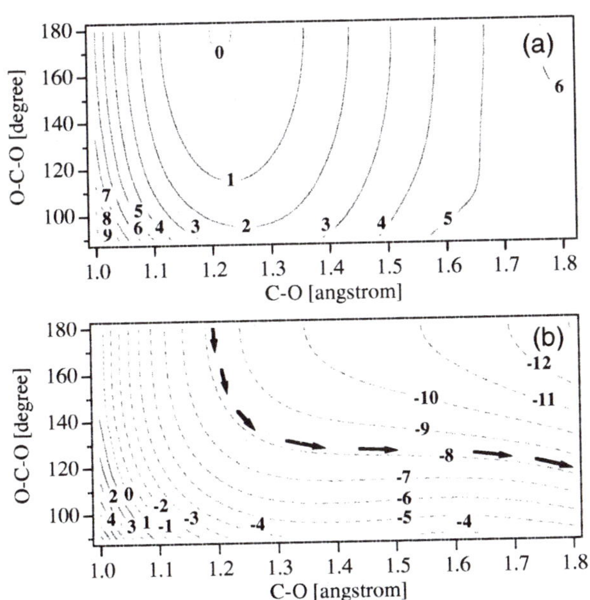

Figure 5. 2D potential surfaces of the lowest singlet adiabatic state of CO_2^{2+} at (a) $\varepsilon(t) = 0$ and (b) 0.2 a.u. The heights of contour lines are indicated in units of eV.

mapping the total wave function onto field-following adiabatic states defined as eigenfunctions of the instantaneous electronic Hamiltonian (which gives the populations or nuclear wave functions of adiabatic states). In a high-intensity and low-frequency regime, only a limited number of adiabatic states participate in the dynamics of bound electrons and the subsequent ionization process. The nuclear motion initiated by field-induced electronic motion can be described by using the time-dependent potentials of adiabatic states.

From the characteristic features of electronic dynamics of H_2^+ and H_2 in intense laser fields, we have extracted the simple electrostatic view that each atom in a molecule is charged by field-induced electron transfer and ionization proceeds via the most unstable atomic site. We have proposed to extend this idea to multi-electron polyatomic molecules. The properties of adiabatic states of multi-electron molecules can be calculated by *ab initio* MO methods. While the adiabatic potentials calculated by an MO method are used to evaluate the nuclear dynamics, the charge distributions on individual atomic sites are used to estimate the ionization probability. Applying the adiabatic state analysis to CO_2 in an intense field, we have revealed that field-induced bond stretching accompanied by a large amplitude bending motion occurs in the CO_2^{2+} stage. This approach is simple but it has wide applicability in predicting the *electronic* and *nuclear* dynamics of polyatomic molecules in intense laser fields.

In intense fields, the electronic wave function changes dramatically on an attosecond scale. Laser control in this regime will therefore bring about ultrafast (intermolecular and intramolecular) electron transfer on a massive scale. The change in charge distribution by an intense field also varies the atomic site that tunnel ionization proceeds from, which changes fragment species due to resultant Coulomb explosions. It may be possible to break a specific bond by controlling Coulomb explosions. Bond-selective cleavage of an HOD molecule can be designed by utilizing the fact that the speed of bond softening of the O-H bond is faster than that of the O-D bond. It is also interesting to control the symmetric and asymmetric pathways of bond dissociation of CO_2. The nuclear dynamics can be manipulated by controlling time-dependent adiabatic surfaces, i.e., by controlling the electronic dynamics (*38*).

This work was supported in part by the Development of High-Density Optical Pulse Generation and Advanced Material Control Techniques and by a grant-in-aid for scientific research from the Ministry of Education, Science and Culture, Japan (12640484). We would like to express our appreciation to Professors K. Yamanouchi and A. Hishikawa for discussion of their work. H.K. would like to thank Professors A. D. Bandrauk, T. T. Nguen-Dang, D. Mathur, H. Sakai, Y. Fujimura and Dr. I. Kawata for their valuable discussions.

References

1. Gavrila, M., Ed. *Atoms in Intense Fields;* Academic Press: NewYork, 1992. Eberly, J. H.; Javanainen, J.; Rzazewski, K. *Phys. Rep.* **1991**, *204*, 331. Lewenstein, M.; Kulander, K. C.; Schafer, K. J.; Bucksbaum, P. H. *Phys. Rev.* A **1995**, *51*, 1495.

2. Keldysh, L. V. *Sov. Phys. JETP* **1965**, *20*, 1307.
3. Muth-Böhm, J.; Becker, A.; Faisal, F. H. M. *Phys. Rev. Lett.* **2000**, *85*, 2280. Augst, S.; Meyerhofer, D. D.; Strickland, D.; Chin, S. L. *J. Opt. Soc. Am. B* **1991**, *8*, 858. Krainov, V. P.; Reiss, H. R.; Smirnov, B. M. *Radiative Processes in Atomic Physics*, Wiley: New York, 1997
4. Corkum, P. B. *Phys. Rev. Lett.* **1993**, *71*, 1994.
5. Walsh, T. D. G.; Ilkov, F. A.; Chin, S. L. *J. Phys. B* **1997**, *30*, 2167.
6. DeWitt, M. J.; Levis, R. J. *Phys. Rev. Lett.* **1998**, 81, 5101. DeWitt, M. J.; Levis, R. J. *J. Chem. Phys.* **1998**, 108, 7739.
7. Posthumus, J. H.; Giles, A. J.; Thompson, M. R.; Codling, K. *J. Phys. B* **1996**, *29*, 5811. Constant, E.; Stapelfelt, H.; Corkum, P. B. *Phys. Rev. Lett.* **1996**, *76*, 4140. Codling, K.; Frasinski, L. J. *J. Phys. B* **1993**, *26*, 783. Schmidt, M.; Normand, D.; Cornaggia C. *Phys. Rev. A* **1994**, *50*, 5037.
8. Zuo, T.; Bandrauk, A. D. *Phys. Rev. A* **1993**, *48*, 3837.
9. Bandrauk, A. D. *Comments At. Mol. Phys. D* **1999**, *1*, 97.
10. Ivanov, M. Y.; Seidemann, T.; Corkum, P. B. *Phys. Rev. A* **1996**, *54*, 1541.
11. Kawata, I.; Kono H. *J. Chem. Phys.* **1999**, *111*, 9498.
12. Kawata, I.; Kono, H.; Fujimura, Y. *J. Chem. Phys.* **1999**, *110*, 11152.
13. Harumiya, K.; Kawata, I.; Kono, H.; Fujimura, Y. *J. Chem. Phys.* **2000**, *113*, 8953.
14. Chelkowski, S.; Zuo, T.; Atabek, O; Bandrauk, A. D. *Phys. Rev. A*, **1995**, *52*, 2977.
15. Kono, H.; Kawata, I. In *Advances in Multi-Photon Processes and Spectroscopy*; Gordon, R. J.; Fujimura, Y. Eds.; World Scientific: Singapore, 2001; Vol. 14, p. 165.
16. Saenz, A. *Phys. Rev. A* **2000**, *61*, 051402 (R).
17. Kawata, I.; Kono, H.; Fujimura, Y.; Bandrauk, A. D. *Phys. Rev. A* **2000**, *62*, 031401(R). Kawata, I.; Bandrauk, A. D.; Kono, H.; Fujimura, Y. *Laser Phys.* **2001**, *11*, 181.
18. Mittleman, M. H. *Introduction to the Theory of Laser-Atom Interactions (2nd.)*; Prenum: New York, 1993.
19. Mulliken, R. S. *J. Chem. Phys.* **1939**, *7*, 20.
20. Yamaki, D.; Kitagawa Y.; Nagao, H.; Nakano, M.;Yoshioka, Y.; Yamaguchi, K. *Int. J. Quant. Chem.* **1999**, *75*, 645.
21. Bandrauk, A. D. In *The Physics of Electronic and Atomic Collisions*; Itikawa, Y. *et al.* Eds.; AIP Conf. Proc. 500: New York, 1999; p.102.
22. Izawa, M.; Harumiya, K.; Kono, H.; Fujimura, Y. to be published.
23. Yu, H.; Zuo, T.; Bandrauk, A. D. *Phys. Rev. A* **1996**, *54*, 3290.
24. Wolniewicz, L. *J. Chem. Phys.* **1993**, *99*, 1851.
25. Sanderson, J. H.; El-Zein, A.; Bryan, W. A.; Newell, W. R.; Langley, A. J.; Taday, P. F. *Phys. Rev. A* **1999**, *59*, R2567.
26. Hishikawa, A.; Iwamae, A.; Yamanouchi, K. *J. Chem. Phys.* **1999**, *111*, 8871.

27. Cornaggia, C. *Phys. Rev. A* **1993**, *54*, R2555.
28. Hishikawa, A.; Iwamae, A.; Yamanouchi, K. *Phys. Rev. Lett.* **1999**; *83*; 1127.
29. Kono, H.; Koseki, S.; Shiota, M.; Fujimura, Y. *J. Phys. Chem. A* **2001**, *105*, 5627.
30. Schmidt, M. W.; Gordon, M. S. *Ann. Rev. Phys. Chem.* **1998**, *49*, 233.
31. Krishnan, R.; Binkley, J. S.; Seeger, R.; Pople, J. A. *J. Chem. Phys.* **1980**, *72*, 650.
32. Schmidt, M. W. *et al. J. Comp. Chem.* **1993**, *14*, 1347-1363.
33. Friedrich, B.; Herschbach, D. *Phys. Rev. Lett.* **1995**, *74*, 4623. Larsen, J. J.; Sakai, H.; Safvan, C. P.; Wendt-Larsen, Ida; Stapelfeldt, H. *J. Chem. Phys.* **1999**, *111*, 7774. Schmidt, M.; Dobosz, S.; Meynadier, P.; D′Oliveira, P.; Normand, D.; Charron, E.; Suzor-Weiner, A. *Phys. Rev. A* **1999**, *60*, 4706.
34. Origoso, J.; Rodríguez, M.; Gupta, M.; Friedrich, B. *J. Chem. Phys.* **1999**, *110*, 3870. Banerjee, S.; Mathur, D.; Kumar, G. R. *Phys. Rev. A* **1999**, *63*, 045401 (R).
35. Posthumus, J. H.; Plumridge, J.; Frasinski, L. J. ; Codling, K.; Langley, A. J.; Tady, P. F. *J. Phys. B* **1998**, *31*, L985. Ellert, C.; Corkum, P. B. *Phys. Rev. A* **1999**, *59*, R3170. Banerjee, S.; Kumar, G. R.; Mathur, D. *Phys. Rev. A* **1999**, *60*, R3369.
36. Spackman, M. A. *J. Phys. Chem.* **1989**, *93*, 7594.
37. Sekino, H.; Bartlett, R. *J. Chem. Phys.* **1993**, *98*, 3022.
38. Abou-Rachid, H.; Nguyen-Dang, T.T.; Atabek, O. *J. Chem. Phys.* **1999**, *110*, 4737.

Alignment and Manipulation

Manipulating Molecules via Combined Electrostatic and Pulsed Nonresonant Laser Fields

Long Cai and Bretislav Friedrich

Department of Chemistry and Chemical Biology, Harvard University, 12 Oxford Street, Cambridge, MA 02138

Any polar molecule can be strongly oriented by the combination of a weak electrostatic field ($10^1 - 10^4$ V/cm) with a nonresonant laser field ($10^8 - 10^{12}$ W/cm^2). The resulting pseudo-first-order Stark effect causes the molecule - whether it is linear or asymmetric - to librate about the field vector with an angular amplitude of, typically, less than $\pm 30°$. A pulsed nonresonant laser field provides the requisite intensity but introduces a time dependence. We show that even for a short-pulse, nonadiabatic interaction, the orientation due to the electrostatic field alone can be greatly enhanced by the pulsed laser field. We describe both the stationary states and the rotational wavepackets created, respectively, by the adiabatic and nonadiabatic interaction with the combined fields as a function of the duration and intensity of the laser pulse, the strength of the electrostatic field, and the tilt angle between the two fields. The orientation and alignment of the molecular axis can he used to tailor spectroscopic transition as well as to control or restrict molecular rotation and translation.

The means to align or orient the molecular axis have been augmented during the past decade by new techniques based on the hybridization of rotational states of molecules with external static electric [1], magnetic [2], or radiative fields [3]-[4]. The hybridization is due to the dipole potential, either permanent (for a polar or paramagnetic molecule in a static field) or induced (for a nonspherical polarizable molecule in a nonresonant radiative field), and gives rise to directional, *pendular* states in which the molecular axis is confined to librate over a limited angular range about the external field vector. Since low rotational states undergo field hybridization more easily than the higher ones, it is advantageous, in order to foster orientation or alignment, to cool the molecules rotationally before they interact with the field. However, the nonresonant polarizability interaction with a laser field is often strong enough to hybridize all rotational states even in a thermal ensemble, which supersedes the requirement that the molecules be cold. Pendular states have been instrumental in a host of recent developments in areas such as stereodynamics of elementary collisions [1], spectroscopy [5]-[7], photodissociation dynamics [8]-[11], molecular focusing [12],[13], and slowing [14] and trapping [15] of molecules.

In a static electric field the axis of a polar molecule is oriented, whereas the axis of a nonspherical molecule in a nonresonant radiative field is aligned with respect to the field vector. As usual, here axial anisotropy is designated *orientation* if it behaves like a single-headed arrow and *alignment* if it behaves like a double-headed arrow.

The orientation of a polar molecule in an electrostatic field can be dramatically enhanced by a combined action with a nonresonant laser field [16],[17] or with a static magnetic field [18],[19]. Here we deal with the former combination which enhances the orientation due to a pseudo-first-order Stark effect. This makes any molecule - whether it is linear or asymmetric - to act almost like a symmetric top.

The pseudo-first-order Stark effect in the combined fields takes place when two nearly degenerate states of opposite parity get coupled by the electric dipole interaction. The near-degeneracy in a nonresonant laser field arises as a consequence of the induced dipole interaction which produces a double-well potential, governed by the anisotropy of molecular polarizability and the laser intensity. The pendular energy levels thus occur as tunneling doublets, whose splitting decreases exponentially with the square root of the strength of the interaction (or of the laser intensity) [20].

The enhancement of the orientation can also be qualitatively understood in terms of the pendular amplitude of the molecular axis. This is limited to a narrow range by the strong polarizability interaction. Since the body-fixed permanent electric dipole moment is coupled to the molecular axis, its angular amplitude is limited by the laser field as well; the electrostatic field then just defines a preferred direction for the oscillating electric dipole. Thus often a very weak static electric field can convert second-order alignment by a laser into a strong first-order orientation. The effect occurs for any polar molecules, as only an anisotropic polarizability is required.

Nonresonant radiative fields of the requisite strength can be attained by focusing a pulsed laser beam. For most small diatomic molecules, laser intensities ranging between $10^8 - 10^{12}$ W/cm^2 suffice to generate an alignment that corresponds to an angular amplitude of the molecular axis of about $\pm 30°$ for the lowest molecular state [3]. However, such radiative fields are time dependent, delivered as pulses (with, typically, Gaussian time profiles). Depending on the width of the time profile (the pulse duration), τ, we distinguish two limiting cases of the interaction of the radiative field with the molecule [21]: (a) If $\tau \gtrsim 5\, \hbar/B$ (where B is the rotational constant of the molecule), the interaction is adiabatic and the molecule behaves as if the radiative field were static at any instant. The states thereby created are the stationary *pendular states* [3] (see also below); (b) For $\tau \lesssim \hbar/B$, the time evolution is nonadiabatic and so the molecule ends up in a *rotational wavepacket*. The wavepacket is comprised of a finite number of free-rotor states and thus may recur after the pulse has passed - giving rise to alignment under field-free conditions, as pointed out by Bandrauk [22] and later analyzed by others [21],[23].

Here we examine the time evolution of the states created by the combined fields, describe the directionality of the eigenstates, and show that even for a nonadiabatic interaction with a short pulse, the orientation due to the electrostatic field alone can be greatly enhanced by the pulsed laser field. We map out the rotational wavepackets created by the nonadiabatic interaction with the combined fields as a function of the duration and intensity of the laser pulse, the strength of the electrostatic field, and the tilt angle between the two fields. We find an analytic solution to the time-dependent Schrödinger equation in the short-pulse limit. This agrees quantitatively with our computations and indicates that the recurring rotational wavepackets form under an impulsive transfer of action from the radiative field to the molecule.

Theory

We consider a $^1\Sigma$ rotor molecule with a permanent dipole moment μ along the molecular axis and polarizability components $\alpha_\parallel$ and $\alpha_\perp$ parallel and perpendicular to the molecular axis. The molecule is subject to a combination of a static electric field, ε_S, with a pulsed plane-polarized nonresonant laser field, ε_L, see Figure 1. The fields ε_S and ε_L can be tilted with respect to one another by an angle β. Collinear fields ($\beta = 0$ or $180°$) exhibit an azimuthal symmetry, and so the permanent and induced dipole potentials involve just the polar angle θ between the molecular axis and the field direction. Consequently, the projection, M, of the molecular angular momentum $\mathbf{J}$ on the field direction, is then a "good" quantum number. We limit consideration to a pulsed plane wave radiation of frequency ν and time profile $g(t)$ such that

$$\varepsilon_L^2(t) = \frac{8\pi}{c} I_0 g(t) \cos^2(2\pi \nu t) \tag{1}$$

where I_0 is the peak intensity. We assume the oscillation frequency ν to be far removed from any molecular resonance and much higher than either

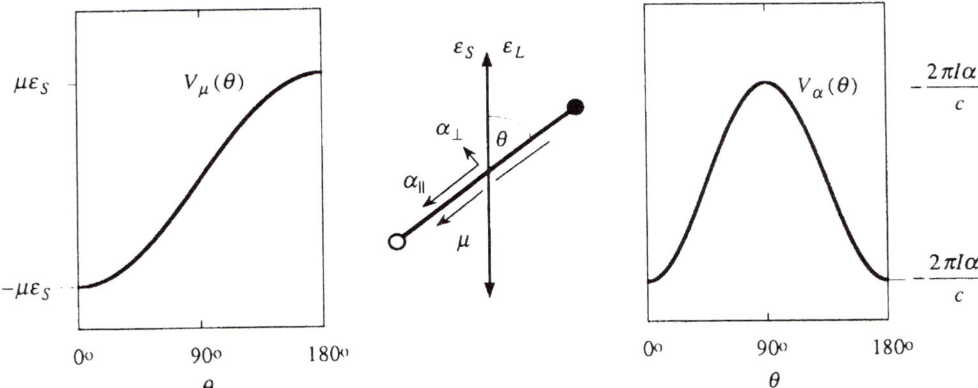

Figure 1: The induced dipole potential, $V_\alpha(\theta) = -\frac{2\pi I}{c}[(\alpha_{||} - \alpha_\perp)\cos^2\theta + \alpha_\perp]$, and the permanent dipole potential, $V_\mu(\theta) = -\mu\varepsilon_S\cos\theta$, with the quantities defined in the diagram in the center. For $V_\alpha(\theta)$, the pair of potential minima in the polar regions is separated by an equatorial barrier. Note that $V_\alpha(0°) = V_\alpha(180°) = -\frac{2\pi I\alpha_{||}}{c}$ and $V_\alpha(90°) = -\frac{2\pi I\alpha_\perp}{c}$. The $V_\mu(\theta)$ potential has a single minimum, $V_\alpha(0°) = -\mu\varepsilon_S$, and a maximum, $V_\mu(180°) = \mu\varepsilon_S$.

τ^{-1} or the rotational periods. The resulting effective Hamiltonian, $H(t)$, is thus averaged over the rapid oscillations. This cancels the interaction between μ and ε_L (see also ref. [24]) and reduces the time dependence of ε_L to that of the time profile,

$$\langle \varepsilon_L^2(t) \rangle = \frac{4\pi}{c} I_0 g(t) \tag{2}$$

With the dimensionless interaction parameters given by

$$\begin{aligned}
\omega_{||,\perp}(t) &= \omega_{||,\perp} g(t) \\
\omega_{||,\perp} &\equiv \frac{2\pi\alpha_{||,\perp} I}{Bc} \\
\Delta\omega &\equiv \omega_{||} - \omega_{\perp} \\
\Delta\omega(t) &= \omega_{||}(t) - \omega_{\perp}(t) \equiv \Delta\omega g(t)
\end{aligned} \tag{3}$$

the Hamiltonian becomes

$$H(t) = B\left[\mathbf{J}^2 - \Delta\omega(t)\cos^2\theta - \omega_{\perp}(t) - \omega\cos\theta \right] \tag{4}$$

If the tilt angle β between the field directions is nonzero, the relation

$$\cos\theta_L = \cos\beta\cos\theta + \sin\beta\sin\theta\cos\phi \tag{5}$$

is employed in Hamiltonian (4), with $\theta, \phi \equiv \theta_S, \phi_S$. In what follows we concentrate on the collinear case and just mention, in Results and Discussion, some of the effects that can be attained with $\beta = 90°$.

The time-dependent Schrödinger equation corresponding to Hamiltonian (4) can be cast in a dimensionless form

$$i\frac{\hbar}{B}\frac{\partial\psi(t)}{\partial t} = \frac{H(t)}{B}\psi(t) \tag{6}$$

and the wavefunctions expanded in terms of a series in field-free rotor wave functions $|JM\rangle \equiv Y_{JM}$ (pertaining to eigenenergies E_J)

$$\psi(\Delta\omega(t), \omega) = \sum_J d_J(\Delta\omega(t), \omega)|J, M\rangle \exp\left[-\frac{iE_J t}{\hbar}\right] \tag{7}$$

whose time-dependent coefficients, $d_J(\Delta\omega(t), \omega) \equiv d_J(t)$, solely determine the solutions at given values of the dimensionless parameters $\Delta\omega(t)$ and ω and at given initial conditions (in the interaction representation). The expansion (or *hybridization*) coefficients $d_J(t)$ can be found from the differential equations

$$i\frac{\hbar}{B}\dot{d}_J(t) = -\sum_{J'} d_{J'}(t)\langle J, M|\Delta\omega(t)\cos^2\theta + \omega_{\perp}(t) + \omega\cos\theta|J', M\rangle$$

$$\times \exp\left[-\frac{i(E_{J'} - E_J)t}{\hbar}\right] \tag{8}$$

Note that the $\cos^2\theta$ and $\cos\theta$ operators connect states that differ in J by $0, \pm2$ and ±1, respectively. In what follows, we consider the molecule to be in the ground rotational state, $|0,0\rangle \equiv Y_{0,0}$, before switching on any of the fields (*i.e.*, for $\omega_{\|,\perp}(t=0)=0$ and $\omega=0$). We take the pulse shape function to be a Gaussian,

$$g(t) = \exp(-\frac{t^2}{\sigma^2}) \tag{9}$$

characterized by a full width at half maximum, $\tau = 2\,(\ln 2)^{1/2}\,\sigma \approx 1.67\sigma$.

Results and Discussion

We examine the effect of both the time extent (pulse duration τ) and magnitude ($\Delta\omega$, proportional to the laser intensity I) of the laser pulse on the solutions of Eq. (8) in the long and short-pulse limits. We chose a permanent dipole interaction characterized by $\omega = 1$, which is attainable for most small ground-state polar molecules with electrostatic fields on the order of up to 10 kV/cm; *e.g.*, $\varepsilon_S = 730$ V/cm for ground-state KCl. The chosen range of $\Delta\omega$ includes values of up to 1000, which is attainable for some of the above molecules with nonresonant laser intensities of less than 5×10^{12} W/cm^2; *e.g.*, $\Delta\omega = 240$ for KCl at 10^{12} W/cm^2. However, strong effects are in place at much lower $\Delta\omega$ values (see below). Note that $\sigma = \hbar/B$ corresponds to 5.3 ps for $B = 1$ cm^{-1}.

Long-Pulse (Adiabatic) Limit

For $\sigma \to \infty$, the time profile $g(t) \to 1$ and Hamiltonian (4) becomes

$$\frac{H(t)}{B} = \mathbf{J}^2 - \left(\Delta\omega\cos^2\theta + \omega_\perp\right) - \omega\cos\theta \tag{10}$$

Its stationary solutions (*pendular states*) [16],[17]

$$\Psi\left(\Delta\omega\right) = \sum_J d_J\left(\Delta\omega,\omega\right)|J,M\rangle \equiv |\tilde{J},M;\Delta\omega,\omega\rangle \tag{11}$$

pertain to eigenvalues

$$\lambda_{\tilde{J},M} = E_{\tilde{J},M}/B + \omega_\perp \tag{12}$$

For $\Delta\omega = \omega = 0$, the eigenproperties become those of a field-free rotor; the eigenfunctions then coincide with spherical harmonics, and the eigenvalues become $\lambda_{\tilde{J},M} \to E_{\tilde{J},M}/B = J(J+1)$. The eigenstates can thus be labeled by M and the nominal value $\tilde{J}$, designating the angular momentum of the field-free rotor state that adiabatically correlates with the high-field hybrid function. In the high-field limit, $\Delta\omega \to \infty$ and/or $\omega \to \infty$, the range of θ is confined near a potential minimum and Eq. (10) reduces to that for a two-dimensional angular harmonic oscillator (harmonic librator).

Figure 2 shows the energy levels $E_{\tilde{J},M}/B$ as a function of the $\Delta\omega$ parameter for $\omega = 0$ (upper panel) and $\omega = 10$ (lower panel). The anisotropic $\cos^2\theta$ potential produces a double well which splits the pendular states bound by these wells into tunneling doublets, see $e.g.$, the $\tilde{J}, M = 0, 0$ and $1, 0$ or $2, 1$ and $1, 1$ pairs. The members of a given tunneling doublet have definite but opposite parities. For large $\Delta\omega$, these levels become nearly degenerate (see upper panel, $\Delta\omega > 15$). As in any double-well potential, the energy difference between the paired levels corresponds to the frequency for tunneling between the wells. This shrinks exponentially with the square root of the laser intensity [20]. Introducing the permanent dipole interaction, proportional to $\cos\theta$, connects these nearly degenerate states of opposite parity and thereby creates a pseudo-first order Stark effect ($cf.$ the large splitting between the members of a given tunneling doublet in the lower panel). Thus, if $\Delta\omega$ is large, even a very weak static field can produce quite strong orientation, with $\langle\cos\theta\rangle$ large and positive for the $0, 0$ state and equally large but negative for the $1, 0$ state, see Figure 3. The sense and magnitude of the orientation of the two members of a tunneling doublet follow from the Hellmann-Feynman theorem, $\langle\cos\theta\rangle = -\partial(E_{\tilde{J},M}/B)/\partial\omega$, applied to the energy shifts. In effect, a strong polarizability interaction is capable of making any polar molecule behave in a static field as if it were almost a symmetric top.

Recently, a direct experimental evidence for the adiabatic enhancement of orientation in the combined fields has been found by Baumfalk et $al.$ [25]. In their experiment, HXeI was subjected to a 5 ns laser pulse of 2.4×10^{12} W/cm^2 collinear with an electrostatic field of 4 V/cm, yielding $\tau \approx 25$ $\hbar/B$, $\Delta\omega = 50$, and $\omega = 0.016$. The observed "complete" orientation of HXeI corresponds to $\langle\cos\theta\rangle \approx 0.91$ for the pendular ground state, $cf.$ Fig. 3.

Short-Pulse (Nonadiabatic) Limit

In the short-pulse limit, $\sigma \to 0$, the time evolution of the initial wavefunction $|\tilde{0}, 0; 0, \omega\rangle$ under the Hamiltonian $H(t)$ is approximated, up to order σ, by a propagator $S(t) = \exp\left[-\frac{i}{\hbar}\int H(t')dt'\right]$, which yields an approximate wavefunction

$$\tilde{\psi}(t) = S(t)|\tilde{0}, 0; 0, \omega\rangle \tag{13}$$

$$= \exp\left[-\frac{i}{\hbar}B\left[t\mathbf{J}^2 - t\omega\cos\theta - \Delta\omega\cos^2\theta G(t)\right]\right]|\tilde{0}, 0; 0, \omega\rangle \tag{14}$$

with $G(t) = \int_{-\infty}^{t} g(t')dt' = \frac{1}{2}\pi^{1/2}\sigma\left[1 + \mathrm{erf}(t/\sigma)\right]$, see also ref. [26]. For $t \gtrsim 12\sigma \equiv t_a$, $G(t_a)$ is of the order σ and the effect of $\mathbf{J}^2$ can be neglected.

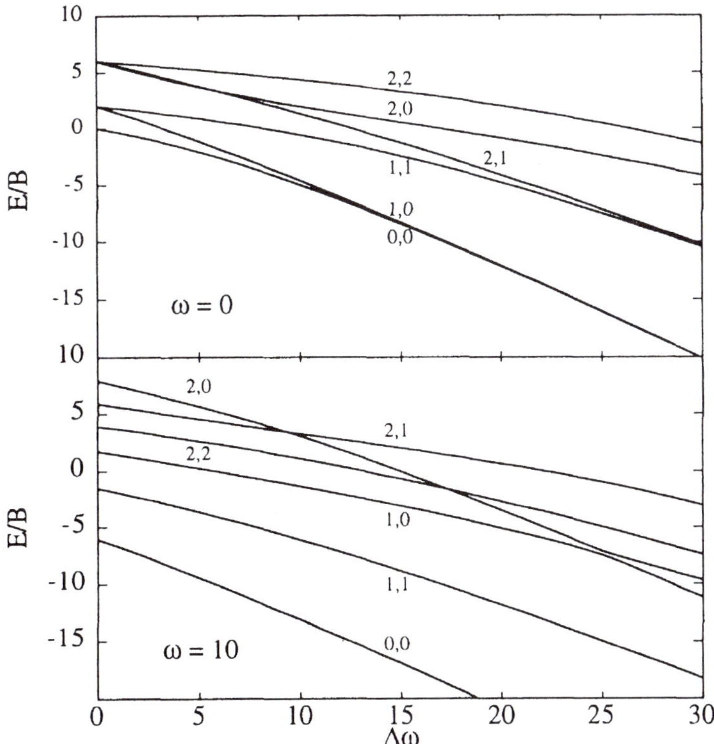

Figure 2: Energy levels (in units of the rotational constant B) of the lowest states of linear molecules as a function of $\Delta\omega$ for $\omega = 0$ (upper panel) and $\omega = 10$ (lower panel) at $\beta = 0°$ or $180°$. The $\Delta\omega$ and ω parameters measure, respectively, the strengths of the laser and static fields. The $\tilde{J}, \tilde{M} = 1, 0$ and $0, 0$ and $2, 1$ and $1, 1$ pairs of states form tunneling doublets of opposite parity. These split as a result of coupling by the electrostatic field.

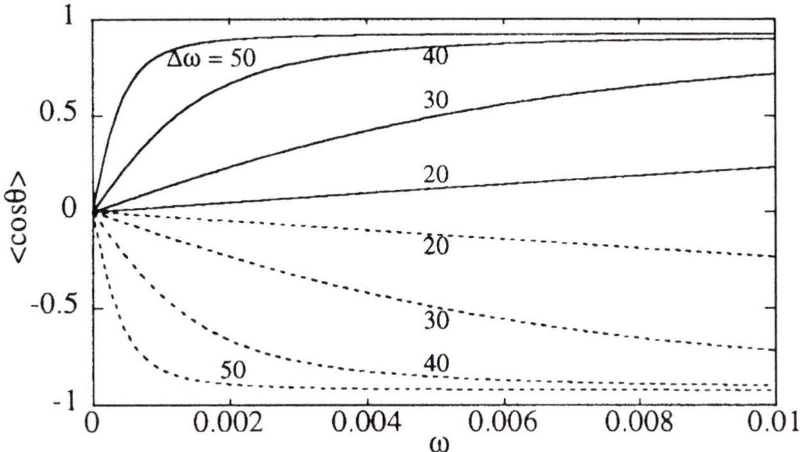

Figure 3: Effect of combined action of collinear fields ($\beta = 0°$ or $180°$) on the expectation value $\langle \cos\theta \rangle$ of orientation cosine, for the $\tilde{J}, \tilde{M} = 0, 0$ (full curves) and $1, 0$ states (dashed curves). Curves show variation with ω, the strength of the permanent dipole interaction (here very weak), for several values of $\Delta\omega$, the strength of the induced dipole interaction. For $\omega = 0$, and $\Delta\omega > 10$, the $0, 0$ and $1, 0$ states are components of a tunneling doublet and become nearly degenerate as $\Delta\omega$ increases. When $\omega > 0$ these levels acquire large effective dipole moments, opposite in sign. Thereby the molecular axis becomes oriented, in the $0, 0$ state parallel to ε_S, in the $1, 0$ state antiparallel.

As a result, the approximate wavefunction becomes

$$\bar{\psi}\left(t \geq t_a\right) \tag{15}$$

$$\approx \exp\left[-\frac{i}{\hbar}Bt\left(\mathbf{J}^2 - \omega\cos\theta\right)\right]\exp\left[\frac{iB\pi^{1/2}\sigma\Delta\omega}{2\hbar}\cos^2\theta\right]|\tilde{0},0;0,\omega\rangle$$

The expansion coefficients in the non-adiabatic limit are obtained by the transformation

$$d_{\tilde{J}}(t \geq t_a) \approx 2\pi\int\langle\tilde{J},0;0,\omega|\,\bar{\psi}\left(t \geq t_a\right)\sin\theta d\theta \equiv \bar{d}_{\tilde{J}}\left(t_a\right)$$

$$= 2\pi\langle\tilde{J},0;0,\omega|\exp\left[\frac{iB\pi^{1/2}\sigma\Delta\omega}{2\hbar}\cos^2\theta\right]|\tilde{0},0;0,\omega\rangle \tag{16}$$

For the purely nonresonant laser interaction (*i.e.*, $\omega = 0$), we obtain

$$\bar{\psi}\left(t_a\right) \approx \exp\left[\frac{iB}{\hbar}\pi^{1/2}\sigma\Delta\omega\cos^2\theta\right]Y_{0,0} \tag{17}$$

which yields the expansion coefficients

$$\bar{d}_J\left(t_a\right) \approx 2\pi\int\bar{\psi}\left(t_a\right)Y_{J,0}\sin\theta d\theta \tag{18}$$

The expansion coefficients were computed by numerically solving eq. (8) and used to evaluate $\langle\cos^2\theta\rangle$ and $\langle\cos\theta\rangle$ that characterize, respectively, the alignment and orientation of the molecular axis (alignment and orientation cosines).

Figure 4 shows the dependence of the alignment and orientation cosines on time (expressed in units of σ) for different values of σ and fixed $\Delta\omega = 100$ and $\omega = 1$. At $\sigma = 0.01\ \hbar/B$ (upper panel), the pulse creates a rotational wavepacket whose alignment and orientation lag behind the pulse shape function and recur with a period determined by σ and $\Delta\omega$ (non-adiabatic behavior). On the time scale of τ, the orientation closely follows the alignment as the permanent electric dipole moment is confined to an angular range preordained by the dominant induced dipole interaction with the laser field. At $\sigma = 1\ \hbar/B$ (lower panel), the time dependence of $\langle\cos^2\theta\rangle$ and $\langle\cos\theta\rangle$ mimics the pulse shape function and by the time the pulse is over, the alignment and orientation of the initial wavefunction, $|\tilde{0},0;0,\omega\rangle$, are nearly recovered (see the nearly stationary oscillations in $\langle\cos^2\theta\rangle$ and $\langle\cos\theta\rangle$, characteristic of the eigenstates). Again, such a full recovery would correspond to the adiabatic limit when the molecule ends up in its initial eigenstate. The middle panel of Fig. 4 shows a case when the pulse establishes a phase relationship among the components of the wavepacket that *happens* to suppress the recurrences of both the alignment and orientation.

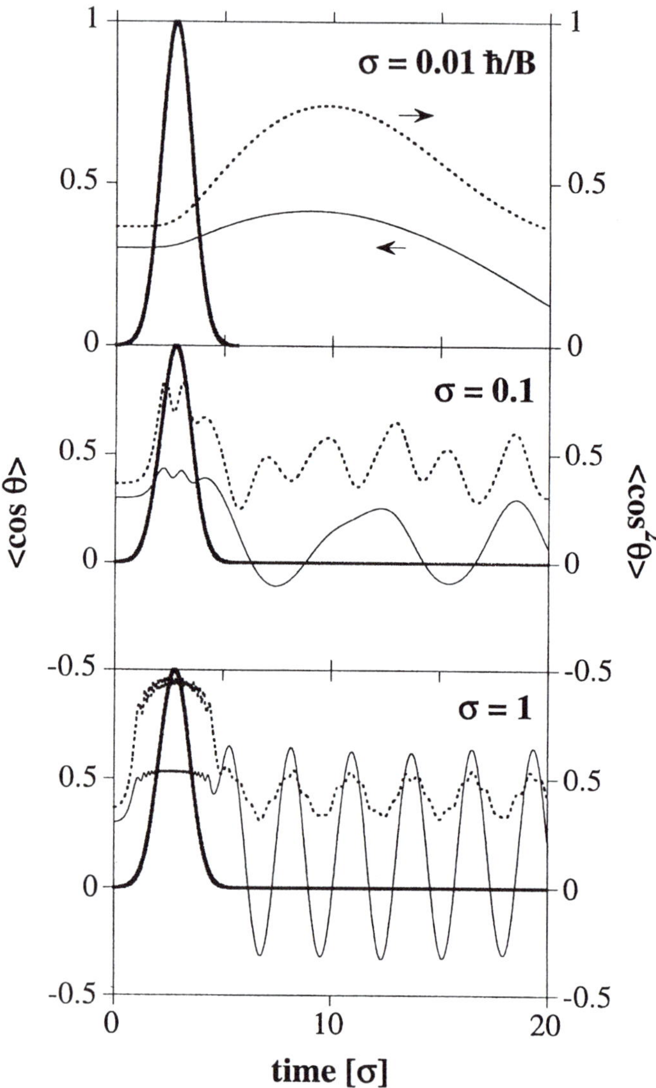

Figure 4: Dependence of the alignment (dotted lines) and orientation (full lines) cosines on time (expressed in terms of σ) for different pulse durations τ and fixed $\Delta\omega = 100$ and $\omega = 1$. The grey curves show the pulse shape function.

Such an anomalous behavior reflects the accidental phase matching among the eigenstates at the end of the pulse.

An intriguing feature of the rotational wavepacket created by a strongly non-adiabatic pulse in the combined fields is revealed when another, second pulse is sent in. If the first pulse creates a wide wavepacket (comprised of many $|\tilde{J}, 0; 0, \omega\rangle$ states), the second pulse, while on, can either restore. suppress or reverse the orientation the wavepacket had during the first pulse. The outcome depends on the delay between the two pulses. We found that, in effect, the second pulse can juggle the wavepacket between the two equivalent wells of the $\cos^2 \theta$ potential. The well centered at $0°$ overlaps with the attractive part of the $\cos \theta$ potential while the well centered at $180°$ overlaps with its repulsive branch. Therefore, when the wavepacket falls into either of the $\cos^2 \theta$ minima, it becomes localized, resulting in a net orientation. A near hit of the repulsive barrier between the minima spreads the wavepacket about equally between them, which cancels the average orientation. The alignment is left essentially intact by the second pulse. For a delay of the second pulse that corresponds to a full rephasing of the wavepacket, the orientation is the same as that created during a single pulse with the same σ and $\Delta\omega$. The case of a resonant two-pulse interaction has been analyzed in ref. [27].

Figure 5 shows the expansion coefficients $d_{\tilde{J}}(t_a)$ for $\tilde{J} \le 3$ as a function of the induced dipole interaction parameter $\Delta\omega$ for a fixed permanent dipole interaction parameter $\omega = 1$ and for different σ. The expansion coefficients pertain to a time t_a after the nonresonant laser pulse (which is centered at $t = 6\sigma$). For $t \ge t_a$ the $d_{\tilde{J}}(t)$'s again evolve, within $\approx 1\%$, to values which they maintain "ever after" and so the $d_{\tilde{J}}(t_a)$ coefficients can be used to reconstruct the wavefunction $\psi(t)$ at any time $t \ge t_a$.

The upper panel of Fig. 5 shows the $d_{\tilde{J}}(t_a)$ coefficients for $\sigma = 0.01$ $\hbar/B$. This corresponds to a strongly non-adiabatic regime, when many of the $d_{\tilde{J}}(t_a)$ coefficients contribute to the wavefunction after the pulse passed over (up to $\tilde{J} = 8$ at $\Delta\omega = 900$).

In the upper panel of Fig. 5 we also plot the approximate $\bar{d}_{\tilde{J}}(t_a)$ coefficients obtained in the non-adiabatic limit, eq. (16). The agreement with the exact results is excellent, just within a few per cent. This implies that the interaction at $\sigma = 0.01 \ \hbar/B$ is governed, over a wide range of field strengths, by an impulsive transfer of action, A, from the radiative field to the molecule: since the angular momentum is

$$L = B \int \frac{d}{d\theta} \left[\Delta\omega(t) \cos^2 \theta + \omega_\perp(t) \right] dt = -B\Delta\omega G(t) \sin 2\theta \qquad (19)$$

the action

$$A = \int L d\theta = B\Delta\omega G(t) \cos^2 \theta \qquad (20)$$

Therefore, the operator that creates the rotational wavepacket in the nonadiabatic limit, eq. (15), is indeed $\exp[iA/\hbar]$. Note that the $d_{\tilde{J}}(t)$ coeffi-

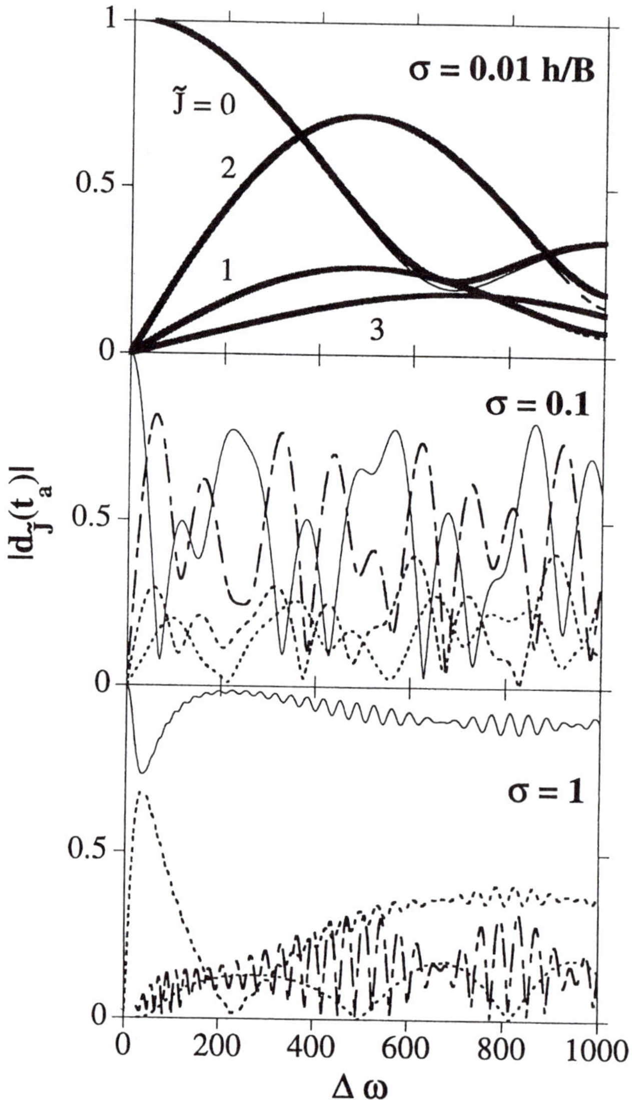

Figure 5: Expansion coefficients $d_{\tilde{J}}(t_a)$ of the time-dependent wavefunction for $\tilde{J}$ up to 3 as a function of the induced dipole interaction parameter $\Delta\omega$ for a fixed permanent dipole interaction parameter $\omega = 1$ and for different pulse durations τ. The grey lines in the upper panel show the approximate $\bar{d}_{\tilde{J}}(t_a)$ coefficients obtained in the non-adiabatic limit, eq. (16).

cients in the non-adiabatic limit scale, at any given time, according to the value of the product $\sigma \Delta\omega$.

The lower panel of Fig. 5 shows all the contributing $d_{\tilde{J}}(t_a)$ coefficients for a pulse with $\sigma = 1 \ \hbar/B$. One can see that at small $\Delta\omega$, the wavefunction consists mainly of a single contribution, namely the $|\tilde{0}, 0; 0, \omega\rangle$ initial state. This is a signature of an adiabatic behavior when the wavefunction follows the radiative field as if it were static at any field strength along the pulse profile. As $\Delta\omega$ increases, deviations from the adiabatic behavior become evident in the augmented contributions from higher $|\tilde{J}, 0; 0, \omega\rangle$ states. The middle panel of Fig. 5 pertains to the intermediate case characterized by $\sigma = 0.1\hbar/B$ when the behavior falls within neither limit and the numerical calculation is the only guide.

Figure 6 shows time averages of the alignment cosine

$$\langle\langle\cos^2\theta\rangle\rangle = \sum_{\tilde{J}} |d_{\tilde{J}}(t_a)|^2 \langle\tilde{J}| \cos^2\theta|\tilde{J}\rangle \tag{21}$$

and of the orientation cosine

$$\langle\langle\cos\theta\rangle\rangle = \sum_{\tilde{J}} |d_{\tilde{J}}(t_a)|^2 \langle\tilde{J}| \cos\theta|\tilde{J}\rangle \tag{22}$$

as functions of the laser pulse duration σ at fixed values of the induced dipole interaction parameter $\Delta\omega = 100, 400$, and 900 and for a fixed value of $\omega = 1$. One can see that both the alignment and orientation persist in the absence of the pulsed laser field. However, the recurrences of alignment and orientation qualitatively differ. While the alignment remains essentially constant, the orientation cosine exhibits large oscillations (between -0.1 and $+0.3$) and a tendency to increase towards the adiabatic limit (where it would reach a value of ≈ 0.3 corresponding to the purely electrostatic interaction). Apart from the case of small $\sigma\Delta\omega$ (for $\sigma \leq 0.05 \ \hbar/B$ in the upper panel pertaining to $\Delta\omega = 100$), the average orientation is actually suppressed at low σ. This is a consequence of the alignment and orientation of the contributing states $|\tilde{J}, 0; 0, \omega\rangle$. Since the states are all aligned more or less along the radiative field, their mixing does not affect the average alignment $\langle\langle\cos^2\theta\rangle\rangle$. On the other hand, the orientation of the $|\tilde{J}, 0; 0, \omega\rangle$ states with $\tilde{J} \geq 1$ is opposite to the one of the $|\tilde{0}, 0; 0, \omega\rangle$ state (which is along the static field). Therefore, as more of the states with $\tilde{J} \geq 1$ contribute to the wavefunction, the average orientation $\langle\langle\cos\theta\rangle\rangle$ decreases. This happens, as one can see in the upper panel of Fig. 4, in the non-adiabatic limit for $\sigma = 0.01 \ \hbar/B$ and $\Delta\omega \geq 300$. On the other hand, the range of $\Delta\omega \leq 100$ for $\sigma = 0.01 \ \hbar/B$ is particularly advantageous for maintaining a large average orientation after the passage of the pulse, since the wavefunction is dominated by the $|\tilde{0}, 0; 0, \omega\rangle$ state oriented along the static field. Thus a desirable alignment or orientation cosine can be obtained by dialing the appropriate values of the σ and $\Delta\omega$ parameters. At choice values of σ and $\Delta\omega$, the alignment can reach $\langle\langle\cos^2\theta\rangle\rangle \gtrsim 0.5$, corresponding to an angular amplitude of less than $\pm 45°$. The alignment

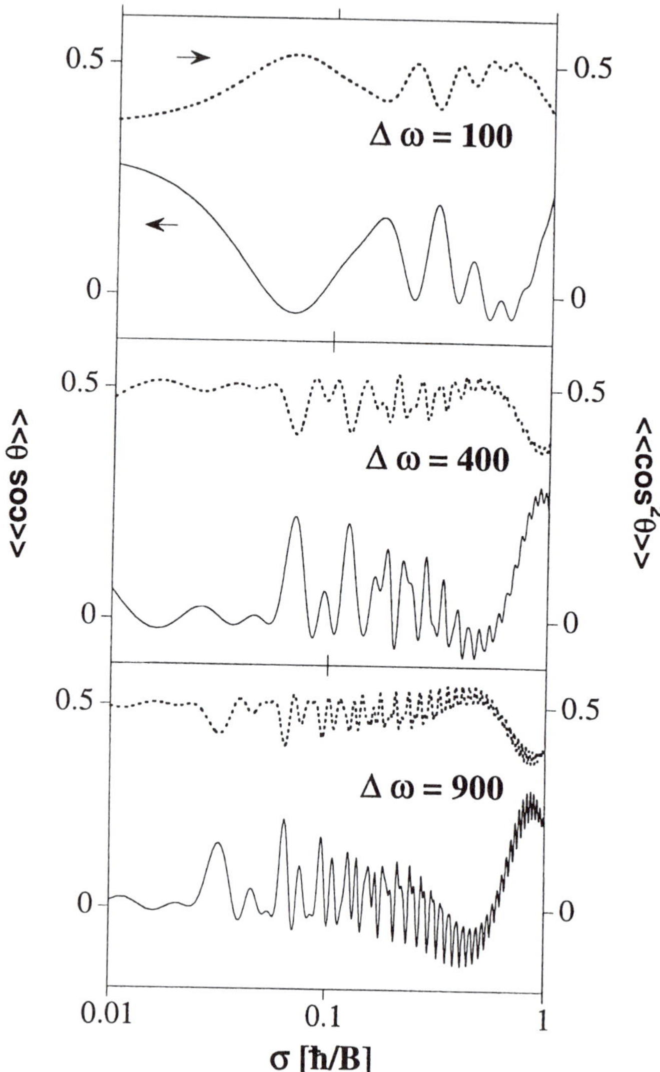

Figure 6: Time averages of the alignment (dotted lines) and orientation (full lines) cosines as functions of the laser pulse duration τ at fixed values of the induced dipole interaction parameter $\Delta\omega = 100$, 400, and 900 and for a fixed value of $\omega = 1$.

tapers off for pulses with $\sigma \rightarrow 1 \ \hbar/B$ as the adiabatic limit is approached. Note that in the adiabatic case, the field-free molecular axis distribution is isotropic, characterized by $\langle\langle\cos\theta\rangle\rangle = 0$ and $\langle\langle\cos^2\theta\rangle\rangle = \frac{1}{3}$.

If ε_S and ε_L are not collinear, the molecular axis can be localized with respect to ϕ as well as θ, since M states as well as J states undergo hybridization. For $\beta = 90°$, rather than cooperating in augmenting the hybridization along a common direction, the perpendicular fields are competing, each trying to have the pendular dipole favor a direction orthogonal to the other field. This can be used to eliminate "wrong way orientation" which otherwise occurs for "low-field seeking" states. For such a state with $\langle\cos\theta\rangle$ negative in a static field, even when ω remains small the orientation can be made positive by subjecting the molecule to a substantial $\Delta\omega$ delivered by an orthogonal laser field. For sufficiently large $\Delta\omega$, a positive $\langle\cos\theta\rangle$ is attained for all ω, ensuring "right-way orientation." Currently, uncertainties imposed by the presence of states with $\langle\cos\theta\rangle$ negative have handicapped interpretation of experiments employing a static field alone to produce oriented pendular molecules [28],[29]. The dependence of $\langle\langle\cos^2\theta\rangle\rangle$ on the tilt angle β is weak, $\langle\langle\cos\theta\rangle\rangle$ remains essentially constant (with respect to the electrostatic field).

Conclusions

The strong induced dipole interaction, easily attainable with pulsed nonresonant laser fields, can be used to enhance a weak permanent dipole orientation during the laser pulse. Such an enhancement is most dramatic if the interaction of the laser field with the molecule is adiabatic. Under suitable conditions, a sizable orientation can recur after the passage of the pulse. The recurring wavepackets afford field-free alignment of the molecular axis. The recurrences become most prevalent if formed under an impulsive transfer of action from the radiative field to the molecule. The effect should find application whenever molecular orientation and alignment are desirable, and should be taken into account in all schemes that make use of pulsed radiative fields and their combination with electrostatic fields.

Acknowledgment

The two-field eigenproblem was originally worked out, on the suggestion of André Bandrauk (Université de Sherbrook), by Dudley Herschbach and B.F. We are grateful to Dudley Herschbach for discussions and encouragement and to Jotin Marango for help with the calculations. L.C. thanks the Harvard College Research Fund for a grant. The support for this work has been provided by the National Science Foundation.

References

[1] Loesch, H.J. *Annu. Rev. Phys. Chem.* **1995**, *46*, 555 and work cited therein.

[2] Slenczka, A.; Friedrich, B.; Herschbach, D. *Phys. Rev. Lett.* **1994**, *72*, 1806.

[3] Friedrich, B.; Herschbach, D. *Phys. Rev. Lett.* **1995**, *74*, 4623; *J. Phys. Chem.* **1995**, *99*, 15686.

[4] Dion, C.M.; Keller, A.; Atabek, O.; Bandrauk, A.D. *Phys. Rev. A* **1999**, *59*, 1382.

[5] Slenczka, A. *Chem. Eur. J.* **1997**, *5* , 1136 and work cited therein.

[6] Kim, W.; Felker, P. *J. Chem. Phys.* **1996**, *104*, 1147; *J. Chem. Phys.* **1997**, *107*, 2193; *J. Chem. Phys.* **1998**, *108*, 6763.

[7] Malvaldi, M.; Persico, M.; Van Leuven, P. *J. Chem. Phys.* **1999**, *111*, 9560.

[8] Block, P.; Bohac, E.; Miller, R.E. *Phys. Rev. Lett.* **1992**, *68*, 1303.

[9] Bhardwaj, V. R.; Safvan, C. P.; Vijayalaksmi, K.; Mathur, D. *J. Phys. B* **1997**, *30*, 3821 and work cited therein.

[10] Karczmarek, J.; Wright, J.; Corkum, P., Ivanov, M. *Phys. Rev. Lett.* **1999**, *82*, 3420.

[11] Sakai, H.; Safvan, C.P.; Larsen, J.J.; Hilligsøe, K.M.; Held, K.; Stapelfeldt, H. *J. Chem. Phys.* **1999**, *110*, 10235.

[12] Seideman, T. *J. Chem. Phys.* **1995**, *103*, 7887; *Phys. Rev. A* **1997**, *56*, R17.

[13] Stapelfeldt, H.; Sakai, H.; Constant, E.; Corkum, P. B. *Phys. Rev. Lett.* **1997**, *79*, 2787.

[14] Friedrich, B. *Phys. Rev. A* **2000**, *61*, 025403.

[15] Takekoshi, T.; Patterson, B.M.; Knize, R.J. *Phys. Rev. Lett.* **1998**, *81*, 5109.

[16] Friedrich, B.; Herschbach, D. *J. Chem. Phys.* **1999**, *111*, 6157.

[17] Friedrich, B.; Herschbach, D. *J. Phys. Chem.* **1999**, *103*, 10280.

[18] Boca, A.; Friedrich, B. *J. Chem. Phys.* **2000** , *112* , 3609.

[19] Friedrich, B.; Herschbach, D. *Phys. Chem. Chem. Phys.* **2000**, *2*, 419.

[20] Friedrich, B.; Herschbach, D. *Z. Phys. D* **1996**, *36*, 221.

[21] Ortigoso, J.; Rodriguez, M.; Gupta, M.; Friedrich, B. *J. Chem. Phys.* **1999**, *110*, 3870.

[22] Bandrauk, A.D.; Claveau, L. *J. Phys. Chem.* **1989**, *93*, 107 and references cited therein.

[23] Seideman, T. *Phys. Rev. Lett.* **1999** , *83*, 4971.

[24] Keller, A.; Dion, C.M.; Atabek, O. *Phys. Rev. A* **2000**, *61*, 023409.

[25] Baumfalk, R.; Nahler, N.H.; Buck, U. *J. Chem. Phys.* **2001**, *114*, 4755.

[26] Henriksen, N.E. *Chem. Phys. Lett.* **1999**, *312*, 196.

[27] Dion, C.M.; Bandrauk, A.D.; Atabek, O.; Keller, A.; Umeda, H.; Fujimura, Y. *Chem. Phys. Lett.* **1999**, *302*, 215.

[28] Friedrich, B.; Herschbach, D.R.; Rost, J.-M.; Rubahn, H.-G.; Renger, M.; Verbeek, M. *J. Chem. Soc. Faraday Trans.* **1993**, *89*, 1539.

[29] van Leuken, J.J.; Bulthuis, J.; Stolte, S.; Loesch, H.J. *J. Phys. Chem.* **1995**, *99*, 13582.

Chapter 20

Prospects for All-Optical Alignment and Quantum State Control of Nonpolar Molecules

A. M. Lyyra[1], F. C. Spano[2], J. Qi[1,3], and T. Kirova[1]

[1]Physics Department, and [2]Chemistry Department, Temple University, Philadelphia, PA 19122
[3]Physics Department, University of Connecticut, Storrs, CT 06269

We have demonstrated Autler-Townes splitting in molecular lithium using continuous wave triple resonance spectroscopy. The Autler-Townes split double resonance lineshape consists of a superposition of several narrower twin peaks, one for each $|M_J|$. The splitting is proportional to $|M_J|$ enabling separation of the individual M_J peaks. Thus, excitation of an individual magnetic sublevel can be used to align the angular momentum of nonpolar molecules all-optically. We also propose a novel quantum control scheme based on using a coupling laser to create dressed states of controllable molecular character. Preliminary calculations are presented for variable singlet triplet character.

Preparing gas-phase molecules with pre-selected initial reactant quantum states and orientation is one of the main goals in chemical reaction dynamics. In addition, initial state selectivity enhances the viability of quantum control schemes which seek to direct the course of chemical reactions (*1-5*). We emphasize here state-selectivity combined with a*ll-optical control of molecular angular momentum alignment and quantum state character.*

Traditional methods to align polar molecules include application of strong dc electric fields (pendular states) *(6)*, focussing by an electric hexapole field *(7)* and optical pumping via polarized light *(8)*. Here, we describe a novel all-optical method of aligning and orienting the angular momentum of nonpolar gas-phase molecules based on the M_J -dependent Autler-Townes (AT) splitting *(9)* using a coupling laser to lift the magnetic sublevel degeneracy of a given target level. The AT split levels are actually dressed by the coupling laser photons. The so-called dressed states project onto *both* molecular eigenstates connected by the coupling laser. Angular momentum alignment can be achieved with a linearly polarized coupling laser whereas orientation requires a circularly polarized coupling laser [*10*].

We have also begun to investigate how the dressed states approach can be exploited for quantum state control. One application is the creation of *controllable* linear combinations of singlet and triplet states – or dressed gateway intermediate levels to enhance access to otherwise 'dark' triplet states in multiple resonance spectroscopy. Such states can also be used for *direct comparison of singlet and triplet state reactivities in reaction dynamics experiments, since the singlet and triplet character of the initial wavefunction can easily be controlled by a laser.*

Alignment and Orientation

Autler-Townes splitting can be used to produce electronically excited molecules with a pre-selected angular momentum alignment (specified $|M_J|$) or orientation specified M_J). AT splitting causes a laser induced fluorescence line from an excited level in an atom or a molecule to split into two peaks when a secondary or coupling laser is applied at or near resonance with a transition between the excited state and the coupling field lower level. This splitting is proportional to the product of the coupling laser field strength and the transition dipole moment. AT splitting in molecules has thus far been observed in the multi-photon ionization spectrum of H_2(*11*), N_2(*12*), CO(*13*) and benzene (*14*), using pulsed laser excitation to overcome Doppler broadening. The present authors were the first to measure *sub-Doppler* AT splitting in the OODR spectrum of a molecular gas (Li_2) using cw lasers (*15*). In this manner multiphoton ionization inherent to pulsed laser excitation from the ground state thermal population is eliminated.

Several groups have devised *all-optical* schemes based on the $|M|$-dependent ac Stark effect to achieve partial alignment in small gas-phase molecules. All have used two lasers. Linskens et al. (*16*) used a strong coupling laser to ac Stark shift the magnetic substates in an electronic ground state of SF_6. Subsequent two-photon absorption to a level with angular momentum quantum number J reveals a complicated multi-peaked lineshape due to the interference of several two-photon pathways. The central peak arises primarily (85%) from molecules with $|M| = J$. Neuhauser and Neusser (*14*) used a strong coupling laser to ac Stark split an intermediate electronic state of benzene with angular momentum quantum number J. Tuning the probe laser while monitoring two-photon ionization reveals the unresolved AT splitting of the intermediate state. For an R branch coupling transition there is also an unshifted peak due to molecules with $|M| = J$ which are not affected by the coupling laser. Rudert, et. al.(*17*) used ac Stark shifting of the Raman pump beam to partially resolve the $|M|$-dependent contributions to the Raman lineshape for a transition in acetylene. In our work we utilize three lasers - pump and probe lasers to affect a sub-Doppler OODR excitation to an upper (target) level, and the coupling laser for AT splitting the upper level. Since it is very unlikely that any two of the lasers are simultaneously resonant with more than one transition there are no complicating interference effects which can arise in two-photon absorption or ionization. Our scheme can potentially resolve the entire M-dependence of the target level making it possible to selectively align or orient molecules in *any* prescribed excited state with sub-Doppler resolution.

Our experimental scheme for alignment and orientation is based on the M_J-dependent AT splitting of the magnetic substates within a molecular level. The splitting depends on M_J through the orientation dependence of the transition dipole moment. Fig. 1a depicts the M_J-dependent AT splitting in an excited level 3 (with angular momentum quantum number J_3) induced by a coupling laser, L3, which is resonant with the transition between level 3 and a second intermediate level 4 (with angular momentum quantum number J_4). The states within the split pair are actually dressed states – eigenstates of the mixed molecule-laser Hamiltonian. When the coupling laser is linearly polarized the splitting is the same for both M_J and $- M_J$; for circular polarization all $2J_3+1$ components have unique AT splittings. The preparation of an excited molecule with a pre-selected angular momentum alignment (orientation) is accomplished by tuning the probe laser L2 in an OODR excitation sequence into resonance with the $|M_J|^{th}$ (M_J^{th}) dressed state.

The degree of alignment or orientation depends on how well resolved the various M-dependent (From here after we drop the subscipt on M_J). AT lineshapes are. A given lineshape consists of two peaks, each of width Γ, separated by the AT splitting. The composite AT spectrum is resolved when the difference between the splittings of the Mth and $(M+1)$th lineshapes exceeds Γ. For the coupling field with polarization α ($\alpha=0,\pm1$), the AT splitting of the Mth substate of level 3 is given by the orientation-dependent Rabi frequency,

$$\Omega_M^\alpha = [\sigma_{J_3 J_4, M}^\alpha]^{1/2} \, \mu_e \, |<v|v'>| \, E_3 / \hbar \tag{1}$$

where E_3 is the coupling field amplitude, μ_e is the electronic transition dipole moment for the transition between levels 3 and 4, and $|<v|v'>|^2$ is the Franck-Condon (FC) factor for the same transition. $\sigma_{JJ'M}^\alpha$ carries the orientation dependence of the $J \rightarrow J'$ oscillator strength. Since $\sigma_{J_3 J_4 M}^0$ depends only on $|M|$, molecular alignment results from a linearly polarized coupling laser. However, because $\sigma_{J_3 J_4 M}^{\pm 1}$ is not symmetric in M circularly polarized light can lead to orientation, i.e. the ability to pre-select any value of M. Eq.(1) shows that greater spectral resolution is obtained by increasing the power of the coupling laser.

Quantum State Control

Laser control and manipulation of atomic and molecular quantum states using lasers falls into two categories: Brumer and Shapiro (*1*) pioneered coherent control in the frequency domain, based on quantum mechanical interference between competing pathways. Tannor and Rice (*2*) extended the approach to the time domain by controlling the dynamics of bound wave packets with properly timed pulse sequences. Experimental demonstrations of coherent control have been made in both time and frequency domains.

We describe here a novel quantum control scheme within the frequency domain based on the use of a coupling laser to create dressed states of controllable molecular (i.e., singlet, triplet) character. The five level diagram in Figure 1b shows the basic excitation scheme. Note that since the pump and probe lasers are resonant the method does not depend on interfering pathways as in the Brumer and Shapiro method. *The goal is to maximize upper level triplet state production by starting with a singlet ground state (S1).* When the coupling laser is absent T2 production is practically zero because of the nearly pure singlet and triplet characters of the two intermediate levels. In this scheme the coupling laser induces AT splitting of S2. As the coupling laser power is increased the lower component of the split pair is pushed closer to resonance with T1. (The T1 level is not affected by the coupling laser due to the electric dipole selection rules). Once the dressed S2 level is resonant with T1 the weak spin-orbit interaction creates dressed window states of roughly equal singlet and triplet character (*18*). Such states can be used as intermediate levels in an OODR scheme gain access to the originally "dark" triplet state T2.

The Lithium dimer spin-orbit interaction that causes "natural" perturbation window levels is very weak (0.1 cm^{-1}). Therefore, only a handful of such perturbed pairs have been found in the $A^1\Sigma_u^+$ and $b^3\Pi_u$ states, a fact that makes the proposed optically induced window level scheme particularly attractive. Supporting calculations based on a numerical solution to the density matrix equations of motion for a five level system interacting with three lasers are presented here.

M-dependent Autler-Townes Splitting

We have reported sub-Doppler AT splitting for two ro-vibrational levels within the $G^1\Pi_g$ electronic excited state of $^7Li_2$ using three linearly polarized lasers. We were successful in achieving partial alignment ($|M|$ selectivity) starting from a J_1=4 ground state. However, we could not resolve the M-dependence of the AT splitting when starting from the J_1=11 ground state because of the power limitations of our cw dye lasers.(*15*) The AT line shapes were shown to be in excellent agreement with a density matrix calculation involving a four level system interacting with three laser fields.

We excited lithium dimers within a five-arm stainless steel heat pipe with three Coherent 699-29 dye lasers (0.5 MHz bandwidth). The pump laser (L1), counter-propagating probe laser (L2), and copropagating coupling laser (L3) were aligned coaxially and were linearly polarized in a common direction. The upper level population was monitored through side-arm fluorescence detection utilizing a photomultiplier tube. With this arrangement we observed AT splitting of the $G^1\Pi_g$ (v=12, J_3, f parity) levels with J_3 = 12 and J_3= 3. Figures 2 and 3 show the resulting spectra obtained by holding the pump and coupling lasers on resonance while scanning the probe laser. Figure 2a shows the usual OODR spectrum for the $G^1\Pi_g$ level. With the coupling laser turned on the OODR signal is split into two unresolved symmetric peaks. Figures 2b-d show the spectra corresponding to three different values of the coupling laser power for J_3 = 12 . The peaks are significantly diminished in height relative to the OODR peak and are several times broader. The splitting is proportional to the square root of the coupling laser power as is also observed in the atomic case.

Fig. 3 shows the AT splitting of the $G^1\Pi_g$ (v=12, J_3=3, f parity) level. Unlike the case with J_3=12, the $|M|$-dependent AT splitting is at least partially resolved. For J_3=3 there are three sets of twin peaks corresponding to $|M|$ = 1,

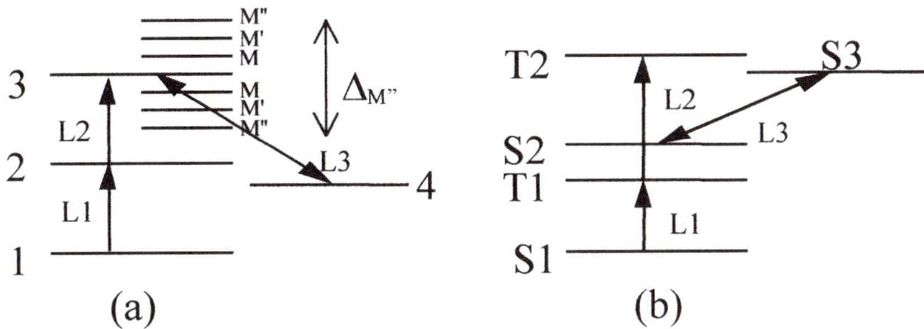

Figure 1. (a) Four level system and excitation scheme used to obtain *M*-dependent AT splitting of the level 3. (b) Five level system and excitation scheme used to move level S2 into resonance with T1 using the coupling laser.

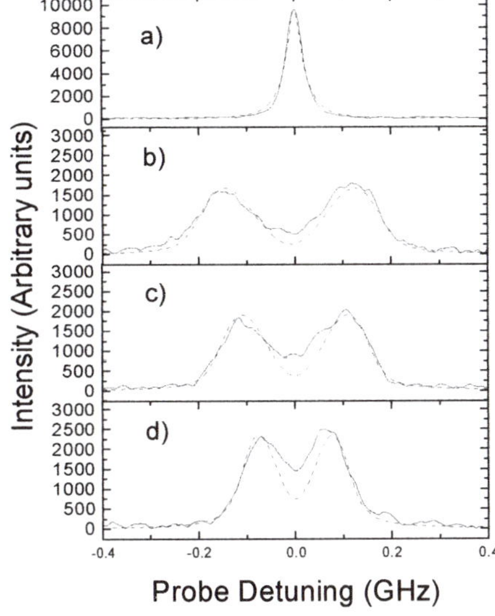

Figure 2. Violet/ UV fluorescence from the $G^1\Pi_g$ (v=12, J=12, f) level as a function of the probe detuning. (a) OODR spectrum with the coupling field off, (b) - (d): AT spectra for a resonant coupling field with a power of 300 mW (b) 200 mW (c), and 100 mW (d). The solid lines indicate experimental spectra while the dashed lines are theoretical simulations based on Eq. (2).

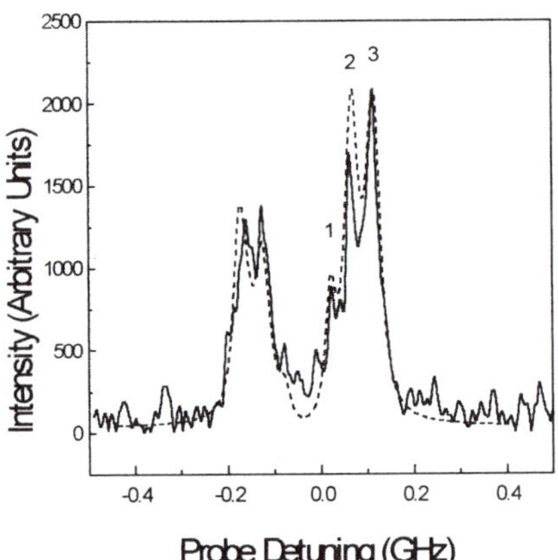

Figure. 3. The AT spectrum of the $G^1\Pi_g$ (v=12, J=3,f) level taken using single channel detection. The coupling laser is slightly detuned with a power of 300 mW. The peaks are labeled according to the value of |M|. The solid line indicates the experimental spectra while the dashed line is a theoretical simulation based on Eq. (2).

2, and 3. The $|M| = 2$ and 3 peaks are clearly distinguished, whereas the $|M| = 1$ peak falls below the noise level.

In order to understand the nature of the AT line shapes in Figures 2 and 3, we solved the density matrix equations of motion for a four level system interacting with pump, probe, and coupling lasers (*10*) (see Figure 1a). The pump and probe are treated perturbatively and the coupling field exactly. The signal $S(\Delta_2)$ is proportional to the steady-state upper level population as a function of probe detuning, Δ_2. It can be expressed as,

$$S(\Delta_2) = \sum_M \int_0^\infty r\,dr < \rho_{33}^M(r, v_z) >_D \qquad (2)$$

where $\rho_{33}^M(r, v_z)$ is the steady-state upper level population for a molecule with orientation M, radial position r relative to the common axis of the three laser beams, and velocity v_z. $<...>_D$ indicates the Doppler average. Under steady state conditions and assuming that the pump laser is tuned to the center of the Doppler broadened line transition, $\rho_{33}^M(r, v_z)$ is given by:

$$\rho_{33}^M(r, v_z) = \frac{\Omega_{1,M}^2(r)\Omega_{2,M}^2(r)\rho_{11}^0}{8A_M\gamma_3(\widetilde{\Delta}_1^2 + \gamma_{12}^2)} \times$$

$$\mathrm{Im}\left\{ \frac{(\widetilde{\Delta}_1 + \widetilde{\Delta}_2 - \widetilde{\Delta}_3 - i\gamma_{14}) - \dfrac{\gamma_4\Omega_{3,M}^2(r)(\widetilde{\Delta}_3 - i\gamma_{34})}{2\Omega_{3,M}^2(r)\gamma_{34} + 4\gamma_4(\widetilde{\Delta}_3^2 + \gamma_{34}^2)}}{(\widetilde{\Delta}_1 + \widetilde{\Delta}_2 - i\gamma_{13})(\widetilde{\Delta}_1 + \widetilde{\Delta}_2 - \widetilde{\Delta}_3 - i\gamma_{14}) - (\Omega_{3,M}(r)/2)^2} \right.$$

$$\left. + \frac{2\gamma_{12}^c}{\gamma_2} \frac{\widetilde{\Delta}_2 - \widetilde{\Delta}_3 - i\gamma_{24} - \dfrac{\gamma_4\Omega_{3,M}^2(r)(\widetilde{\Delta}_3 - i\gamma_{34})}{2\Omega_{3,M}^2(r)\gamma_{34} + 4\gamma_4(\widetilde{\Delta}_3^2 + \gamma_{34}^2)}}{(\widetilde{\Delta}_2 - i\gamma_{23})(\widetilde{\Delta}_2 - \widetilde{\Delta}_3 - i\gamma_{24}) - (\Omega_{3,M}(r)/2)^2} \right\} \qquad (3)$$

with $A_M = 1 - \dfrac{\Gamma_{34}}{\gamma_3} + \dfrac{\gamma_4}{\gamma_3}\dfrac{\Omega_{3,M}^2(r)\gamma_{34} + 2\Gamma_{34}(\widetilde{\Delta}_3^2 + \gamma_{34}^2)}{\Omega_{3,M}^2(r)\gamma_{34} + 2\gamma_4(\widetilde{\Delta}_3^2 + \gamma_{34}^2)}$

In Eq.(3), the Doppler-shifted detunings of the pump, probe and coupling lasers from the molecular transition frequencies $\omega_{ij} \equiv (\varepsilon_i - \varepsilon_j)/\hbar$ are,

$$\widetilde{\Delta}_1 \equiv \Delta_1 + k_1 v_z, \quad \Delta_1 \equiv \omega_{21} - \omega_1$$

$$\widetilde{\Delta}_2 \equiv \Delta_2 + k_2 v_z \, , \quad \Delta_2 \equiv \omega_{32} - \omega_2$$
$$\widetilde{\Delta}_3 \equiv \Delta_3 + k_3 v_z \, , \quad \Delta_3 \equiv \omega_{34} - \omega_3 \qquad (4)$$

respectively.

The Rabi frequencies of all three lasers depend on both the molecule's orientation as well as its distance from the beam axis. When the ith laser is tuned near resonance with the i, $i+1$ transition, the Rabi frequency is given by,

$$\Omega_{i,M}(r) \equiv (\mu^M_{i,i+1} E_i / \hbar) \exp[-(r/w_i)^2] \qquad (5)$$

where E_i and w_i are the (on-axis) electric field and spot size respectively of the ith laser and $\mu^M_{i,i+1}$ is the M-dependent transition dipole moment between levels i and $i+1$ (10). In addition, γ_i in Eq. (3) is the damping rate of the ith level, including radiative as well as collisional contributions. The decay rate of the coherence between levels i and j ($\neq i$) is $\gamma_{ij} = (\gamma_i + \gamma_j)/2 + \gamma^c_{ij}$, where γ^c_{ij} is the pure dephasing contribution induced by phase interrupting collisions. Γ_{34} is the damping rate of the upper level to the intermediate level, $A^1\Sigma_u^+(v=14, J=12)$. When the coupling field is zero Eq.(3) reduces to the OODR line shape expression derived previously (10, 15).

In order to compare with the experiments of Figures 2 and 3 we evaluated Eq. (2) using expressions for $\mu^M_{i,i+1}$ corresponding to an R transition for the pump laser and Q transitions for both the probe and coupling lasers (10). The intermediate level lifetime $\gamma_2^{-1}(\approx \gamma_4^{-1})$ was set to the experimentally determined value of 17 nsec , which includes an 18.8 nsec natural lifetime extrapolated to 17 nsec at 200 mTorr to include collisional redistribution (10). Since information about the various dephasing rates in Li_2 is not available, we assumed a simplified model where $\gamma^c_{i1} = \gamma^c$ and $\gamma^c_{ij} = \gamma^c_{i1} + \gamma^c_{j1}$ ($i,j \neq 1$). We performed our calculations by first locating values of γ_3 and γ^c which, when inserted into Equations (2) and (3), best reproduced (i) the experimental OODR line shape in Figure 2a when $E_3=0$, and (ii) the experimental AT spectrum for the highest coupling field power (Figure 2b), with E_3 adjusted to reproduce the experimental AT splitting. The values $\gamma^c = 32$ MHz and $\gamma_3 = 2\gamma_2$ gave the dashed curves in Figures 2a and 2b. Calculations at lower coupling field powers (dashed curves in Figures 2c and 2d) were conducted without any additional parameter changes or renormalizations. In Figure 3 we used $\gamma^c = 10$ MHz with all others unchanged from the $J_3 = 12$ case. The slight asymmetry of the peaks is due to detuning of the coupling field.

Figures 2 and 3 show our theoretical AT line shapes to be in excellent agreement with experiment. The broad unresolved twin peaks in Figures 2b-d reflect the distribution of $|M|$-dependent Rabi frequencies. The lineshape consists of a superposition of eleven narrow twin peaks, one for each non-zero value of $|M|$ for $J_3 = 12$, except $|M| = J$ since these states are not excited in the R(11) pump transition. The splitting for each set is approximately the on-axis Rabi frequency, obtained by inserting the transition dipole moment factors into Eq.(1):

$$\Omega_{3,m}(r=0)/2\pi \approx (|m|/\sqrt{J(J+1)})\mu_\perp |<v|v'>| E_3/h \qquad (6)$$

where the linearity in $|M|$ derives from the rotational selection rule for a Q transition. Our calculations show that the individual magnetic sublevel lines are approximately 10-15 MHz broader than the OODR line due to the variation of the Rabi frequency over the Gaussian transverse profile. The peak of the composite AT line shape for Fig. 2 coincides with that for the $|M| = 8$ component. Hence, for the Q(12) transition the overall AT splitting is $\Delta_{AT} \approx 0.64\mu_\perp |<v|v'>| E_3/h$. This experimental AT splitting with known E_3 and FC factor yields a value of 2.4 au for the $\mu_\perp$.

Eq.(1) shows that in order to resolve the individual AT line shapes for each value of M (and hence achieve better alignment and orientation) one can either 1) increase the coupling field intensity, 2) find a coupling transition with greater transition dipole moment or FC factor, or 3) reduce the line broadening from collisions by utilizing a molecular beam. Because all $\sigma^\alpha_{JJ_1M}$ in Eq.(1) scale as J^{-1} the ability to resolve the M-dependent AT line shapes decreases with increasing J. This is the reason why we were able to partially resolve the splittings for $J_3=3$, but not for $J_3=12$, in our previous study (15).

The most effective and controllable way to resolve the M-dependent components of the composite AT lineshape is to increase the power of the coupling field *while maintaining a spot size several times larger than the pump and probe laser spot sizes* . This last condition ensures that the linewidth of each peak in a split pair for a given value of M is not increased through transverse profile broadening (TPB). TPB is an inhomogeneous broadening mechanism arising from the radially dependent Rabi frequency. TPB is liminated and the best resolution is therefore obtained for a flat-top transverse profile. However, Gaussian profiles can be used as long as the spot size of the coupling laser is several times larger than the pump and probe laser spot sizes. This ensures that molecules within the pump volume experience a uniform coupling field.

In what follows we determine theoretically through the use of Equations (2) and (3) the coupling field power necessary to achieve magnetic substate resolution of the previously studied AT line shape of the $G^1\Pi_g$ $J_3=12$ level in diatomic lithium (see Figure 2). In Ref. 15 we induced AT splittings as high as 300 MHz using a 300 mW linearly polarized coupling laser focused to a spot size of 450 μm (see Figure 2b). In order to maintain the same number of detected photons as in Ref. 15 (and hence preserve the signal to noise ratio) we assume in the present calculation the same parameters for the pump and probe lasers; respective spot sizes of 220 and 360 μm with powers of less than 10 mW each to avoid saturation broadening. All molecular parameters are identical to those used in the simulations in Fig. 2.

Fig. 4 shows the calculated AT spectrum for various coupling laser powers and for various coupling laser spot sizes. It is apparent that doubling the spot size to 900 μm over the spot size used in Ref.15 leads to a significant increase in resolution, almost to the optimum achieved with a flat-top profile. The power needed to obtain full resolution of all components is 120 W for the 900 μm spot-size case, which is about 400 times greater than the maximum output power of our cw Argon Ion pumped dye lasers. However, this power is modest compared to the powers available with typical pulsed dye lasers. A 120 W, 30 nsec long (square) pulse has only 3.6 μJ of energy. A build-up cavity could be also used with the advantage of maintaining the narrow bandwidth of the coupling field of 1MHz.

Creation of Dressed Singlet-Triplet Gateway States

AT splitting can also be used to achieve quantum control by creating dressed states with controllable admixtures of selected molecular eigenstates. We propose a demonstration of this control scheme by creating singlet-triplet window levels in molecular Lithium with a strong coupling laser. Such levels can serve as intermediate levels in the OODR excitation of high lying triplet levels, as shown schematically in Fig. 1b. Li_2 is an excellent candidate for such experiments because of its weak spin-orbit coupling (<0.2 cm^{-1}). The proposed scheme may be considered as an all-optical alternative to anticrossing spectroscopy (19) in which strong magnetic fields are used to tune triplet substates into resonance with singlet states. Initially, the singlet state S2 and triplet level T1 differ in energy by an amount much greater than the spin-orbit coupling so that they are of predominantly singlet and triplet character, respectively. The coupling field shifts S2 into resonance with T1 via AT splitting in order to promote mixing. The admixures are in fact dressed states; the coupling field creates dressed eigenstates consisting of linear combinations of the states |S2,n>, |T1,n> and |S3,n-1> where n is the number of coupling

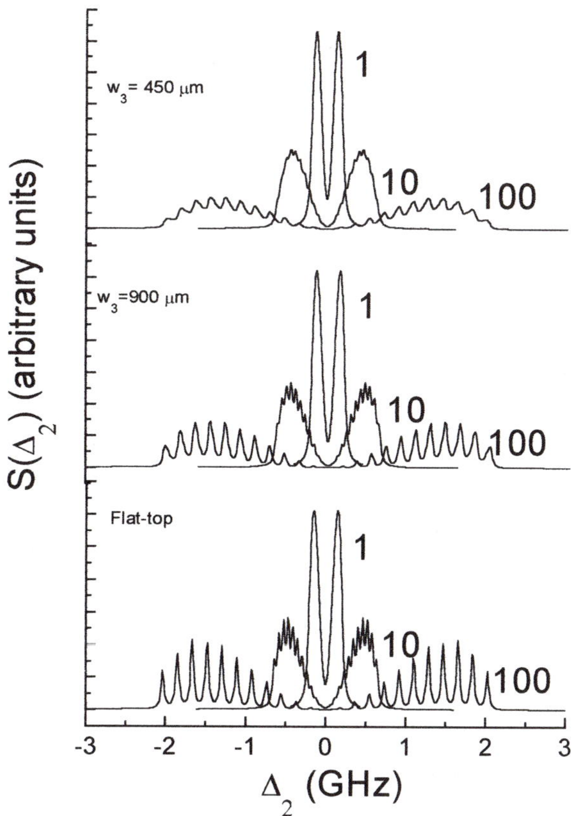

Figure 4. Calculated AT spectra for three coupling laser spot sizes. The numbers refer to the relative on-axis coupling laser *intensities*. For the 450 um spot size the on-axis intensity of unity corresponds to a power of 0.3W. For the flat-top profile we took a radius of 1000 μm.

laser photons. With the coupling field off the excited triplet state (T2) production is very low because of the $\Delta S = 0$ selection rule. With the coupling field tuned to the S2-S3 transition the S2 state undergoes AT splitting (the T1 state is unaffected). If the coupling field Rabi frequency Ω_3 is equal to the T1-S2 splitting Δ then the state $2^{-1/2}\{|S2,n> - |S3,n-1>\}$ (the lower member of the AT split pair) becomes resonant with $|T1,n>$. In this case the small spin orbit coupling will create the eigenstates, $2^{-1}\{|S2,n> - |S3,n-1>\} \pm 2^{-1/2}\,|T1,n>$. Either of these states can serve as window states for the enhanced OODR excitation of the upper triplet level T2.

To test this idea we solved the density matrix equations of motion for the five level system in Fig.1b interacting with pump, probe, and coupling lasers. S2 was predominantly singlet (99.98%) and T1 predominantly triplet (99.98%). These mixing coefficients reflect the condition that the energy difference (Δ) between these states is much greater than the spin-orbit coupling. Fig. 5a shows the T2 population as a function of the coupling field Rabi frequency. The pump laser is tuned to the $S1 \rightarrow T1$ transition and the probe is tuned to the $T1 \rightarrow T2$ transition. Note that T2 production peaks when the Rabi frequency Ω_3 is equal to the S2-T1 splitting Δ. Fig. 5b shows the T2 population as a function of pump detuning (while maintaining the two-photon resonance condition, $\omega_1 + \omega_2 = \omega_{T2,S1}$). T2 production peaks when the pump is nearly resonant with the $S1 \rightarrow T1$ transition. In all calculations the Rabi frequencies of the pump and probe were taken to be equal to the assumed uniform relaxation rates of all the levels. The maximum rate of triplet state production in Figure 5a is almost two-orders of magnitude greater than T2 production in the absence of the coupling laser. The theory used for these calculations applies to the case of a singlet state S2 with $J = 0$. For levels with larger J values several light-induced M-dependent S2-T1 resonances are expected at different coupling field powers. The theory that includes the M-dependency of the AT splitting will be presented in a forthcoming paper.

To test this idea using diatomic lithium requires locating a singlet-triplet pair with Δ of the order of 1 cm^{-1} to ensure, that the states are only slightly affected by the spin-orbit coupling, which is about 0.1 cm^{-1}. Several candidates have been identified. For example the $^7Li_2$ $A^1\Sigma_u^+$ v=27, J=1 and $b^3\Pi_u$ v=23, N=2 levels are 0.627cm^{-1} apart. In addition to using these optically induced window levels to augment the very few existing Li_2 perturbation window levels, these optically induced mixed states could also be used to measure differences in singlet vs. triplet reactivities. Such an experiment would shed light on the *fundamental question of the effect of molecular spin on reactivity*, since all else is unchanged with the exception of the singlet-triplet character of wavefunction, which is varied.

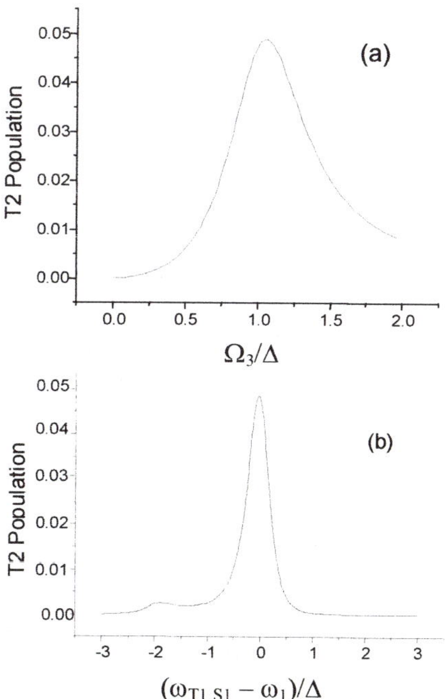

Figure 5. a) T2 population as a function of coupling laser power. b) T2 population as a function of pump laser detuning

Conclusions

We have demonstrated Autler-Townes splitting in molecular lithium using continuous wave triple resonance spectroscopy. The Autler-Townes split double resonance lineshape consists of a superposition of several narrower twin peaks, one for each $|M_J|$. The splitting is proportional to $|M_J|$ enabling separation of the individual M_J peaks. Thus, excitation of an individual magnetic sublevel can be used to align the angular momentum of nonpolar molecules all-optically. We have also proposed a novel quantum control scheme based on using a coupling laser to create dressed states of controllable molecular character. We have presented preliminary calculations for variable singlet and triplet character dressed states.

Acknowledgements

We gratefully acknowledge support for this work from the National Science Foundation through grant PHY 9983533. We also wish to thank our colleagues Prof. L. Li, Prof. L. M. Narducci, Prof. J. Huennekens, Dr. G. Lazarov and Jenny Magnes and Angelos Lazoudis for their contributions to this work.

References

1. Brumer, P.; Shapiro, M. Accts. Chem. Res. **22**, 407-413 (1989).
2. Tannor, D. J; Rice, S. A. Adv. Chem. Phys. **70**, 441-523 (1988).
3. Gordon R. J.; Rice, S. A. Annual Review of Physical Chemistry, **48**, 601-641 (1997).
4. Zhu, L.; Kleiman, V.; Li, X.; Lu, S. P.; Trentelman, K.; and Gordon, R. J. Science **270**, 77-80 (1995)
5. Gordon,R.J.; Zhu, L.C.; Seideman,T. Accounts Chem. Res., 32, 1007-1016 (1999).
6. Friedrich, B.; Slenczka, A.; and Hershbach, D. R. Chem. Phys. Letters **221,** 333-340 (1994).
7. Cho,V. A.; and Bernstein, R. B.; J. Phys. Chem. **95**, 8129-8136 (1999).
8. Bergemann,K.; edited by G. Scoles, Oxford University Press, New York, 293-344 (1988).

9.   Autler S. H.; Townes, C. H. Phys. Rev. **100**, 703-722 (1955)

10. Spano, F.C. J. Chem. Phys **114**, 276-288 (2001)

11. Quesada,M. A.; Lau, A. M. F.; Parker,D. H.; D. W. Chandler. Phys. Rev. **36**, 4107-110 (1987).

12. Girard, Sitz,G. O.; Zare,R.N.; Billy,N.; and Vique,J. J.Chem. Phys. **97**, 26-41 (1992).

13. Xu,S.; . Sha, G.;Jing,B.; Chen X.; Zhang,D.; J.Chem. Phys., **100**, 6122-6124(1994).

14. Neuhauser, R.; Neusser, H. J.; J. Chem. Phys. **103,** 5362-5365 (1995).

15. Qi, J.; Lazarov, G.; Li, L.; Narducci, L.M.; Lyyra, A.M.; and Spano, F.C. Phys. Rev. Lett. 83, 288-291 (1999).

16. Liskens, A. F.; Dam, N, Reuss, J.; and Sartakov, B. J. Chem. Phys. **101,** 9384-9394 (1994).

17. Rudert, A. D.; Martin,J.; Zacharias H.; and Halpern J. B. Chem. Phys. Lett. **294**, 381-386 (1998).

18. Li, L., Zhu, Q.,Lyyra, A.M., Whang, T,-J.,Stwalley, W.C.,Field, R. W, and Alexander, M.H. J. Chem. Phys. **97**, 7211-7219(1992)

19. Dupre, D.; Jost, R.; Lombardi, M.; Green, P.G.;Abramson. E.; Field, R.W. Chemical Physics **152**, 293-318(1991)

Chapter 21

Alignment of Neutral Molecules by a Strong Nonresonant Linearly Polarized Laser Field

Hirofumi Sakai[1], Jakob Juul Larsen[2], C. P. Safvan[2], Ida Wendt-Larsen[3], Henrik Stapelfeldt[3], and Tsuneto Kanai[4]

[1]Department of Physics, Graduate School of Science, The University of Tokyo, 7–3–1, Hongo, Bunkyo–ku, Tokyo 113–0033, Japan
[2]Institute of Physics and Astronomy, [3]Department of Chemistry, University of Århus, DK–8000 Århus C, Denmark
[4]Department of Physics, Faculty of Science, The University of Tokyo, 7–3–1, Hongo, Bunkyo-ku, Tokyo 113–0033, Japan

We demonstrate that a strong nonresonant nanosecond laser pulse can be used to align neutral molecules. Our technique, applicable to nonpolar as well as polar molecules, relies on the anisotropic interaction between the strong laser field and the induced dipole moment of the molecules. The degree of alignment is enhanced by increasing the laser intensity or by lowering the initial rotational temperature of the molecules. We measure the alignment by detecting the instantaneous direction of the photodissociated fragments with ion imaging techniques. We have also investigated the possibility of molecular orientation using an asymmetric potential created by the superposition of strong, nonresonant, two-color ($\omega + 2\omega$) laser fields. The time-independent Schrödinger equations are numerically solved in an adiabatic regime. Our results suggest that our approach can be used to orient polar molecules if they are rotationally cold.

1. Experimental demonstration of molecular alignment by a strong nonresonant linearly-polarized laser field[*]

1-1. Introduction

Controlling and manipulating the external degree of freedom of atoms and molecules is one of the intriguing subjects in modern physics and chemistry. This is evident in atomic physics where lasers provide powerful means to control the velocity and position of neutral atoms (*1*), and in studies of condensed matter where scanning probe microscopes enable selection and manipulation of individual atoms and molecules (*2*). Chemists have aimed to control the spatial orientation of molecules because it enables them to study orientational effects in chemical reaction dynamics (*3*). This kind of control can be achieved by orienting certain symmetric top molecules with electric hexapole fields (*3-5*), or by orienting molecules possessing large ratios of the permanent dipole moment to the rotational constant with strong static electric fields (*3,6,7*). The advantage of these methods is that they can arrange polar molecules in a "head-versus-tail" order (orientation). However, there are many molecules, most notably nonpolar molecules, to which they do not apply.

Friedrich and Herschbach suggested exploiting the anisotropic polarizability interaction of an intense nonresonant laser field with the induced dipole moment of molecules (*8,9*). The interaction creates a potential minimum for the molecules along the polarization axis of the laser field, forcing them to librate over a limited angle range instead of rotating freely with random spatial directions. Since the interaction between the laser field and a permanent dipole moment averages out to zero, the molecules are aligned rather than oriented. This new technique can be regarded as a modern version of the optical Kerr effect (*10*) with the special emphasis changed from the study of electro-optic properties of liquids to alignment of isolated molecules in gas phase. Furthermore, it indicates an important new direction in molecular science because the use of strong nonresonant laser fields, in addition to alignment presented in this article, may transfer many of the successful applications of laser-induced manipulation of atoms to molecules such as trapping (*8,9,11,12*) and focusing (*13,14*). The convincing evidence of laser-induced alignment was first reported by Kim and Felker (*15,16*). Our demonstration (*17-19*) provides further experimental evidence of laser-induced alignment by using a strong linearly-polarized, infrared, nanosecond laser pulse to align a sample of neutral molecules and femtosecond laser-induced photodissociation to measure the degree of alignment. In particular, we emphasize the general applicability of the scheme by aligning I_2, ICl, CS_2, CH_3I, and C_6H_5I molecules.

Fortunately, our demonstration could trigger active experimental studies in this filed. In this section, we summarize the technical essence of our laser-induced molecular alignment, discuss important implications of our work and review some of the related experimental progresses.

1-2. Experimental

A schematic of the experimental setup is shown in Fig. 1. To determine the degree of alignment of the molecules we measure their instantaneous direction while the alignment laser is still on. This is necessary because the molecules evolve back into field-free rotational states when the alignment field is switched off. In the case of iodine, the measurement is performed by photodissociating the molecules (20) during the alignment pulse with a femtosecond "pump" pulse and observing the direction of the photofragments with a two-dimensional detector. Since the detector is only sensitive to ions, it is necessary to ionize the photofrafragments without changing their direction. This is done by firing a second, intense femtosecond "probe" pulse after the dissociation is complete. In the case of other molecules we use one intense femtosecond pulse to produce multiply-charged ions of the molecules.

A pulsed supersonic beam, formed by expanding the molecules in 760 Torr of argon or helium through a 0.5-mm nozzle, is crossed at $90°$ by the focused laser beams. The partial pressure of the molecular gas is adjusted to a few mbar. The alignment pulses come from a pulsed single-longitudinal-mode Nd:YAG laser ($\lambda = 1064$ nm) with a pulse duration of ~ 3.5 ns (full width at half maximum). The alignment pulse is focused to a spot size, $\omega_0^{YAG} \sim 42$ μm, yielding a maximum intensity of $\sim 1.4 \times 10^{12}$ W/cm^2. The pump pulses used in

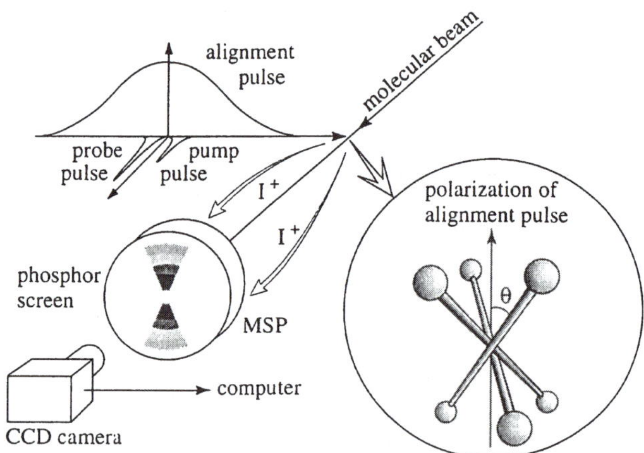

Figure 1. A schematic of the experimental setup. The polarization of the lasers are indicated by the arrows (upper left-hand side). The inset shows the aligned molecules making an angle θ with the polarization axis of the alignment laser. (Reproduced with permission from reference 17.)

the iodine experiment originate from an optical parametric amplifier and are centered at 688 nm. The pulse length is 100 fs and the peak intensity is ~ 3.0 x 10^{12} W/cm^2. The pulse energy is kept below saturation of the dissociating transition (*21*). The beam waist of the probe pulse ($\omega_0^{probe} \sim 32$ µm) is adjusted to be smaller than ω_0^{YAG} to ensure that we probe only those molecules that have been exposed to the alignment field. The pulses originate from the Ti:sapphire laser system and are centered at 800 nm. The pulse width is 100 fs and the peak intensity is ~ 7.0 x 10^{13} W/cm^2 using linear polarization. When probing alignment with single pulse induced dissociative ionization we use circularly polarized pulses with a peak intensity of ~ 8.0 x 10^{13} W/cm^2. The laser pulses are spatially overlapped using dichroic mirrors and focused, by a 30-cm focal length lens, into the vacuum chamber, where they cross a molecular beam at 90°

The 1 kHz femtosecond laser system serves as the master clock in the present experiment. Since the maximum repetition rate of the molecular beam and the charge-coupled device (CCD) camera is 20 Hz, a homemade box triggers on every 50th femtosecond laser pulse to be used in the experiment. The time difference between the two femtosecond pulses is fixed by an optical delay line and the alignment pulse is electronically synchronized to the femtosecond pulses with a precision better than ±0.5 ns. The ionized fragments are accelerated by a static field toward an ion detector positioned on-axis with the molecular beam (*22*). The detector consists of a microsphere plate (MSP) backed by a phosphor screen. Each ion detected leads to emission of a localized flash of light which is recorded by a CCD camera. Fast electronic gating of the CCD camera allows us to record mass and charge selected ion images.

1-3. Alignment of iodine and other molecules

We measure the spatial direction of the I$_2$ molecules by dissociating them using a femtosecond pump pulse as mentioned above. The pump pulse is resonant with the perpendicular $A \; ^3\Pi_{1u} \leftarrow X \; ^1\Sigma_g^+$ transition and a dissociative wave packet with a final internuclear speed of 8.9 Å/ps is created. The velocity vector of the atomic iodine fragments is a direct measure of the initial direction of the molecule. At an optical delay of 200 ps the iodine atoms are separated by almost 2000 Å so the Coulomb energy released if both fragments are ionized is negligible compared to the dissociation energy. Hereby, we ensure that the velocity of all I$^+$ ions originating from ionization of the dissociated molecules is the same.

The pump pulses are polarized perpendicular to the detector surface and it is therefore mainly molecules with their internuclear axis parallel to the detector surface that are dissociated. As a result, the ionized fragments, all having the same speed, will be detected at the same radial position on the detector but at different angular positions, corresponding to the initial direction of the parent molecule. The two-dimensional ion images of I$^+$ ions, shown in Fig. 2, include the basic experimental information. The circularly symmetric ion image recorded when only the femtosecond pulses are used [Fig. 2(A)] exhibit a prominent ring with a diameter, d, of 7.4 mm and a more diffuse distribution in the outer part of the image. The ring results from molecules photodissociated by the pump pulse and subsequently ionized by the probe pulse. The observation of a circularly symmetric ring shows that the initial direction of the molecules is random. The deviations from this uniform angular distribution are used to identify alignment as discussed later. The I^+ ions detected away from the well-defined photodissociation channel originate from fragmentation of multiply charged molecular ions formed when the intense probe pulse interacts with those I_2 molecules that did not undergo excitation by the pump pulse (21).

When the alignment pulse, polarized perpendicular to the molecular beam axis, is applied together with the femtosecond pulses the angular distribution of

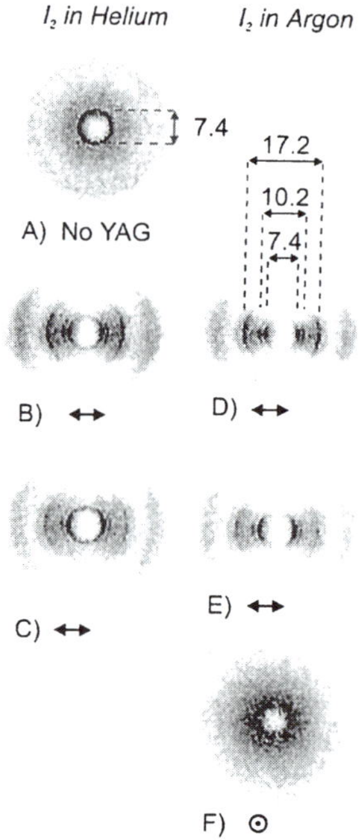

Figure 2. Ion images of I^+ fragments recorded for I_2 seeded in He (left images) or in Ar (right images), and for different intensities of the YAG pulse. (A) $I_{YAG} = 0$. (B), (D), (F) $I_{YAG} = 1.4 \times 10^{12}$ W/cm^2. (C), (E) $I_{YAG} = 5.0 \times 10^{11}$ W/cm^2. The femtosecond pulses are polarized along the molecular beam axis and the polarization direction of the YAG is shown at each image. The length scale in (A) and (D) are given in millimeters. In all images the central portion (radius ≤ 2.5 mm) of the detector is not active. (Reproduced with permission from reference 17.)

the photodissociation channel at $d = 7.4$ mm changes dramatically and becomes localized around the polarization direction of the alignment pulse. This is illustrated in Figs. 2(B)-2(E) for different intensities of the YAG pulse, I_{YAG}, and for I_2 seeded in Ar or in He. The pump pulse is synchronized to the maximum intensity of the alignment pulse. We interpret the localization of the angular distribution as a result of the I_2 molecules being aligned along the electric field vector of the alignment pulse.

An obvious test of this interpretation is to record an ion image with the YAG polarization parallel rather than perpendicular to the molecular beam axis. The result is shown in Fig. 2(F) where all experimental conditions are identical to those in Fig. 2(D) except that the polarization of the YAG pulse is rotated by 90°. The complete disappearance of the photodissociation channel at $d = 7.4$ mm confirms that the direction of the molecules are now confined along the direction of the YAG polarization, perpendicular to the plane from which dissociation is possible. Instead, the aligned molecules are multiply ionized by the intense probe pulse and their subsequent fragmentation produces a circularly symmetric ion image with a radial distribution strongly peaked in the center of the detector. (Note that the detector has a small blind spot in the center.)

The two important parameters determining the degree of alignment are the initial rotational temperature of the I_2 molecules and the intensity of the YAG pulse. The rotational temperature is controlled by using either Ar or He in the supersonic expansion because the Ar carrier gas produces I_2 molecules with lower rotational temperature than does the He carrier gas (23). In agreement with theory (8,9), we observe more localized angular distribution, i.e. stronger alignment, when Ar is used for a fixed intensity of the YAG pulse [compare Figs. 2(B) and 2(D), or 2(C) and 2(E)]. Similarly, we observe that the alignment becomes more pronounced if the intensity of the YAG pulse is increased and the initial rotational temperature is kept constant. This is seen by comparing Figs. 2(B) and 2(C) (I_2 in He), or Figs. 2(D) and 2(E) (I_2 in Ar). The appearance of two additional pairs of half rings with diameters ~ 10.2 and ~ 17.2 mm is discussed in Refs. 17-19.

We can quantify the degree of alignment from the angular distribution shown in Fig. 2. For each set of (I_{YAG}, Ar/He) we have determined the average value of $\cos^2\theta$, $<<\cos^2\theta>>$, where θ is the angle between the YAG polarization axis and the internuclear axis. A detailed procedure to calculate the parameter $<<\cos^2\theta>>$ is described in Ref. 18. The results are shown in Fig. 3. We also calculated $<<\cos^2\theta>>$, based on the theory of Refs. 8 and 9, as a function of I_{YAG} and for different rotational temperatures, T_{rot} (24). As seen in Fig. 3, good agreement between the experimental results and the calculations (full curves) are obtained for $T_{rot} = 3.5$ K (I_2 in Ar) and $T_{rot} = 7$ K (I_2 in He). Such rotational temperatures are consistent with the results from measurements of T_{rot} on a molecular beam (23) similar to that reported here. A more precise comparison between the measured and the calculated values of $<<\cos^2\theta>>$ requires an independent determination of T_{rot} in our experiment.

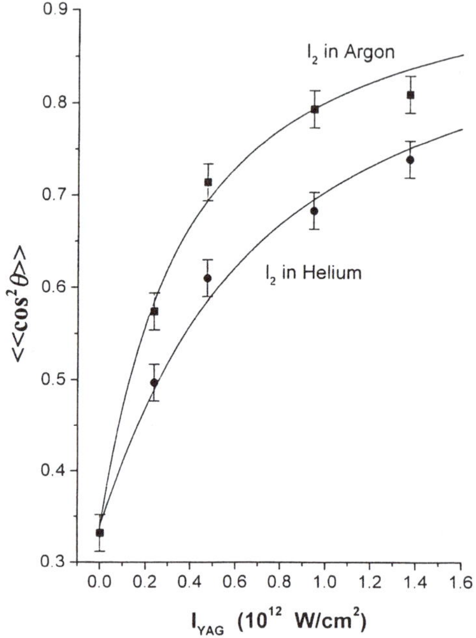

Figure 3. The observed value of $<<\cos^2\theta>>$ for different I_{YAG} and for I_2 in Ar (squares) and in He (circles). The full curves are the calculated values of $<<\cos^2\theta>>$. Note that for a sample of randomly oriented molecules the theoretical value of $<<\cos^2\theta>>$ is 1/3. (Reproduced with permission from reference 17.)

The initiation and termination of the alignment when the YAG pulse is turned on and subsequently switched off can be followed by changing the delay of the pump pulse with respect to the YAG pulse. It is confirmed that the degree of alignment, i.e. $<<\cos^2\theta>>$, follows the intensity of the YAG pulse, ensuring that the present experiments were performed in an adiabatic regime.

To demonstrate that laser induced alignment is generally applicable to molecules other than I_2 we carried out experiments on C_6H_5I, ICl, CS_2, and CH_3I. The alignment is probed with circularly polarized femtosecond pulses, centered at 800 nm, inducing dissociative ionization, and by recording the ionized fragments (I^+ or S^+). This is a convenient and generally applicable method to confirm if the molecules are aligned. We observe that for all these molecules the angular distributions of the ionized fragments become more localized around the polarization axis of the YAG pulse as I_{YAG} is increased. This is a qualitative manifestation of alignment similar to what we observe for iodine. Detailed discussions about the alignment of these molecules are given in Refs. 18 and 19.

1-4. Concluding remarks

It should be noticed that the laser used for alignment is nonresonant (*25*) with respect to both the electronic and vibrational structure of the molecule because excitation of any of these internal degrees of freedom can lead to dissociation or ionization before alignment. In our experiment the YAG pulse does not cause any observable dissociation or ionization. Generally, electronic resonances can be avoided by employing infrared radiation and keeping the intensity below the value where tunnel ionization becomes significant (*26*). To avoid vibrational excitation it is necessary to inspect carefully the infrared absorption spectrum of the particular molecule. If the wavelength of the YAG laser accidentally coincides with a vibrational (overtone) transition and thereby causes significant dissociation, it is possible to align the molecule using a different wavelength produced by nonlinear optics methods or by employing a different laser like a CO_2 laser.

An experiment making use of laser aligned molecules must be carried out in the presence of the alignment field because the pendular states evolve back into field-free states when the field is turned off. The presence of the intense laser field can influence some measurements, e.g., electronic states will be Stark shifted or additional photons can be absorbed as discussed in Refs. 17-19. In the near future it might prove experimentally possible to use shorter, nonadiabatic pulses to produce field-free alignment by recurrences of rotational wave packets as discussed by Ortigoso *et al.* (*27*).

Our demonstration of molecular alignment has importance for a number of areas in molecular science. (i) Controlling the alignment of the molecular geometry enables enhancement or suppression of photochemical events that depend on the angle between the polarization axis of the incident light and a specific molecular axis. (ii) The duration of the YAG pulse is sufficiently long that picosecond or femtosecond time-resolved studies can be performed on a sample of aligned molecules. For example, this is important for femtosecond timed Coulomb explosion studies (*28*). (iii) Spectroscopy of pendular states as recently demonstrated by Kim and Felker in Raman experiments (*15,16*). (iv) In addition to alignment, laser-induced dipole forces have lately been employed to deflect neutral molecules (*13,14*). The use of nonresonant intense laser fields should allow many aspects of atomic optics to be transferred to molecules such as trapping, wave guiding, and focusing. In our recent paper (*18*), some potential applications using the technique of aligning neutral molecules are discussed. They are (a) Selective dissociation, (b) Alignment of molecular plane, (c) Control of strong laser field ionization, (d) Applications in chemical control, (e) Dissociative rotational cooling, and (f) Steric effects in bimolecular reactions.

Selective dissociation and alignment of molecular plane have been already demonstrated. Using a sample of iodine molecules aligned by the present technique, Larsen *et al.* controlled the branching ratio of the I + I and I + I* photodissociation channels by a factor of 26 (*29*). The selective photodissociation was achieved by irradiating the aligned molecules with femtosecond pulses polarized parallel or perpendicular to the polarization axis of the alignment laser. They have also demonstrated that an intense, elliptically polarized, nonresonant laser field can simultaneously force all three axes of a molecule to align along given axes fixed in space, thus inhibiting the free rotation in all three Euler angles (*30*). 3,4-dibromothiophene molecules were three dimensionally aligned with an elliptically polarized nanosecond laser pulse. On the other hand, Corkum and his coworkers proposed (*31*) and demonstrated (*32*) an optical centrifuge for molecules. Using a combination of two counterrotating circularly polarized beams, they accelerated the rate of polarization rotation from 0 to 6 THz in 50 ps, spinning chlorine molecules from near rest up to angular momentum states $J \sim 420$. Their data show that the molecular bond is broken and the molecule dissociates at the highest spinning rate.

2. Numerical simulations of molecular orientation using strong, nonresonant, two-color laser fields[**]

2-1. Introduction

As demonstrated in the previous section, we have already succeeded in aligning neutral molecules by a strong linearly-polarized laser field (*17-19*). The realization of molecular orientation to arrange polar molecules in a "head-versus-tail" order is a next challenging subject. The molecular orientation should greatly expand the range of applications in stereodynamical studies of chemical reactions (*3*). Experimentally, a hexapole field (*4,5,33*) and/or a brute force field (*6,34*) can be used to orient molecules. Since a hexapole field acts as a selector, the resulting molecular density is as low as or slightly higher than the background density. Such a low molecular density hinders an efficient data acquisition. On the other hand, the degree of orientation and alignment achieved by the brute force field is lower than that by our laser-induced technique (*18,34*). Furthermore, Stark effects by the strong electrostatic field may be undesirable in some precise spectroscopic studies.

Vrakking and Stolte (*35*) proposed coherent control of molecular orientation by two-color phase-locked laser excitation. In their scheme, hexapole state-selected NO molecules have to be prepared first. Since their technique utilizes the transition dipole moment, resonant excitation processes are required to form an oriented superposition of both negative and positive ground-state parity levels. Consequently, a remarkable feature of their scheme is that even with nanosecond laser pulses the orientation displays oscillatory behavior related to the number of Rabi oscillations. Therefore, the time domain of experiments on a sample of oriented molecules achieved by their approach is substantially restricted to the femtosecond region as they addressed. Dion *et al.* (*36*) studied the orientation dynamics of HCN in the presence of a superposition of intense, linearly-polarized infrared laser pulses of frequency ω and 2ω. In their approach, 2ω is resonant with a $0 \rightarrow 1$ vibrational transition and the duration of the laser pulse is 1.7 ps. It means that their process is not adiabatic.

Here we propose an alternative and versatile approach to orient neutral polar molecules. The basic idea is to use an asymmetric electric field (i.e. an asymmetric potential) created by the superposition of intense, *nonresonant*, two-color laser fields. The key is to include anisotropic hyperpolarizability interaction as well as polarizability interaction. We solve the time-independent Schrödinger equation numerically in an adiabatic regime where the orientation proceeds slowly compared to the rotational period of the molecule (8,9). We calculate the orientation parameter, i.e. the ensemble average of $\cos\theta$, $<<\cos\theta>>$, and the alignment parameter, i.e. the ensemble average of $\cos^2\theta$, $<<\cos^2\theta>>$, where θ is the angle between the polarization direction of the laser field and the molecular axis. We evaluate their laser intensity dependence and the time evolution during the laser pulse. The extension of our approach to a nonadiabatic regime (27) can have a potential to realize a field-free orientation, which will be discussed below.

2-2. Numerical method

The form of the electric field is expressed by

$$E(t) = E_0(t)[\cos(\omega t) + \gamma \cos(2\omega t + \phi)], \tag{1}$$

where γ and ϕ are the relative field strength and phase, respectively. For simplicity, we consider a rigid rotor subject to radiation with an asymmetric field given by Eq. (1). The Schrödinger equation is

$$[BJ^2 + W_\mu(\theta) + W_{pol}(\theta) + W_{hyp}(\theta)]\Psi(\theta) = E\Psi(\theta), \tag{2}$$

with B rotational constant, J^2 squared angular momentum operator, and E eigenenergy. The interaction potentials $W_\mu(\theta)$, $W_{pol}(\theta)$, and $W_{hyp}(\theta)$ are based on the permanent dipole moment along the internuclear axis, the polarizability, and the hyperpolarizability, respectivlely. They are

$$W_\mu(\theta) = -\mu E \cos\theta, \tag{3}$$

$$W_{pol}(\theta) = -\frac{1}{2}[(\alpha_\| - \alpha_\perp)\cos^2\theta + \alpha_\perp]E^2 \tag{4}$$

$$= -\frac{1}{2}(\Delta\alpha\cos^2\theta + \alpha_\perp)E^2, \tag{5}$$

$$\Delta\alpha \equiv \alpha_\| - \alpha_\perp, \tag{6}$$

$$W_{hyp}(\theta) = -\frac{1}{6}[(\beta_{\parallel} - 3\beta_{\perp})\cos^3\theta + 3\beta_{\perp}\cos\theta]E^3, \tag{7}$$

with θ the angle between the polarization axis and the molecular axis, μ a permanent dipole moment along the internuclear axis, $\alpha_{\parallel}$ and $\alpha_{\perp}$ polarizability components, parallel and perpendicular to the axis, respectively, and $\beta_{\parallel}$ and $\beta_{\perp}$ hyperpolarizability components, parallel and perpendicular to the axis, respectively. We note that for the laser frequencies $\omega/2\pi$ much greater than the reciprocal of its pulse duration τ_p, $\omega/2\pi \gg \tau_p^{-1}$, the permanent dipole interaction $W_\mu(\theta)$ averaged over τ_p becomes zero. In order to maximize the asymmetry of the potential, we have to maximize $< E^3 >= (3/4)\gamma E_0^{3}\cos\phi$ ($<...>$ shows the average over τ_p). We choose $\phi = 0$ to achieve maximal asymmetry and $\gamma = 1$, which is limited by the practical reason. Actually, higher intensity of shorter-wavelength 2ω pulse could promote the ionization of molecules. The resulting electric field is

$$E(t) = E_0(t)[\cos(\omega t) + \cos(2\omega t)]. \tag{8}$$

The total interaction energy W_{total} averaged over τ_p is

$$W_{total} = W_{pol} + W_{hyp} \tag{9}$$

$$= -\frac{1}{2}\alpha_{\perp}E_0^{2} - \frac{3}{8}\beta_{\perp}E_0^{3}\cos\theta - \frac{1}{2}\Delta\alpha E_0^{2}\cos^2\theta - \frac{1}{8}(\beta_{\parallel} - 3\beta_{\perp})E_0^{3}\cos^3\theta. \tag{10}$$

Dividing the both hand sides of Eq. (10) by $-B$ and introducing parameters a_0, a_1, c^2, and a_3 for the future convenience, we obtain

$$-\frac{W_{total}}{B} = \frac{\alpha_{\perp}E_0^{2}}{2B} + \frac{3\beta_{\perp}E_0^{3}}{8B}\cos\theta + \frac{\Delta\alpha E_0^{2}}{2B}\cos^2\theta + \frac{(\beta_{\parallel} - 3\beta_{\perp})E_0^{3}}{8B}\cos^3\theta \tag{11}$$

$$\equiv a_0 + a_1\cos\theta + c^2\cos^2\theta + a_3\cos^3\theta. \tag{12}$$

The Schrödinger equation to be solved is

$$\left[\frac{d}{dz}\left[(1-z^2)\frac{d}{dz}\right] - \frac{M^2}{1-z^2} + \lambda_{\tilde{J},M} + a_1 z + c^2 z^2 + a_3 z^3\right]S_{\tilde{J},M}(\theta) = 0, \tag{13}$$

where $z = \cos\theta$. The eigenvalues $\lambda_{\tilde{J},M}$ and eigenfunctions $S_{\tilde{J},M}(\theta)$ can be numerically computed. The eigenvalues $\lambda_{\tilde{J},M}$ are related to the eigenenergies by $\lambda_{\tilde{J},M} = a_0 + E_{\tilde{J},M}/B$.

The orientation parameters $<\cos\theta>_{J,M}$ and the alignment parameters $<\cos^2\theta>_{J,M}$ can be evaluated by means of the Hellmann-Feynman theorem. These parameters are averaged over all possible initial rotational states of a molecular gas to obtain their ensemble averages $<<\cos\theta>>$ and $<<\cos^2\theta>>$, when we assume a Boltzmann distribution as an initial rotational states. Further detailed description of the numerical method is given in our recent paper (*37*).

2-3. Results and discussions

We consider a FCN molecule as a sample in the following discussion. We assume that an orientation pulse is a spatially and temporally Gaussian pulse whose pulse width (full width at half maximum) is 5 ns and that we probe the region of 0.8 x (spot size of the focused laser beam). The spatial averaging is taken into account when we calculate the orientation parameter $<<\cos\theta>>$ and the alignment parameter $<<\cos^2\theta>>$.

We calculated eigenfunctions for the lowest six states as functions of angle θ. It is noted that the eigenfunction for the lowest state is highly asymmetric and strongly localized around $\theta = 0$. It means that it is essential for the molecular orientation to produce rotationally cold molecules so that all of the molecules sit in the lowest state.

The dependence of the orientation parameter $<<\cos\theta>>$ and the alignment parameter $<<\cos^2\theta>>$ on the laser intensity is presented for some rotational temperatures T_{rot} in Fig. 4 (It should be noted that for a sample of randomly oriented molecules the theoretical value of $<<\cos^2\theta>>$ is 1/3). The positive $<<\cos\theta>>$ means that the F-atom is in the region of $0 < \theta < \pi/2$ according to the definition of polarizability and hyperpolarizability (*38*). It can be seen that significant orientation (and alignment) is realized at a lower rotational temperature of around 1 K, which can be achieved by the molecular beam technique (*39*). Figure 4 shows that a very low rotational temperature is more crucial for orientation than for alignment. We notice that significant orientation and alignment are realized at relatively low laser intensities of $\sim 10^{12}$ W/cm^2. Although hyperpolarizabilities are essential for the molecular orientation, some calculations on model molecules assuming appropriate parameters suggest that the relatively big difference between polarizability components as well as the very low rotational temperature (1 K) is responsible for the good response to the laser field. At the peak intensity of 1.4 x 10^{12} W/cm^2, the difference between the two minima in the asymmetric potential is estimated to be 68 μeV.

For FCN molecules in a static field of 100 kV/cm and at $T_{rot} = 4$ K, the $<<\cos\theta>>$ is calculated to be 0.37 (*40*), which is more than twice the value given by our approach at the peak intensity of 1.4 x 10^{12} W/cm^2 and the same rotational temperature. However, a very low rotational temperature makes our approach more advantageous. Actually, the $<<\cos\theta>>$ and $<<\cos^2\theta>>$ are estimated to be 0.52 and 0.78, respectively, at 1.4 x 10^{12} W/cm^2 and $T_{rot} = 1$ K.

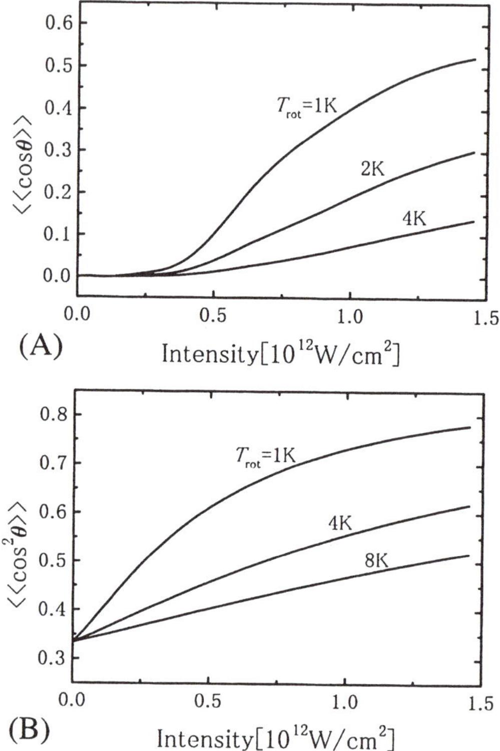

Figure 4. The dependence of $\langle\langle\cos\theta\rangle\rangle$ (A) and $\langle\langle\cos^2\theta\rangle\rangle$ (B) on the peak intensity for some rotational temperatures. Note that for a sample of randomly oriented molecules the theoretical value of $\langle\langle\cos^2\theta\rangle\rangle$ is 1/3.

Furthermore, since the ionization potential of FCN is 13.3 eV (*41*), we could increase the peak intensity at least up to ~ 10^{13} W/cm^2 without any significant ionization of FCN molecules (*26*). It means that even higher degree of orientation $\langle\langle\cos\theta\rangle\rangle$ can be achieved. In this sense, our approach is superior to brute force orientation (*34*).

The time evolutions of $\langle\langle\cos\theta\rangle\rangle$ and $\langle\langle\cos^2\theta\rangle\rangle$ during the laser pulse are also simulated. We used the peak intensity of 1.4 x 10^{12} W/cm^2 and T_{rot} = 1 K. In an adiabatic regime, the time evolutions of the parameters reflect the temporal profile of the orientation pulse and their peaks coincide with that of the laser pulse (*17-19,27,42*) The duration of the orientation and alignment is sufficiently long that picosecond or femtosecond time-resolved experiments can be performed on a sample of oriented molecules.

We performed similar simulation on a model molecule assuming appropriate parameters and confirmed that it could be oriented. This ensures that the parameters for FCN molecules are not special. It should be emphasized that molecules for which we know all the necessary parameters for the simulation are fairy limited.

As mentioned in Section 1, Larsen *et al.* demonstrated that an intense, elliptically polarized, nonresonant laser field can simultaneously force all three axes of a molecule to align along given axes fixed in space (*30*). It might be possible to realize perfect three-dimensional orientation of asymmetric top molecules if our approach for the molecular orientation is combined with their three-dimensional alignment technique.

As mentioned in the introduction, a field-free orientation could be realized by employing our approach in a nonadiabatic regime. Actually, we have already confirmed its possibilities by solving the time-dependent Schrödinger equations (*42*). The results will be presented elsewhere in the near future. Friedrich and Herschbach (*43,44*) have discussed enhanced orientation of polar molecules by combined electrostatic and nonresonant induced dipole forces. Although this is indeed an alternative and useful approach, it is impossible to realize a field-free orientation even in a nonadiabatic regime (*45*) because an electrostatic field is still applied.

2-4. Conclusion

We examined the possibility of molecular orientation using an asymmetric potential created by the superposition of two-color laser fields. We numerically solved the time-independent Schrödinger equation in an adiabatic regime and evaluated the orientation parameter $<<\cos\theta>>$ and the alignment parameter $<<\cos^2\theta>>$. We calculated their laser intensity dependence and time evolution during the laser pulse. Although molecules have to be rotationally cold, we have found that they could be oriented by our approach. For FCN molecules, $<<\cos\theta>>$ and $<<\cos^2\theta>>$ were estimated to be 0.52 and 0.78, respectively, at the peak intensity of 1.4×10^{12} W/cm^2, where $T_{rot} = 1$ K was assumed. Although we believe our approach is novel and useful, the most appropriate method should be complementarily chosen among a variety of approaches.

[*] H. Sakai, J. J. Larsen, C. P. Safvan, I. Wendt-Larsen, and H. Stapelfeldt contributed to the work of Section 1.

[**] T. Kanai and H. Sakai contributed to the work of Section 2.

References and notes

1. For a review, see Adams, C. S.; Riis, E. *Prog. Quantum Electron.* **1997**, *21*, 1.
2. Stroscio, J. A.; Eigler, D. M. *Science* **1991**, *254*, 1319.

3. *J. Phys. Chem. A* **1997**, *101,* No. 41, special issue on Stereodynamics of Chemical Reactions.

4. Cho, V. A.; Bernstein, R. B. *J. Phys. Chem.* **1991**, *95,* 8129.

5. Parker, D. H.; Bernstein, R. B. *Annu. Rev. Phys. Chem.* **1989**, *40,* 561.

6. Friedrich, B.; Herschbach, D. R. *Nature (London)* **1991**, *353,* 412.

7. Loesch, H. L.; Remscheid, A. *J. Phys. Chem.* **1991**, *95,* 8194.

8. Friedrich, B.; Herscbach, D. *Phys. Rev. Lett.* **1995**, *74,* 4623.

9. Friedrich, B.; Herscbach, D. *J. Phys. Chem.* **1995**, *99,* 15686.

10. Williams, J. H. *Adv. Chem. Phys.* **1993**, *85,* 361.

11. Weinstein J. D. *et al. Nature (London)* **1998**, *395,* 148.

12. Takekoshi, T.; Patterson, B. M.; Knize, R. J. *Phys. Rev. Lett.* **1998**, *81,* 5105.

13. Stapelfeldt, H.; Sakai, H.; Constant, E.; Corkum, P. B. *Phys. Rev. Lett.* **1997**, *79, 2787.*

14. Sakai, H.; Tarasevitch, A.; Danilov, J.; Stapelfeldt, H.; Yip, R. W.; Ellert, C.; Constant, E.; Corkum, P. B. *Phys. Rev. A* **1998**, *57, 2794.*

15. Kim, W.; Felker, P. M. *J. Chem. Phys.* **1996**, *104,* 1147.

16. Kim, W.; Felker, P. M. *J. Chem. Phys.* **1998**, *108,* 6763.

17. Sakai, H.; Safvan, C. P.; Larsen, J. J.; Hilligsøe, K. M.; Hald, K.; Stapelfeldt, H. *J. Chem. Phys.* **1999**, *110,* 10235.

18. Larsen, J. J.; Sakai, H.; Safvan, C. P.; Wendt-Larsen, I.; Stapelfeldt, H. *J. Chem. Phys.* **1999**, *111,* 7774.

19. Sakai, H.; Larsen, J. J.; Safvan, C. P.; Wendt-Larsen, I.; Hilligsøe, K. M.; Hald, K.; Stapelfeldt, H. *Advances in multiphoton processes and spectroscopy,* World Scientific, 2000; Vol. 14, pp 135-150.

20. Photodissociation as a probe of the molecular orientation has been reported by other groups, see, for example, Bazalgette, G. *et al. J. Phys. Chem. A* **1998**, *102,* 1098.

21. Larsen, J. J. *et al.* J. *Chem. Phys.* **1998**, *109,* 8857.

22. Eppink, T. J. B.; Parker, D. H. *Rev. Sci. Instrum.* **1997**, *68,* 3477.

23. McClelland, G. M.; Saenger, K. L.; Valentini, J. J.; Herschbach, D. R. *J. Phys. Chem.* **1979**, *83,* 947.

24. The calculated values of $\langle\langle\cos^2\theta\rangle\rangle$ are averaged over the different intensity regions of the focal volume of the YAG pulse.

25. A different method for achieving alignment, based on a resonant laser field, has been suggested: Seideman, T. *J. Chem. Phys.* **1995**, *103,* 7887.

26. Dietrich, P.; Corkum, P. B.; *J. Chem. Phys.* **1992**, *97,* 3187.

27. Ortigoso, J.; Rodríguez, M.; Gupta, M.; Friedrich, B. *J. Chem. Phys.* **1999**, *110,* 3870.

28. Ellert, Ch. *et al. Philos. Trans. R. Soc. London, Ser. A* **1998**, *356,* 329.

29. Larsen, J. J.; Wendt-Larsen, I.; Stapelfeldt, H. *Phys. Rev. Lett.* **1999**, *83,* 1123.

30. Larsen, J. J.; Hald, K.; Bjerre, N.; Stapelfeldt, H.; Seideman, T. *Phys. Rev. Lett.* **2000**, *85,* 2470.

31. Karczmarek, J.; Wright, J.; Corkum, P.; Ivanov, M. *Phys. Rev. Lett.* **1999**, *82,* 3420.

32. Villeneuve, D. M.; Aseyev, S. A.; Dietrich, P.; Spanner, M.; Ivanov, M. Yu.; Corkum, P. B. *Phys. Rev. Lett.* **2000**, *85,* 542.

33. See, for example, Kasai, T.; Fukawa, T.; Matsunami, T.; Che, D.-C.; Ohashi, K.; Fukunishi, Y.; Ohoyama, H.; Kuwata, K. *Rev. Sci. Instrum.* **1993,** *64,* 1150 and references therein.

34. Li, H.; Franks, K. J.; Hanson, R. J.; Kong, W. *J. Phys. Chem. A* **1998,** *102,* 8084.

35. Vrakking, M. J. J.; Stolte, S. *Chem. Phys. Lett.* **1997,** *271,* 209.

36. Dion, M.; Bandrauk, A. D.; Atabek, O.; Keller, A.; Umeda, H.; Fujimura, Y. *Chem. Phys. Lett.* **1999,** *302,* 215.

37. Kanai, T.; Sakai, H. "Numerical simulations of molecular orientation using strong, nonresonant, two-color laser fields," submitted.

38. Maroulis, G.; Pouchan, C.; *Chem. Phys.* **1997,** *215,* 67.

39. *Atomic and Molecular Beam Methods;* Scoles, G., Ed.; Oxford University Press, Oxford, 1988; Vol. I.

40. Referee's comment.

41. Radzig, A. A.; Smirnov, B. M. *Reference Data on Atoms, Molecules, and Ions;* Springer-Verlag, 1985.

42. Hayakawa, T.; Sakai, H. in preparation.

43. Friedrich, B.; Herschbach, D. *J. Chem. Phys.* **1999,** *111,* 6157.

44. Friedrich, B.; Herschbach, D. *J. Phys. Chem. A* **1999,** *103,* 10280.

45. Cai, L.; Marango, J.; Friedrich, B. *Phys. Rev. Lett.* **2001,** *86,* 775.

Chapter 22

Propensity of Molecules to Spatially Align In Linearly–Polarized, Intense Light Fields

D. Mathur, S. Banerjee, and G. Ravindra Kumar

Tata Institute of Fundamental Research, Homi Bhabha Road, Mumbai 400 005, India

The propensity of molecules to spatially align in intense, linearly-polarized light is understood using a classical model that incorporates enhanced ionization and higher-order polarizabilities. It is now possible to predict whether, and under what conditions, a molecule may spatially align on the basis of the parameters: laser pulse duration, peak intensity and the ratio of molecular polarizability to moment of inertia.

Introduction

Spatial alignment of isolated molecules is a subset of one of the central endeavors of physicists and chemists, namely to control the external degrees of freedom of molecules at the microscopic level. The polarizability interaction of an intense, linearly-polarized light field with the induced dipole moment of molecules gives rise to a double-well potential; the resulting angular realignment of molecular axes is akin to the interconversion of left- and right-handed enantiomers that was considered by Hund over 7 decades ago in terms of similar potentials (*1,2*). Spatial alignment of individual molecules can also be considered a special facet of the optical Kerr effect (*3*). Practically, studies of spatial alignment of molecules are interesting because of tantalizing possibilities of pendular-state spectroscopy (*4-6*), coherent control experiments (*7*), and molecular trapping and focusing (*8,9*). Theoretical progress has been achieved with quantal descriptions of alignment (see for instance, *2,5,6,10,11*). Although such treatments have provided new insights, the methodologies adopted are not simple and general access to such methods is restricted. To provide useful guideposts, simpler, albeit not rigorous, treatments might help provide a measure of predictability as to the propensity of molecules to spatially align upon exposure to laser fields of a given intensity and pulse duration. To this end we report here a classical model of the alignment dynamics that is readily usable and of utility.

An experimental signature of light-field-induced spatial alignment is the anisotropic angular distribution that is measured for fragment ions produced in the course of molecular dissociative ionization. Anisotropic ion distributions are characteristic of molecules undergoing dissociative ionization in strong, linearly-polarized light. Although this has been established for a long time, the underlying causes are still being clarified. Early experimental work (*12*) attributed such anisotropies to the dependence of ionization rate on the angle of the molecular axis with the light field vector, **E**. However, later work seemed to show that molecules tend to spatially reorient in a linearly-polarized field, leading to a fragment ion distribution that is strongly peaked in the field direction (*12*). Subsequent examination of the evidence showed that, at least in some cases, the conclusions pertaining to spatial alignment were incorrect, and such processes were not as ubiquitous as was originally thought (*13,14*). It now appears that the extent and nature of the angular anisotropy is determined by two processes: (i) the dependence of the ionization rate on the angle between the molecular axis and **E**, and (ii) field-induced motion of the molecule, leading to the spatial alignment of the most polarizable axis parallel to **E**. Experimentally measured anisotropies that arise from effect (i) are said to be due to "geometric alignment" whereas those due to effect (ii) are said to arise from "dynamic alignment". It is clearly important to be able to identify the significant contributor to experimentally observed anisotropies.

Conventional wisdom dictates that the alignment dynamics are governed by the molecular polarizability, α, which, in turn, is governed by a $\cos^2\theta$ potential. The total angular momentum of each molecule is coupled to the laser field through α. The second-order field-molecule interaction potential, $V(\theta)$, is

$$V(\theta) = \frac{1}{2} E^2 \left[\alpha_{\parallel} \cos^2\theta + \alpha_{\perp} \sin^2\theta\right]$$

(1)

where E is the average field strength and $\alpha_{\parallel}$ and $\alpha_{\perp}$ are, respectively, the polarizability components parallel and perpendicular to the bond axis. The $V(\theta)$ term causes spatial alignment; the field-molecule interaction energy overwhelms the field-free rotational energy (*4*) and induces dynamic alignment. For polyatomic molecules, the spatial properties of α determine the overall alignment dynamics (*15,16*). By way of illustration, Fig. 1 shows some typical

338

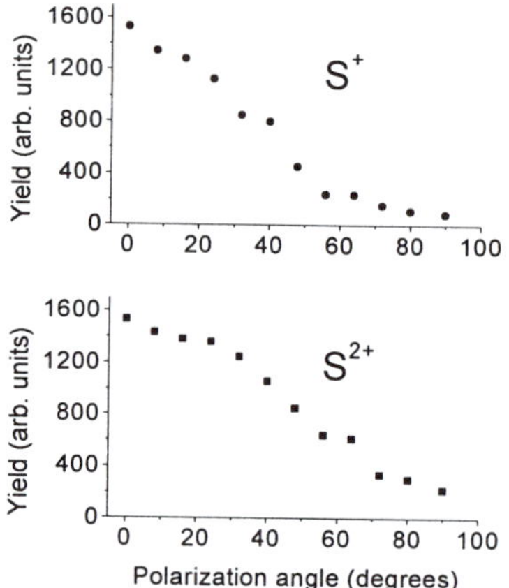

Figure 1. Angular distributions of S^{q+} fragments ions produced upon irradiation of linear CS_2 molecules by 100fs-long pulses (800nm) of intensity 10^{15} W cm^{-2}. 0° denotes alignment of the laser polarization vector parallel to the spectrometer axis. Very similar distributions were obtained when 35ps-long pulses were used, at lower intensities (10^{13} W cm^{-2}). Are the anisotropic distributions a signature of light-field-induced spatial alignment?

angular distributions for S^+ and S^{2+} fragments formed upon irradiation of linear CS_2 by 100fs pulses of 800nm radiation with peak intensity 10^{15} W cm^{-2}. The very sharp fall of ion yield in both instances is very similar to the observations made in experiments conducted using longer (35ps) pulses (*4*). The question arises as to whether the observed anisotropies are due to dynamic or geometric alignment.

In the recent past it has become possible to seek an experimental answer to this question (*14*). In our approach, dynamic and geometric alignment are distinguished by probing the width of the angular distribution over a range of laser intensities. This is equivalent to measuring the ratio of ion yields for orthogonal laser polarizations, A_{pp}, defined as

$$A_{pp} = \frac{A_{\parallel}}{A_{\perp}}, \tag{2}$$

where $A_{\parallel}$ is the ion yield for laser polarization parallel to the detector axis and $A_{\perp}$ that perpendicular to the detector axis. The method is based on the fact that when spatial alignment does occur, increase in the laser intensity should lead to an even larger torque on the molecule, leading to a decrease in the width of the angular distribution. Thus, if w denotes the width of the angular distribution of a particular fragment channel, then for the case of dynamic alignment,

$$\tau \propto \sqrt{I}, \quad w \approx \frac{1}{\tau} \quad \Rightarrow \frac{\partial w}{\partial I} < 0. \tag{3}$$

A contrary situation occurs for the case of geometric alignment. In this the probability of ionization at 90° increases as a function of intensity. The angular width also increases as higher intensities are accessed. Thus

$$\frac{\partial w}{\partial I} > 0. \tag{4}$$

For our experiments these conditions translate to A_{pp} increasing with intensity for the case of dynamic alignment, and decreasing for geometric alignment.

Figure 2 depicts the measured variation with laser intensity of the ratio of $S^{q+}_{\parallel}/S^{q+}_{\perp}$. In the case of geometric alignment, we expect the $\perp$-component to enhance with laser intensity. Consequently, the $S^{+}_{\parallel}/S^{+}_{\perp}$ ratio should fall as the

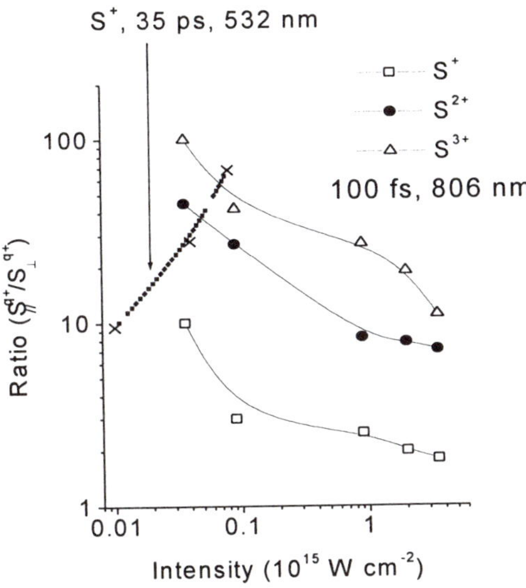

Figure 2. A simple experimental test to distinguish between dynamic and geometric alignment. 35ps pulses result in the former whereas 100fs pulses give rise to the latter, even though both produce anisotropic angular distributions (see text, and ref.14).

laser intensity increases. Our 100fs results indeed indicate this: significant falls occur in the $S^+_\parallel/S^+_\perp$, $S^{2+}_\parallel/S^{2+}_\perp$ and $S^{3+}_\parallel/S^{3+}_\perp$ ratios as the laser intensity is increased from 10^{13} to 10^{15} W cm^{-2}. Geometric alignment clearly dominates in the 100fs case. Upon using 35ps duration laser pulses (from an Nd:YAG laser, at $10^{13} - 10^{14}$ W cm^{-2}), the *opposite* occurs. The $S^+_\parallel/S^+_\perp$ ratio now increases with laser intensity. Similar observations were also made for S^{2+} ions. Dynamic alignment of CS_2 clearly occurs when longer-duration (35ps) laser pulses are used, even though the intensity accessed is lower than in the femtosecond experiments.

To place our results in perspective we note that work carried out by others has clearly demonstrated that 100fs light pulses of peak intensities 10^{15} W cm^{-2} induce significant spatial reorientation in light molecules like H_2 and N_2. However, heavy molecules like I_2 do not align. CS_2 is an intermediate case, and it also does not align in 100fs fields, even though anisotropic angular distributions are measured. We note that for many other molecules, anisotropy of the angular distribution of the fragment ions has been used to infer that they align with the light polarization vector. Such evidence is obviously not sufficient.

It is, therefore, very important to establish, both on the basis of the properties of the molecule under investigation and on the characteristics of the laser light that is used, the extent of spatial alignment that occurs. To this end a simple model with predictive power is clearly necessary.

Classical Model for Alignment

A comprehensive study was carried out by us in order to develop a model that might be used to predict the propensity of molecules to spatially align on the basis of parameters that are readily accessible (*17*). In any given analysis of spatial alignment three factors play a crucial role: (i) the peak intensity of the laser pulse, (ii) its temporal duration, and (iii) the ratio of the molecular polarizability (ground or excited state) to the moment of inertia ($A=\alpha/I$). In fact, the classification of molecules as light or heavy with regard to alignment should be on the basis of A and not the mass alone. At high enough intensities, nonlinear effects become significant. Molecular hyperpolarizabilites may then be expected to contribute substantially, although their role in the alignment dynamics has hitherto not been explicitly considered. It is also well established that field ionization of molecules is ubiquitous with short pulse lasers. The important implication of the field ionization model relevant to alignment is the breakup of the molecule at a critical distance, R_c, that is larger than the equilibrium internuclear separation, R_e, whose precise temporal location depends on the laser pulse. The stretching of the internuclear axis increases the moment of inertia and leads to a slowing down of reorientation; dissociation at R_c implies that the molecule will dissociate before the peak intensity is reached, except for ultrashort light pulses.

The process of alignment is modeled by considering a rotor in a time dependent **E** field (17). The interaction Hamiltonian is given by $H_I = -\mu.E$, where

$$\mu = \mu_0 + \frac{1}{2}\alpha\mathbf{E} + \frac{1}{6}\beta\mathbf{EE} + \frac{1}{24}\gamma\mathbf{EEE} + \dots \tag{5}$$

The Lagrangian for this system, considering only linear polarizabilty, is:

$$L = \frac{1}{2}I\left[\left(\frac{d\theta}{dt}\right)^2 + \sin^2\theta\left(\frac{d\phi}{dt}\right)^2\right] + \frac{1}{2}E(t)^2\left[\alpha_{\parallel}\cos^2\theta + \alpha_{\perp}\sin^2\theta\right], \tag{6}$$

where θ is the angle made by the molecular axis with E. The equation of motion is then given by

$$\frac{d^2\theta}{dt^2} = -\frac{\alpha_{eff}}{2I}E(t)^2\sin 2\theta - \frac{2}{r}\left(\frac{dr}{dt}\right)\left(\frac{d\theta}{dt}\right), \tag{7}$$

where α_{eff} is the effective polarizability. In the low-field limit this is the linear polarizability ($\alpha_{eff} = \alpha_{\parallel}-\alpha_{\perp}$). The modification of this in high fields is discussed later. To keep our calculations as realistic as possible we have used a Gaussian laser pulse, $E=E_0\exp[-(t/\tau)^2]$. Spatial variations within the laser beam are not taken into account since it has been shown recently that intensity selective experiments minimize focal volume effects (19).

In eq. (7), the first term causes reorientation while the second term, the so-called damping term, impedes the motion of the molecular axis towards the light field vector, due to the elongation of the molecular axis from R_e to R_c after the removal of one or two electrons by tunnel- or over-the-barrier ionization. The equation is solved used a 4th-order Runge-Kutta algorithm for a range of light field parameters: laser intensity range 10^{12}-10^{15} W cm^{-2}, and pulse durations from 40fs to 2ps. Instead of molecule-specific calculations, the A-parameter has been taken to lie in the interval 2×10^4-6×10^7, covering the entire gamut of light and heavy species, amongst them H_2, N_2, CS_2 and I_2. The initial direction of the molecular axis, θ_0, is taken to be random in space. After the light pulse is switched on, the angular position is calculated as a function of time for various values of θ_0, as shown in Fig. 3(a). From this one obtains a plot of θ_f vs. θ_i, where θ_i is the initial angular position and θ_f is the angular orientation of the

342

molecular axis at a particular instant. This is shown in Fig. 3(b). To obtain the distribution of the molecular axis at any instant of time we note that initially

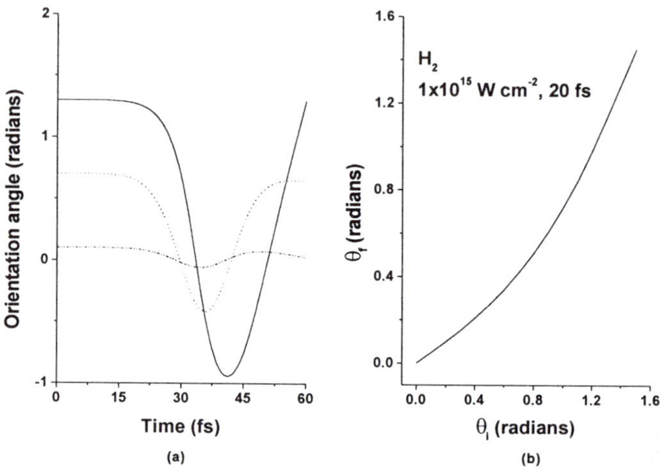

Figure 3. (a) Time evolution of the molecular alignment for various initial orientations at a peak intensity of 10^{15} W cm^{-2} and temporal width of 20 fs. (b) Alignment of the molecular axis at the point of breakup predicted by our model as a function of initial orientation (see text).

$$\Delta N_i = N_i(\theta)\Delta\theta_i, \tag{8}$$

where ΔN_i is the number of molecules in the interval $\Delta\theta_i$ with axis oriented at θ_I, and $N_i(\theta)$ is the number density as a function of orientation. Initially, the ensemble is isotropically distributed in space and N_i is independent of θ. As the ensemble interacts with the light pulse this distribution gets modified. Thus at some instant of time, t,

$$\Delta N(t) = N(t,\theta_f)\Delta\theta_f. \tag{9}$$

Since the number of molecules is preserved, that is, $\Sigma_\theta \Delta N_i = \Sigma_\theta \Delta N(t)$, it follows that the angular distribution, $N(t,\theta)$, is given by the following expression:

$$N(t,\theta) = a\left(\frac{\partial\theta_f}{\partial\theta_i}\right)^{-1}, \qquad (10)$$

where a is a constant of proportionality. Since one is interested only in the relative width, this constant can be set to 1 without loss of generality. The procedure cannot be used when the applied field is so strong that the molecular axis crosses $\theta=0$. In such cases the molecule oscillates about $\theta=0$, and the derivative blows up. A counting method is then used to obtain the angular distribution by interpolating the relation between θ_f and θ_i and then enumerating the number of points which lie in a given angular interval.

The trajectories shown in Fig. 3 correspond to H_2 exposed to 20fs pulses at 10^{15} W cm^{-2}. As is seen from the slope in Fig. 3(b), the extent of reorientation in H_2 is negligible under these conditions. Similar calculations were carried out for a range of parameters as specified above. Figure 4 presents, in pictorial form, a summary of all the results of our calculations. The hatched surface shown in the

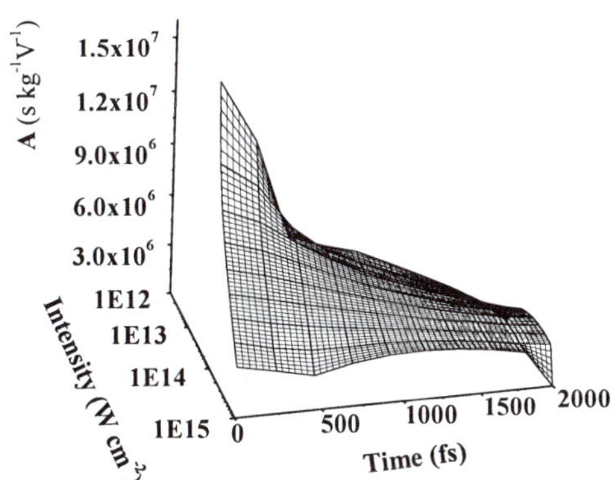

Figure 4. The hatched surface demarcates regions where molecular spatial alignment is significant from those where no significant reorientation occurs. Points lying below the surface correspond to the case of no alignment while all points lying on the surface, and above it, lead to the molecular axis being aligned along the light polarization vector.

figure demarcates regions where molecular spatial alignment is significant from those where no significant reorientation occurs. All points, as defined by the four parameters, that lie above the surface correspond to molecules that are liable to become 'tightly' aligned by the linearly polarized light field, while the opposite holds for points that lie below the surface. Note that we are dealing here only with linear polarizabilities. The nonlinear polarizability components serve only to strengthen the alignment. Thus, the demarcation based on α alone is very rigorous.

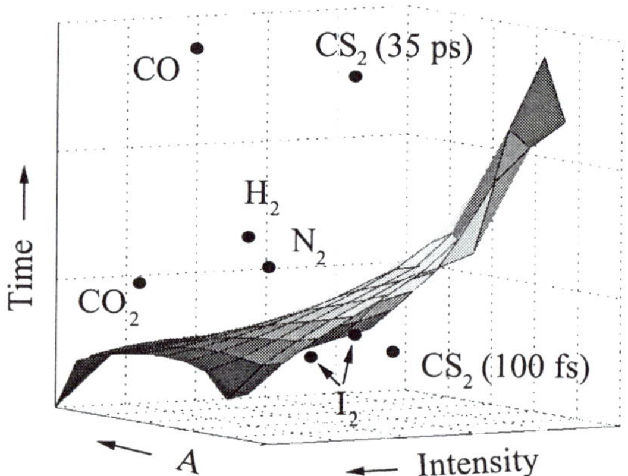

Figure 5. Experimental data for various molecules placed in relation to the surface predicted using our model calculations. Points lying below the surface correspond to the case of no alignment while all points lying on the surface, and above it, lead to the molecular axis being aligned along the light polarization vector. CO [ref. 12], CO$_2$ [ref. 22], H$_2$, N$_2$ [ref. 13], I$_2$ [ref. 23], CS$_2$ (picosecond data, above the surface [ref. 4], femtosecond data, below the surface [ref. 17]). The axes ranges are the same as in Figure 4.

Our calculations so far pertain to the position of the molecular axis. However alignment is deduced from the anisotropy of fragment ions. To make the connection with experimental data we note that the angular distributions that are shown are those that would practically be measured using a spectrometer with a small acceptance angle. Thus, our calculations are especially relevant to the angular distributions of highly charged ions (that possess large kinetic energies). This is also important in the context of the residual angular momentum as the molecule rotates. This is sufficiently large to cause significant rotation on the time scale of the laser pulse but negligible compared to the energy of the fragment ions (typically more than 1 eV). Figure 5 compares the results of our calculation with experimental data for various molecules. The location of each molecule relative to the surface is in precise agreement with known facts as to whether or not alignment occurs. Thus our model is able to summarize a large body of experimental work. In addition, the utility of the surface is obvious for future experiments. New data on alignment can be compared to these results to check for the validity of the model.

Enhanced Ionization

Real ionization dynamics occur through an enhanced ionization (EI) mechanism wherein one or two electrons are removed at the first ionization step. Subsequent to the first ionization step that occurs at R_e, the two residual ions mutually repel each other, leading to an increase in the bond length. This results in one or more Stark-shifted electronic levels rising above the potential barrier that separates the atomic cores, at which point multiple electron ejection occurs, leading to molecular fragmentation (*20*). EI can modify the reorientation rate in two ways. Firstly, as the moment of inertia increases, the magnitude of the first term in Eq. (7) will reduce. In addition, the damping term will come into play, leading to a further decrease in the rate at which the molecule rotates towards the light field vector. Thus, the overall effect of EI will be to impede the process of alignment. It is important to investigate the extent EI might modify our first-order calculations. Since EI parameters are available only for a few molecules, we have carried out these calculations for some standard cases. These can be extended to any other molecule once the relevant parameters are established, either by calculation or experiment.

Figure 6 shows the angular distribution for H_2 for a pulse duration of 40fs at a peak laser intensity of 10^{15} W cm^{-2}, with and without the damping term. It is clear that the reorientation of H_2 is not significantly affected when the damping term is included. There are two major reasons for this. Firstly, A is extremely large and the torque experienced by H_2 is sufficient to induce reorientation despite the presence of an opposing force. Secondly, the fact that the ionization energy (and hence, the appearance intensity) of H_2 is quite high, the damping force only comes into play close to the peak of the laser pulse, by which time the molecular axis is already aligned with the light polarization vector. Interestingly, the width of the angular distribution with damping included is actually smaller than when no damping is present. This arises due to the fact that the angular velocity without damping is larger, causing the molecular axis to execute large amplitude oscillations about $\theta=0$; hence, there will exist instants at which the peak of the angular distribution will shift away from zero. This is reminiscent of pendular states and associated experimental observations reported previously (*4*).

A contrary situation is also depicted in Fig. 6 when linear CS_2 molecules are exposed to 100fs light fields at 10^{15} W cm^{-2}. Here, the lower ionization energy of the molecule and the relatively small value of A, leads to virtually no reorientation of the S-C-S axes with the direction of the **E** field at the point at which dissociation occurs. Strong alignment can be expected if it is assumed that the molecule survives undissociated till the peak of the laser pulse, a fact

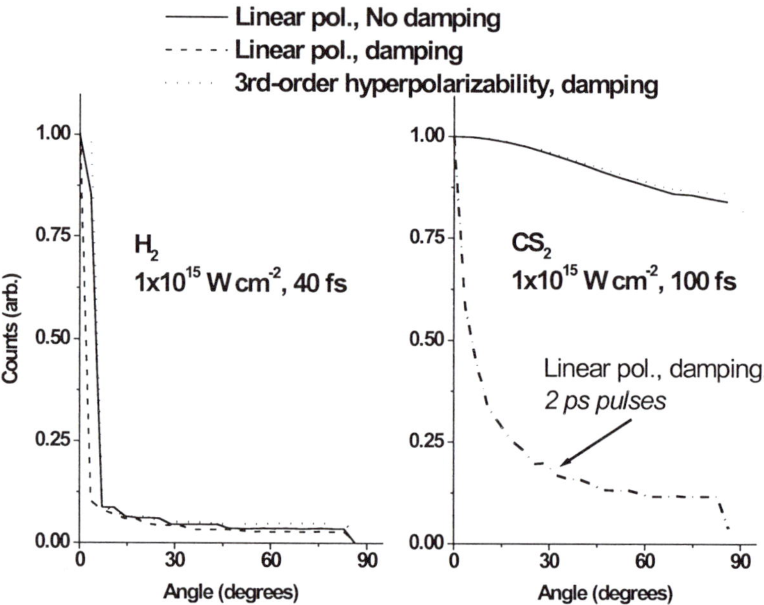

Figure 6. Alignment dynamics calculated for H$_2$ and CS$_2$, with enhanced ionization taken into account.

contrary to experimental observation, and illustrates the essentiality of EI in any such model. Similar calculations have been carried out for other molecules like N_2 and I_2. The situation for N_2 is similar to H_2 because of the similarity in the relevant molecular parameters. In the case of I_2, reorientation of the molecule is not significant even without the damping term. Once the damping term is included there is only a small deflection of the molecular axis.

Hyperpolarizabilities

Hitherto, only the polarization response that arises from the linear term has been considered. To what extent is this justified, especially at intensities in the range of 10^{12} - 10^{15} W cm^{-2}? It is important to note that hyperpolarizabilites are significant only at the highest intensities. This is obvious when we compare the integrals which define the work done by the field on the molecule by each order of the hyperpolarizability. For longer pulses, a model based on linear polarizability is sufficient since the dissociation of the molecule occurs on the rising edge of the pulse. However, as the pulse duration becomes shorter (<50fs), the molecule will survive till the maximum intensity is reached, and the reorientation due to the higher order terms will become comparable to that due to the linear term, and may even exceed it!

To account for hyperpolarizability, approximations need to be made. The magnitude of the second- and higher-order susceptibility tensors is not known in most cases. We consider here the case for H_2, taking into account the third order term due to the electronic response γ_e (21). The polarizability term is modified as follows:

$$\alpha \sin 2\theta E^2 \rightarrow \alpha \sin 2\theta E^2 + \gamma_e \sin 4\theta E^4 + \dots \qquad (11)$$

Figure 7 shows the effect of nonlinearity on the reorientation of the molecular axis for H_2 with a 20fs pulse. The inclusion of only the third-order term leads to a significantly larger reorientation of the H-H axis as compared to the case when only the linear term is considered. As noted previously, such effects will be significant only for very short pulses. This is shown in Fig. 6 for the case of a

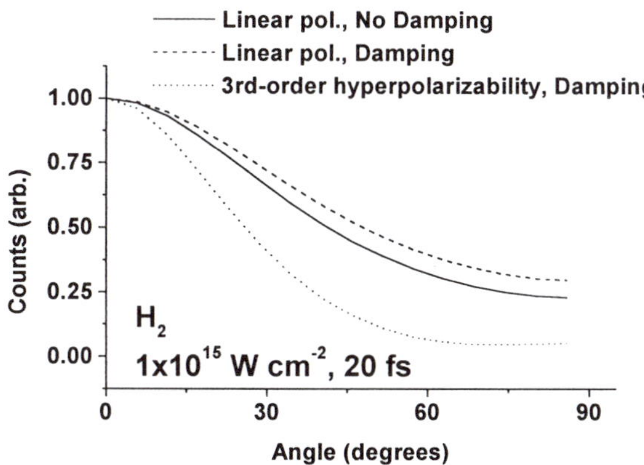

Figure 7. Alignment of the molecular axis of H₂.

40fs pulse wherein the effect of the third-order term is much smaller than that for the 20fs case. Of course, it is obvious that as the laser pulses get shorter even higher-order terms begin to play a significant role, an indication of highly nonperturbative coupling. One can speculate that heavy molecules, like I_2, may align with sufficiently short pulses because of the contribution from higher-order terms. However there are no reports of experiments with ultrashort (<20fs) pulses. It is clearly necessary to test this conjecture experimentally since very little is known about the high-order polarizabilities of almost all molecules.

Summary

We have considered light-field-induced molecular realignment using a classical model. Despite the obvious limitations of classical models, our readily-accessible model supplements high-level quantal treatments (see, for instance *2,5,6,7,10,11,* and references therein) and the results that we obtain are of considerable utility in understanding the existing body of experimental work on spatial alignment of molecules. Using the model, it is also possible to make predictions on whether or not, and under what circumstances, molecules are spatially aligned when subjected to intense, linearly-polarized, short-duration light fields. The rigor of the model is demonstrated. We have also incorporated, for the first time, (i) the role of enhanced ionization in the reorientation of molecules and (ii) the role of hyperpolarizability. It is shown that higher order contributions to the dipole moment are very important for extremely short light pulses. It is predicted that this will lead to alignment of molecules even for sub-50fs pulses that are becoming increasingly accessible to experimentalists.

Acknowledgements

Several colleagues have participated in our experiments on spatial alignment of molecules; it is a pleasure to acknowledge contributions made by V. R. Bhardwaj, K. Vijayalakshmi, C. P. Safvan, R. V. Thomas, and F. A. Rajgara. The Department of Science and Technology is thanked for generous support for our high-intensity femtosecond laser system.

References

1.   Hund, F. *Z. Phys.* **1927**, 43, 805.
2.   Friedrich, B; Herschbach, D. *Z. Phys. D*, **1996**, 36, 221.
3.   Williams, J. H. *Adv. Chem. Phys.*, **1993**, 85, 361.
4.   Kumar, G. R; Gross, P; Safvan, C. P; Rajgara, F. A; Mathur, D. *Phys. Rev. A*, **1996**, 53, 3098.

5. Kim, W; Felker, P. M. *J. Chem. Phys.*, **1998**, 108, 6763.

6. Ortigoso, J; Rodríguez, M; Gupta, M; Friedrich, B. *J. Chem. Phys.*, **1999**, 110, 3870.

7. Charron, E; Giusti-Suzor, A; Mies, F. H. *Phys. Rev. A*, 1994, 49, R641.

8. Weinstein, J. D. *Nature*, 1998, 395, 148.

9. Stapelfeldt, H; Sakai, H; Constant, E; Corkum, P. B. *Phys. Rev. Lett.*, **1997**, 79, 2787.

10. Seideman, T. *J. Chem. Phys.,***1995**, 103, 7887.

11. Friedrich, B; Herschbach, D. *Phys. Rev. Lett.,* **1995**, 74, 4623.

12. Frasinski, L. J; Codling, K; Hatherly, P. A. *Science*, **1989**, 246, 1029.

13. Normand, D; Lompre, L. A; Cornaggia, C. *J Phys. B*, **1992**, 25, L497.

14. Posthumus, J. H; Plumridge, J; Thomas, M. K; Codling, K; Frasinski, L. J. *J Phys. B*, **1998**, 31, L553.

15. Banerjee, S; Ravindra Kumar, G; Mathur, D. *Phys. Rev. A* **1999**, 60, R3369.

16. Bhardwaj, V. R; Safvan, C. P; Vijayalakshmi, K; Mathur, D. *J. Phys. B,* **1997**, 30, 3821.

17. Bhardwaj, V. R; Vijayalakshmi, K; Mathur, D. *Phys. Rev. A,* **1997**, 56, 2455.

18. Banerjee, S; Mathur, D; Ravindra Kumar, G. *Phys. Rev. A* **2001**, 63, 045401.

19. Dion, C. M.; Keller, A; Atabek, O; Bandrauk, A. D. *Phys. Rev. A* **1999**, 59, 1382.

20. Banerjee, S; Ravindra Kumar, G; Mathur, D. *J. Phys. B*, **1999**, 32, L305.

21. Zuo, T; Bandrauk, A. D. *Phys. Rev. A* **1995**, 52, R2511.

22. Shelton, D. P. *Mol. Phys.*, **1987**, 60, 65.

23. Sanderson, J. H; Thomas, R. V; Bryan, W. A; Newell, W. R; Langley, A. J; Taday, P. F. *J Phys. B*, **1998**, 31 L599.

24. Ellert, Ch; Corkum, P. B. *Phys. Rev. A,* **1999**, 59, R3170.

Author Index

See also Channel phase; Double-pulse manipulation of vibrational wavepackets; Four-wave mixing (FWM)

Condensed phase quantum control, fundamental issues, 133

Conjugated polymer, mode coupling, 172–175

Control

closed loop, of quantum systems, 8–13

high harmonic generation, 12

high quality, 3–4

tailored laser fields, 2–3

See also Channel phase; Chirped laser pulses; Four-wave mixing (FWM)

Control field design, perturbation theory regime, 4–5

Cost functionals

design, 6

optimal control theory, 5

Coulomb explosion of molecules, intense laser fields, 260–263

Cresyl violet (CV)

CV doped in poly(vinyl alcohol) (PVA), 175

Duschinsky rotation, 175–183

instantaneous amplitudes and frequencies for ring-breathing mode in CV/PVA and CV/poly(methyl methacrylate), 177, 179f

modulation of adiabatic harmonic potential of CV in vibrational coordinates, 180

transient differential transmittance as function of delay time, 175, 176f

See also Real-time spectroscopy

D

Degenerate final states

optimizing population transfer, 26–30

See also Population transfer

Degrees of freedom, rotational, enantiomers, 44–45

Dipole moment functions, selective enantiomer preparation, 44

Dissipative environments. *See* Quantum control

Dissociation

calculated channel phases, 57, 58f

control, 11–12

HI, 49, 51

possible coupling schemes, 54, 55f

See also Channel phase

Dissociative ionization

above threshold dissociation (ATD) spectra, 224

above threshold ionization (ATI) spectra, 224

asymmetry at phase $\phi=0$, 229–230

asymmetry in calculated ATD spectra in presence of ionization, 230, 232

ATD spectra, 230–234

ATD spectra for phases $\phi=0$ and $\phi=\pi/2$ cases, 232, 234

ATI spectra, 227–230

attempts to control using $\omega + 2\omega$ coherent superposition scheme, 222

backward asymmetry in electron ATI and proton ATD, 224

calculating proton kinetic energy spectra of H_2^+, 225f

coherent laser field superposition, 224

complete dynamic, non-Born-Oppenheimer TDSE, 222, 224

counterintuitive tunneling and intuitive classical regimes, 230

dressed states for $\omega + 2\omega$ dissociation, 233f

drift velocity, 229

electron kinetic energy (ATI) spectrum for H_2^+ in laser field, 223f